3000 800045 88205
St. Louis Community College

D0206076

Disseminating Darwinism

This innovative collection of original essays focuses on the ways in which geography, gender, race, and religion influenced the reception of Darwinism in the English-speaking world of the late nineteenth and early twentieth centuries. Although studies of Darwin and Darwinism have increased dramatically in the past few decades, the way in which various groups and regions responded to Darwinism remains relatively unexplored. Little is known about the role of locale in affecting responses to evolutionary theories. The contributions to this volume collectively illustrate the importance of local social, physical, and religious arrangements, while showing that neither distance from Darwin's home at Down nor size of community greatly influenced how various regions responded to Darwinism. Essays spanning the world from Great Britain and North America to Australia and New Zealand explore the various meanings of Darwinism in these widely separated locales, while other chapters focus on the difference race, religion, and gender made in the debates over evolution.

Ronald L. Numbers is the Hilldale and William Coleman Professor of the History of Science and Medicine at the University of Wisconsin, Madison, a fellow of the American Academy of Arts and Sciences, the former president of the American Society of Church History, and the president of the History of Science Society. He is author or editor of over twenty books, including *The Creationists* (Alfred A. Knopf, 1992) and *Darwinism Comes to America* (Harvard University Press, 1998), and is currently writing a history of science in America for Cambridge University Press. Having formerly served as the editor of *Isis*, the journal of the History of Science Society, he is now coediting (with David C. Lindberg) an eight-volume *Cambridge History of Science*.

John Stenhouse is Senior Lecturer in History at the University of Otago, New Zealand. He is coeditor of *Science and Theology: Questions at the Interface* (T. & T. Clark, 1994). Professor Stenhouse has also been published in numerous journals, including *Journal of the History of Biology, New Zealand Journal of History, Journal of Religious History, Metascience,* and *Pacifica: Australian Theological Studies*.

Disseminating Darwinism: The Role of Place, Race, Religion, and Gender

Edited by
RONALD L. NUMBERS
University of Wisconsin, Madison
And
JOHN STENHOUSE
University of Otago, New Zealand

PUBLISHED BY THE PRESS SYNDICATE OF THE UNIVERSITY OF CAMBRIDGE
The Pitt Building, Trumpington Street, Cambridge, United Kingdom

CAMBRIDGE UNIVERSITY PRESS
The Edinburgh Building, Cambridge CB2 2RU, UK
40 West 20th Street, New York, NY 10011-4211, USA
10 Stamford Road, Oakleigh, Melbourne 3166, Australia
Ruiz de Alarcón 13, 28014 Madrid, Spain
Dock House, The Waterfront, Cape Town 8001, South Africa

http://www.cambridge.org

First published 1999
First paperback edition 2001
Reprinted 2001

Printed in the United States of America

Typeset Palatino 10/12 pt. in LATEX2e [TB]

A catalog record for this book is available from the British Library

Library of Congress Cataloging in Publication data is available

ISBN 0 521 62071 6 hardback
ISBN 0 521 01105 1 paperback

CONTENTS

CONTRIBUTORS

ERIC D. ANDERSON is professor of history at Pacific Union College, Angwin, California. He is the author of *Race and Politics in North Carolina, 1872–1901: The Black Second* (1981), coeditor of *The Facts of Reconstruction: Essays in Honor of John Hope Franklin* (1991), and coauthor with Alfred A. Moss, Jr., of *Dangerous Donations: Northern Philanthropy and Southern Black Education, 1902–1930* (1999).

R. SCOTT APPLEBY is professor of history and director of the Cushwa Center for the Study of American Catholicism at the University of Notre Dame. He is the author of *"Church and Age Unite!" The Modernist Impulse in American Catholicism* (1992), coauthor of *The Glory and the Power: The Fundamentalist Challenge to the Modern World* (1992), coeditor, with Martin E. Marty, of the five volumes of *The Fundamentalism Project* (1991–1995), and author of *The Ambivalence of the Sacred: Religion, Violence, and Reconciliation* (1999).

BARRY W. BUTCHER is lecturer in the social studies of science at Deakin University in Geelong, Australia. In 1992 he completed his Ph.D. at the University of Melbourne, where he wrote his dissertation on "Darwinism and Australia, 1836–1914." He is the author of several articles on the reception of Darwinism in Australia, including "Darwinism, Social Darwinism, and the Australian Aborigines: A Reevaluation," in *Darwin's Laboratory: Evolutionary Theory and Natural History in the Pacific*, ed. Roy MacLeod and Philip F. Rehbock (1994), pp. 371–94. He is currently working on an oral history of Australian science.

MARK R. JORGENSEN is a Ph.D. candidate in the Department of Sociology at the University of Minnesota, where he is completing a dissertation on the role of patent rivalry and litigation in stimulating or inhibiting inventive activity in American industry, and an M.A. candidate in the Program in the History of Science and Technology. Other research interests include the evolution of market structure in the history of computing; the social construction of environmental catastrophism; and risk, ethics, and responsibility in the automobile industry.

SALLY GREGORY KOHLSTEDT is professor of the history of science and director of the Center for Advanced Feminist Studies at the University of Minnesota and former president of the History of Science Society. She is the author of *The Formation of the American Scientific Community: The American Association for the Advancement of Science* (1976) and editor of several books, including *International Science and National Scientific Identity: Australia between Britain and America* (1991), coedited with R. W. Home; *Women, Gender, and Science: New Directions* (1997), coedited with Helen Longino; and *Women and Science: An Isis Reader* (1999). She is a fellow of the American Association for the Advancement of Science and the American Antiquarian Society.

DAVID N. LIVINGSTONE is professor of geography and intellectual history at The Queen's University of Belfast in Northern Ireland. He is the author of *Nathaniel Southgate Shaler and the Culture of American Science* (1987), *Darwin's Forgotten Defenders* (1987), *The Geographical Tradition: Episodes in the History of a Contested Enterprise* (1992), and, with Ronald A. Wells, *Ulster-American Religion* (1999). He is currently writing a book on the spaces of science for Basil Blackwell and coediting the volume on *Modern Science in National and International Context* in the *Cambridge History of Science*. He is a fellow of the British Academy, a member of the Royal Irish Academy, a recipient of the Royal Geographical Society's Back Award, and a recipient of the Royal Scottish Geographical Society's Centenary Medal.

RONALD L. NUMBERS is Hilldale and William Coleman professor of the history of science and medicine at the University of Wisconsin, Madison, and onetime William Evans visiting fellow in religious studies at the University of Otago in Dunedin, New Zealand. He is the author or editor of over twenty books, including *The Creationists* (1992) and *Darwinism Comes to America* (1998). He is currently writing a history of science in America for Cambridge University Press and coediting, with David C. Lindberg, the eight-volume *Cambridge History of Science*. He is a fellow of the American Academy of Arts and Sciences, the president of the American Society of Church History, and the president-elect of the History of Science Society.

JON H. ROBERTS is professor of history at the University of Wisconsin, Stevens Point. In the past he has taught in the history departments of Harvard University (1980–1985), the University of Wisconsin, Madison (1989), and the University of Michigan (1992–1993). He is the author of the prize-winning book *Darwinism and the Divine in America: Protestant Intellectuals and Organic Evolution, 1859–1900* (1988). He is currently writing a history of psychology and religion in the United States since the late nineteenth century.

JOHN STENHOUSE is senior lecturer in history at the University of Otago in Dunedin, New Zealand, and a former fellow in the Institute for Research in the Humanities at the University of Wisconsin, Madison (1996). He has written a number of articles on the history of evolution in New Zealand, including "The Darwinian Enlightenment and New Zealand Politics," in *Darwin's Laboratory: Evolutionary Theory and Natural History in the Pacific*, ed. Roy MacLeod and Philip

F. Rehbock (1994), pp. 395–425. He is coeditor of *Science and Theology: Questions at the Interface* (1994). He is currently writing a history of the impact of evolution on racial ideas and attitudes in New Zealand and an essay on "Missionary Science" for the *Cambridge History of Science* volume on *Modern Science in National and International Context.*

LESTER D. STEPHENS is professor emeritus of history at the University of Georgia. His many publications include *Joseph LeConte: Gentle Prophet of Evolution* (1982), *Ancient Animals and Other Wondrous Things: The Story of Francis Simmons Holmes, Paleontologist and Curator of the Charleston Museum* (1988), and *Science, Race, and Religion in the American South: John Bachman and the Charleston Circle of Naturalists* (1999). For over a decade he edited the History of American Science and Technology Series published by the University of Alabama Press, and in the early 1990s he served on the executive committee of the History of Science Society.

MARC SWETLITZ earned an M.A. degree in Jewish studies from the Jewish Theological Seminary in New York City (1984) and a Ph.D. degree in the history of science from the University of Chicago (1991), where he wrote his dissertation on "Julian Huxley, George Gaylord Simpson, and the Idea of Progress in 20th-Century Evolutionary Biology." He subsequently held postdoctoral fellowships in the history of science at the University of Minnesota, Brandeis University, and the University of Oklahoma before joining Arthur Andersen and Co. in St. Charles, Illinois, where he is an instructional design manager in the Center for Professional Education. His publications include "Responses of American Reform Rabbis to Evolutionary Theory, 1864–1888," in *The Interaction of Scientific and Judaic Cultures in Modern Times*, ed. Yakov Rabkin and Ira Robinson (1995), pp. 103–25.

SUZANNE E. ZELLER is associate professor of history at Wilfrid Laurier University in Waterloo, Ontario, Canada. She is the author of *Inventing Canada: Early Victorian Science and the Idea of a Transcontinental Nation* (1987), *Land of Promise, Promised Land: The Culture of Victorian Science in Canada* (1996), and various articles on the history of Canadian science. With T. H. Levere she is writing on "Science in Canada" for the *Cambridge History of Science* volume on *Modern Science in National and International Context.* Her research interests focus on the history of culture and ideas in colonial societies, especially Victorian science and environmental history.

ACKNOWLEDGMENTS

With one exception, the papers in this volume were first presented at a conference on "Responding to Darwin: New Perspectives on the Darwinian Revolution," held in Dunedin, New Zealand, 12–15 May 1994. (David N. Livingstone, who unfortunately could not attend the conference, has graciously permitted us to include his essay in the published proceedings.) This conference was made possible largely through the generosity of the Division of International Programs of the National Science Foundation (represented by Dr. Charles Wallace). Additional support came from the Royal Society of New Zealand (represented by Mrs. Sue Usher), the Alexander Turnbull Library Endowment Trust in Wellington, the Hocken Library and the Division of Humanities at the University of Otago, Knox College in Dunedin, the Departments of History and Religious Studies at the University of Otago, and the Department of the History of Medicine at the University of Wisconsin, Madison. To all these institutions we are most grateful.

We wish to thank, in addition to the contributors to this volume, the following conference participants: Ruth Barton, University of Auckland; Keith R. Benson, University of Washington; Barbara Brookes, University of Otago; Mark Francis, Canterbury University; Chris Gousmett, Dunedin, New Zealand; Pamela M. Henson, Smithsonian Institution; John Angus Laurent, Griffith University; Peter J. Lineham, Massey University; Roy M. MacLeod, University of Sydney; Peter Matheson, The Theological Hall, Knox College; James R. Moore, The Open University; John Omer-Cooper, University of Otago; Gordon S. Parsonson, University of Otago; Gary Tee, University of Auckland; Stephen P. Weldon, University of Wisconsin, Madison; and Janice E. Wilson, University of Otago. Their critiques and comments, as well as those of two anonymous referees, have significantly improved this volume.

INTRODUCTION

RONALD L. NUMBERS AND JOHN STENHOUSE

In 1859 the English naturalist Charles Darwin, a resident of Down outside of London, published his controversial views on the origin of species. In a landmark book entitled *On the Origin of Species by Means of Natural Selection,* he argued against the conventional notion that God had supernaturally created the original types of plants and animals and in favor of the idea that they had evolved naturally over long periods of time primarily, though not exclusively, by means of random variation and natural selection. News of his heretical views spread rapidly, and before long even the citizens of such remote outposts of British civilization as Dunedin, New Zealand, halfway around the globe from Down and home of the southernmost university in the world, were debating the merits of Darwinism.

The essays in this volume focus specifically on the ways in which geography, gender, race, and religion influenced responses to Darwin. Chronologically, they span the period from the publication of the *Origin* to the 1930s, when Darwin's theory of natural selection finally captured the allegiance of the scientific community. Geographically, they concentrate on the English-speaking world, especially Canada, Australia, New Zealand, and the United States. Although historians of science have been examining Darwin's influence for decades and have produced a number of notable studies, our knowledge of how various groups and regions responded to Darwinism remains spotty. For example, despite the availability of such works as Thomas F. Glick's *Comparative Reception of Darwinism* (1974) and the section "Towards the Comparative Reception of Darwinism" in David Kohn's *Darwinian Heritage* (1985) – neither of which covers Australia, New Zealand, or Canada – we still know relatively little about the role of locale in affecting responses to Darwin.[1]

The case studies in this volume illustrate the importance of local social and religious arrangements in affecting responses to Darwinism, a term, it should be noted, that conjured up markedly different images for different people. The essays show that neither distance from Down nor size of community greatly influenced how regions responded to Darwinism, although the smaller the community the more likely it was that individual personalities would dominate the debates. Institutional maturity also seems to have made some difference. In Canada and Australia, for example, where nonevolutionists frequently continued to occupy scientific chairs established before 1859, evolution entered the universities relatively slowly. In New Zealand, in contrast, which did not establish a university until 1869, evolutionists often occupied scientific chairs from their establishment. This made it easier in principle for evolution to gain a foothold. However, on occasion concern about the vulnerability of youthful institutions led their leaders to shy away from involvement in possibly damaging Darwinian debates.

Local environments, both physical and social, seem to have colored responses to evolution. In Canada, for example, the settlers' struggle to survive in a harsh physical environment predisposed some to see a measure of plausibility in a Darwinian view of nature. The New Zealand environment, though temperate in climate, likewise contributed to a positive view of the doctrine of survival of the fittest. There the main threat came from the social environment, in particular from the indigenous Maori, who stood in the way of white expansion. Faced with this obstacle, some settlers employed evolution as an ideological weapon in their struggle against the Maori.

In the field of Darwinian studies, few topics have received more attention than the responses of the religious. Jon H. Roberts' *Darwinism and the Divine in America* (1988), David N. Livingstone's *Darwin's Forgotten Defenders* (1987), and James R. Moore's *Post-Darwinian Controversies* (1979), to name only three of the most important recent studies, represent only a small fraction of the large body of literature on this subject.[2] Yet the extent to which such factors as geographical location and denominational affiliation made a difference in responding to Darwinism remains unclear. In addition, the responses of Catholics and Jews have remained comparatively unexplored.

In contrast to historians who have seen theological interests as central in determining the responses of the religious, Livingstone stresses the significance of geographical locality. He argues that local conditions noticeably affected the ways in which orthodox Calvinists (mostly Presbyterians) in Princeton, New Jersey; Belfast, Northern Ireland; and Edinburgh, Scotland reacted to Darwinism. For example, John Tyndall's notorious attack on Christianity in a Belfast speech in 1874 tended to sour northern Irish Presbyterians on Darwinism, whereas the absence of such frontal attacks in Princeton left their theologically similar brethren across the Atlantic psychologically less hostile to evolutionary claims. Similarly, John Stenhouse suggests that the introduction of evolution in Dunedin, New Zealand, by Anglicans and Methodists contributed to the somewhat jaundiced response by their Presbyterian rivals.

In evaluating American Protestant responses to Darwinism, Jon H. Roberts emphasizes the ways in which epistemological issues influenced attitudes toward Darwinism. In particular, he stresses the importance of biblical (as opposed to natural theological) concerns among opponents of evolution, even before the fundamentalist controversies of the 1920s. He also downplays the importance of denominational labels, arguing that within mainline Protestantism denominational allegiances have little predictive value in identifying the degree of support for Darwinism – and have no apparent correlation with the mechanism of evolution embraced. Only on the margins of Protestantism – in some Holiness, Lutheran, and Adventist groups, for example – did denominational identity strongly influence attitudes toward evolution. Roberts urges caution in attaching too much weight to such analytical categories as social interest, cultural strain, and geography in explaining opposition to evolution. Although the American South produced fewer Protestant proponents of evolution than the North, the differences between the two regions were less striking than is sometimes alleged.

Ronald L. Numbers and Lester D. Stephens drive home this point in their reexamination of Darwinism in the American South. In contrast to the conventional view of the South as uniquely hostile to evolution, they highlight the extent to which evolution gained a foothold in the region, in both churches and schools. In fact, the outburst of antievolution sentiment in the 1920s sprang to a considerable degree from the extent to which Darwinism had penetrated southern thinking. Even during the height of the movement to ban the teaching of evolution, a majority of southern states repelled efforts to restrict the teaching of evolution in public schools. Significantly, southern opponents of Darwinism, like many critics elsewhere, focused their efforts on curtailing the teaching of *human* evolution rather than organic evolution generally.

The essays on American Catholic and Jewish responses to Darwinism, by R. Scott Appleby and Marc Swetlitz, respectively, illustrate the ways in which both religious identity and national context made a difference. During the late nineteenth century, neither the Catholic nor the Jewish communities in America produced many scientists who could serve as mediators between scientific and religious claims, and both communities were handicapped in participating in the discussion because of their relatively large numbers of non-English-speaking members. Nineteenth-century Catholic intellectuals also lived under the shadow of the Vatican's negative attitude toward developmental theories; thus in the United States (as well as in French Canada, as Zeller suggests), they found it more difficult openly to come to terms with evolution than their Protestant counterparts. The comparatively few Catholics who embraced evolution favored mechanisms that could be harmonized with natural and revealed theology, insisted on the special creation of the human soul, and stressed the correspondence of their views with those of the church fathers. In the United States, Catholic theistic evolutionists freely advanced their views until the mid-1890s, when officials in Rome, fearing that the accommodation of religious beliefs to American culture was progressing too far, tried to silence them.

In the Jewish community, too, the debate over Darwinism occurred as part of a larger dispute between traditionalists and reformers, who often appealed to evolution as a justification for their religious agenda. Although Jewish arguments for and against evolution often paralleled those of Protestants, Jewish rabbis who accepted evolution did not follow the liberal Protestant clergy, who reformulated their theology to emphasize God's immanence.

Two other factors – race and gender – have also largely escaped the attention of historians of science. Although such works as John S. Haller's *Outcasts from Evolution* (1971) have looked at the ways that scientists viewed persons of African descent in the context of evolution, the opinions of African Americans themselves have gone virtually unnoticed, except for passing mention in such books as Alfred A. Moss's *The American Negro Academy* (1981).[3] The same is true for women. Historians and philosophers of science have paid increasing attention to the implications of evolutionary theories for women, but few of them have focused on the ways in which women themselves responded to Darwinism. All we have are bits and pieces of the story found in such works as Nancy G. Slack's essay on nineteenth-century women botanists (1987), Deborah Jean Warner's biography of Graceanna Lewis (1979), and Evelleen Richards' chapter on "Huxley and Woman's Place in Science" (1989).[4]

In his essay on African-American responses to organic evolution, Eric Anderson finds little evidence that black intellectuals in America linked racism to Darwinism. The failure of such African Americans to denounce Darwinism for giving racism scientific legitimacy leads him to question the extent to which evolution served as "the chief scientific authority for racists." (In examining the attitudes of white southerners toward evolution, Numbers and Stephens find few references to racial matters.) Black Americans, Anderson points out, worried far more about the implications of pre-Darwinian polygenism and about predictions of Negro extinction than about evolutionism, and they opposed Darwinism more for religious than for racial reasons. Although racists occasionally commandeered Darwinism to reinforce their ideologies, they possessed ample scientific ammunition without dragging evolution into the argument. Because many of the most virulent racists were antievolutionists, they could scarcely appeal to a theory they rejected. John Stenhouse shows that the Maoris of New Zealand reacted to Darwinism in ways similar to those of black Americans. Neither group saw evolution as a massive ideological threat, in part because they faced far more pressing concerns, such as sheer survival.

Sally Gregory Kohlstedt and Mark R. Jorgensen examine how educated women reacted to Darwin's emphasis on differences between the sexes and how his theories influenced the "irrepressible woman question." Although at first there were few women scientists in a position to challenge Darwin's implied defense of Victorian gender roles, and fewer still who wished to do so, some nineteenth-century feminists, such as Antoinette Brown Blackwell, repudiated Darwinian assumptions about women's physical and mental inferiority, whereas others, such as Charlotte Perkins Gilman, chose to stress the ways in which evolution could benefit women. By the turn of the century a handful of women scientists were reexamining the evidence for intellectual differences between the sexes and finding much of it wanting. "Their research," conclude Kohlstedt and Jorgensen, openly challenged "Darwinian notions of 'woman's nature'" and established "empirical standards for testing individual and sex differences."

In the ways mentioned here, the essays in this volume contribute to refining and extending our knowledge of Darwin's reception in the English-speaking world. Some of them treat topics that historians have previously ignored; others challenge prevailing views. Together, they not only demonstrate how deeply Darwinism penetrated diverse English-language cultures but illuminate the great variety of ways in which place, race, religion, and gender modified Darwin's reception.

Notes

1 Thomas F. Glick, ed., *The Comparative Reception of Darwinism* (Austin: University of Texas Press, 1974); David Kohn, ed., *The Darwinian Heritage* (Princeton, N.J.: Princeton University Press, 1985), Part 3: "Towards the Comparative Reception of Darwinism." In 1988 the University of Chicago Press brought out a new edition of the Glick collection. In 1984 the philosopher of science David L. Hull noted that no one had yet demonstrated a correlation "between the reception of Darwin's theory around the world and the larger characteristics of these societies"; see "Evolutionary Thinking Observed," *Science*, 223 (1984):923–24, quotation on p. 923.

2 James R. Moore, *The Post-Darwinian Controversies: A Study of the Protestant Struggle to Come to Terms with Darwin in Great Britain and America, 1870–1900* (Cambridge: Cambridge University Press, 1979); David N. Livingstone,

Darwin's Forgotten Defenders: The Encounter between Evangelical Theology and Evolutionary Thought (Grand Rapids, Mich.: Eerdmans, 1987); and Jon H. Roberts, *Darwinism and the Divine in America: Protestant Intellectuals and Organic Evolution, 1859–1900* (Madison: University of Wisconsin Press, 1988).

3 John S. Haller, Jr., *Outcasts from Evolution: Scientific Attitudes of Racial Inferiority, 1859–1900* (Urbana: University of Illinois Press, 1971); and Alfred A. Moss, Jr., *The American Negro Academy: Voice of the Talented Tenth* (Baton Rouge: Louisiana State University Press, 1981).

4 Nancy G. Slack, "Nineteenth-Century American Women Botanists: Wives, Widows, and Work," in *Uneasy Careers and Intimate Lives: Women in Science, 1789–1979* (New Brunswick, N.J.: Rutgers University Press, 1987), pp. 77–103; Deborah Jean Warner, *Graceanna Lewis: Scientist and Humanitarian* (Washington, D.C.: Smithsonian Institution, 1979); and Evelleen Richards, "Huxley and Woman's Place in Science: The 'Woman Question' and the Control of Victorian Anthropology," in *History, Humanity and Evolution: Essays for John C. Greene*, ed. Moore (Cambridge: Cambridge University Press, 1989), pp. 253–84. See also Rosaleen Love, "Darwinism and Feminism: The 'Woman Question' in the Life and Work of Olive Schreiner and Charlotte Perkins Gilman," in *The Wider Domain of Evolutionary Thought*, eds. D. R. Oldroyd and I. Langham (Dordrecht: D. Reidel, 1983), pp. 113–31.

1

Science, region, and religion: the reception of Darwinism in Princeton, Belfast, and Edinburgh

DAVID N. LIVINGSTONE

In recent years there has been a remarkable "spatial turn" among students of society and culture. The genealogy of this twist of events is both multifaceted and complex. Among philosophers, social theorists, and historians of science there has been a renewed emphasis on the significance of the local, the specific, the situated. Some philosophers thus argue that what passes as a good reason for believing a claim is different from time to time, and from place to place. Rationality, it turns out, is in large measure situation specific, such that what counts as rational is contingent on the context within which people are located.[1] Good grounds for holding a certain belief are evidently different for a twelfth-century milkmaid, a Renaissance alchemist, and a twentieth-century astrophysicist. Among social theorists there has also been a recovery of spatiality. The importance of the diverse locales within which social life is played out has assumed considerable significance with such writers as Clifford Geertz, Erving Goffman, and Anthony Giddens. In Geertz's telling, for example, law turns out not to be ecumenical but local knowledge – local in terms of place, time, class, issue, and what he terms "accent."[2] For Goffman, the situations facilitating human assemblages – gatherings, social occasions, informal encounters, and so on – furnish agents with those repertoires of structural meaning that they draw upon to constitute communication.[3] In Giddens's case it is because of the routinization of everyday life that he sees human agents as transacting their affairs in a variety of locales – settings of interaction which are themselves frequently zoned to facilitate routine social practices. As he puts it, "space is not an empty dimension along which social

I am greatly indebted to the Rev. William O. Harris, archivist at the Speer Library, Princeton Theologial Seminary, and to Dr. Bradley Gundlach for much assistance with archival queries when I visited the Speer Library. Their generosity and help are deeply appreciated. A modified version of this paper also appears in David N. Livingstone, D. G. Hart, and Mark A. Noll, eds., *Evangelicals and Science in Historical Perspective* (New York: Oxford University Press, 1999).

groupings become structured, but has to be considered in terms of its involvement in the constitution of systems of interaction."[4]

These writers certainly do not exhaust the range of sources contributing to this spatial resurgence. We might, for example, canvas the work of figures such as Michel Foucault and Edward Said, who deploy spatial categories for rather different purposes; but survey is not my intention here. Rather it is to alert us to an increasing acknowledgment of the spatial in cultural life – an awareness that is being increasingly recognized among historians of science.[5] Thus, attention has been called to the role of experimental space in the production of scientific knowledge, the significance of the uneven distribution of scientific information, the diffusion tracks along which scientific ideas and their associated instrumental gadgetry migrate, the management of laboratory space, the power relations exhibited in the transmission of scientific lore from specialist space to public place, the political geography and social topography of scientific subcultures, and the institutionalization and policing of the sites in which the reproduction of scientific cultures is effected. The cumulative effect of these investigations is to draw attention to the local, regional, and national features of science – an enterprise hitherto regarded as prototypically universal.

The implications of these recent moves for my present task are of considerable dimensions. My suspicion is that the project of reconstructing the historical relations between science and religion might similarly benefit from a localizing strategy that seeks to situate responses or encounters in their respective socio-spatial settings. To pursue such a program, I suggest, will inevitably mean abandoning grand narratives that trade in abstract and idealist "isms." It will not do to speak of *the* encounter between evangelicalism and evolutionism or Calvinism and Darwinism. Instead I think we will be better advised to seek to uncover how particular religious communities, in particular space-time settings, developed particular tactics for coping with particular evolutionary theses. Here I want to make a preliminary stab at elucidating the responses of certain conservative Christians to evolutionary claims in a number of different locales during the second half of the nineteenth century. If my argument is in the neighborhood of a correct analysis, the specifics of these situations turn out to be of crucial importance.

1874: Three statements

During 1874, in three different cities, Presbyterians with seriously similar theological commitments issued their judgments on the theories of evolution that were gaining ascendancy in the English-speaking intellectual world. In Edinburgh, Belfast, and Princeton, differing assessments of the new biology were to be heard; in these different situations, varying rhetorical stances were adopted; for in these religious spaces, different circumstances prevailed.

In October of 1874 Robert Rainy, the new principal of New College, Edinburgh – the theological college of the Free Church – delivered his inaugural address. His subject was "Evolution and Theology" and, according to his biographer Simpson, it "attracted considerable attention." "The religious mind of the day," according to Simpson, "was disturbed about Darwinism and apprehensive lest it should affect the foundations of faith; and that a man of Dr. Rainy's known piety and orthodoxy should, from the Principal's chair of the New College, frankly accept the legitimacy of the application of evolution even to man's descent and find it a point on which the theologian 'may be perfectly at ease' reassured many minds."[6] Indeed, Rainy did make it clear that while some found evolution objectionable, he himself did not feel justified "in imputing an irreligious position to the Evolutionist." Accordingly, he insisted that even if "the evolution of all animal life in the world shall be shown to be due to the gradual action of permanent forces and properties of matter" – a claim he himself actually doubted – that would have no bearing on "the argument of the Theist [or on] the mind of a reverent spectator of nature."[7]

Methodologically, Rainy was prepared to allow considerable autonomy to the scientific enterprise, believing that it was "often in its right in keeping to its own path" irrespective of interpretations of scripture. But this certainly did not mean that Christians were never justified in "contesting the ground." To the contrary, evolution could be mobilized for infidel purposes; these simply had to be challenged. As an illustration, he referred to the atomic theory of Democritus – an example which, as we shall see, had a particular poignancy in the winter of 1874. And yet, for all that, Rainy was remarkably sanguine about evolutionary speculation, even to the extent of welcoming evolutionary interrogations of human development. Not that he ruled out divine intervention in human origins; he certainly insisted on "direct Divine interposition."[8] But by decorporealizing the *imago dei*, he found it possible to liberate Christian anthropology from detailed questions over similarities in the human and anthropoid skeletons. Thereby the door was opened to an evolutionary account of the human physical form.

That same winter in Belfast, J. L. Porter, professor of biblical criticism in the General Assembly's College (and later president of Queen's College), delivered the opening address to the Presbyterian faculty and students.[9] Here he ominously spoke of the "evil tendencies of recent scientific theories" – and that of evolution in particular – which threatened to "quench every virtuous thought." The need for theological colleges was thus more urgent than ever, so that "heavenly light is preserved and cherished." What was more, he declared that he was "prepared to show that not a single scientific fact has ever been established" from which the pernicious dogmas of Huxley and Tyndall could be "logically deduced."[10] Within a few weeks, on the last day of November, Porter would pursue this same theme in an address on "Science and Revelation: Their Distinctive Provinces," which inaugurated a series of winter lectures on Science and Religion in Rosemary Street Church in downtown Belfast. The need for a clear-cut boundary line – both in terms of content and methodology – between the provinces of science and theology were of the utmost importance to Porter, and while he was happy to insist that "no theological dogma can annul a fact of science," the question of "crude theories and wild speculations" – in which evolutionists were all too prone to engage – was a different matter. As for Darwin himself, *The Origin of Species* was described as having made empirically "one of the most important contributions to modern science"; in logic, by contrast, it was "an utter failure." The facts, in other words, were welcome; the theory was alien. The problem was that the latter was utterly unsupported by the former. In sum, the book was "not scientific." Darwin was not to be substituted for Paley.[11] In the key Calvinist spaces of Belfast and Edinburgh, different attitudes to evolution theory were already being promulgated.

Earlier in May of that same year, Charles Hodge, arguably the most influential theologian at Princeton Theological Seminary during the first century of its existence, had published his last work, *What Is Darwinism?* It contained Hodge's considered treatment of the Darwinian theory and can appropriately be regarded an extended exercise in definition. Thereby Hodge believed its nature could be ascertained and the lineaments of an appropriate Christian response plotted. Because he was certain that Darwin's use of the word "natural" was "antithetical to supernatural," Hodge insisted that "in using the expression Natural Selection, Mr. Darwin intends to exclude design, or final causes." Here the very essence of the theory lay exposed. That "this natural selection is without design, being conducted by unintelligent physical causes," Hodge explained, was "by far the most important and only distinctive element of his theory." In a nutshell, the denial of design was the very "life and soul of his system" and the single feature that brought "it into conflict not only with Christianity, but with the fundamental principles of natural religion."[12] By this definitional move, Hodge could set the terms of the debate and adjudicate on who was or was not a Darwinian. It plainly meant that those such as Asa Gray who considered themselves Christian Darwinians were either mistaken or just plain mixed up; that label had no meaning. Thus for all his efforts to teleologize Darwinism,

Gray simply was "not a Darwinian."[13] That he was a Christian *evolutionist*, Hodge had no doubt; but that was a different matter. For to Hodge, Darwinism simply was atheism.

Three situations

These key pronouncements, of course, were not elaborated in a vacuum; rather they were broadcast in differing ideological contexts. My suggestion is that in these diverse arenas different issues were facing Presbyterian communities, and that these were crucially significant in conditioning the rhetorical stances that were adopted by a variety of religious commentators on evolutionary theory. Besides this, different voices were being sounded in different ways, and their modes of expression, whether bellicose or irenic, did much to set the tone of the local science-religion encounter. I suggest that such prevailing circumstances – and no doubt there are many more – had an important influence on the style of language that was available to theologically conservative spokesmen pronouncing on evolution, which in turn determined not only what could be *said*, but what could be *heard*, about evolution in these three different localities.

In Edinburgh Robert Rainy attempted to "retain the evangelical heritage of the F[ree] C[hurch] while keeping pace with rapid contemporary developments" – a strategy that increasingly made him a controversial figure. To much of the secular press he became known as "Dr. Misty as well as Dr. Rainy," even though he was named by Gladstone as the greatest living Scotsman. Later in his career as an ecclesiastical statesman and architect of the union of the Free Church and the United Presbyterian Church, he found it necessary to "offer unlimited hospitality to biblical criticism." And if in the case of William Robertson Smith he judged it politic to sacrifice that particular critic, he later protected Marcus Dods, A. B. Bruce, and George Adam Smith in a sequence of heresy trials – judgments which secured the later censure of those retaining the classical Calvinism of the Free Church.[14] Certainly Rainy's forté was ecclesiastical polity rather than intellectual engagement. But what *is* significant is that because, as Drummond and Bulloch put it, he "was circumspect and heedful of public reactions," his endorsement of an evolutionary reading of human descent at New College in 1874 is indicative of a general lack of anxiety about evolution among Scottish Calvinists.[15]

In fact, during the 1870s in Edinburgh, the Darwinian issue paled in significance beside two other intellectual currents assaulting the orthodox Scottish mind. To be sure it was, as Simpson writes, a "period when a somewhat scornful materialism was asserting itself – it was the time of Tyndall's famous British Association speech about 'the promise and potency of matter'," but in Simpson's judgment these issues did not "greatly affect Scottish thought."[16] Of much greater moment was the matter of biblical criticism and the influence of German idealist philosophy.[17] As for the former, matters came to a head in 1876 when the protracted heresy trial concerning William Robertson Smith began; this was litigation which would, in due course, result in his dismissal from the Old Testament chair at the Free Church college in Aberdeen. What had sparked off the matter was Smith's contribution to the ninth edition of the *Encyclopaedia Britannica* – the "canonical expression," according to Alasdair MacIntyre, of the Enlightenment thrust of contemporary Edinburgh high culture.[18] Smith's entries on *Angel* and *Bible* revealed his espousal of the Graf–Wellhausen theory of the Pentateuchal documents.[19] And this, together with the ninth edition's progressivist ethos, threatened to subvert the traditional authority of scripture by promising to provide an architectonic overview of knowledge.

Suspicions about Robertson Smith's orthodoxy gravitated around such issues as his querying of the Mosaic authorship of the Pentateuch and the historicity of Deuteronomy, his understanding of biblical inspiration, and his account of Old Testament prophecy. Initially the General Assembly of 1878 acquitted him of the charges, but within a short space of time he found himself arraigned again on the deuteronomic issue; though admonished by the Moderator, he was not subject to any discipline. A few days after the end of the latest trial, however, his piece for the *Encyclopaedia Britannica* on "Hebrew Language and Literature" appeared in print. While it had obviously been prepared long before the last meeting of the General Assembly, its point of publication gave every impression of a flaunting of ecclesiastical authority. In consequence he was deposed from his chair at Aberdeen, on the recommendation of Robert Rainy, and though his position as a minister of the Free Church was not affected, he refused to accept a salary from that position.[20]

In due course the polymathic Robertson Smith would assume general editorship of the *Encyclopaedia Britannica* and eventually occupy the chair of Arabic at the University of Cambridge. Indeed he had already impressively used Boolean mathematical logic to challenge recent efforts to arithmetize geometry, and later would produce an immensely influential speculative account of *The Religion of the Semites* (1889) in which he urged that a primitive sense of tribal unity was articulated in a ceremonial meal provided for through ritual cannibalism. Revitalization of the tribe's sense of belonging, he believed, was secured – as George Davie pungently expresses it – through "eating the gobbets of throbbing flesh, newly-killed, of their fellow-tribesmen."[21] What with this, and

J. F. Maclennan's extraordinary account in *Ancient Marriage* (1865) of the matriarchal and polyandric origins of civilization, ultimately rooted in the unintended consequences of female infanticide, there seems to have been enough sex and violence among radical Calvinists in Scotland to satisfy even the likes of Freud, though, if Davie is to be believed, even Freud, despite his acknowledged indebtedness to Robertson Smith, found cannibalistic nostalgia just a bit too much. Such theories, as Davie further puts it, "made clear to the world that the Calvinist civilization of Scotland, far from averting its eyes from the 'facts of life,' could be ultra-realistic in regard to the role of sex in history."[22]

In an environment where biblical criticism, speculative prehistory, and idealist philosophy were considered much more threatening than science – after all Lord Kelvin was an Episcopalian of Low Church sympathies who regularly worshipped in a Free Church congregation – the need for a rational defense of Christianity was felt by those such as Lord Gifford, who endowed the celebrated series of lectures on natural theology that bears his name. In many ways this was an attempt to out-rationalize the rationalists; thereby the Gifford lecturers deployed the very tools of those whom they sought to outmaneuver.[23] If intellectual energies *were* to be deployed by conservative Christians in the combating of the new infidels, these were the arenas in which engagement was seen to be required.

The situation was different in Belfast. On Wednesday, 19 August 1874 the *Northern Whig* enthusiastically announced the coming of the Parliament of Science – the British Association (BA) – to Belfast. Ironically, the meeting was being welcomed to the city as a temporary respite from "spinning and weaving, and Orange riots, and ecclesiastical squabbles." Nevertheless, "some hot discussions" were predicted "in the biological section," between advocates of human evolution and those "intellectual people – not to speak of religious people at all – who believe there is a gulf between man and gorilla."[24] The Belfast meeting was to be what Moore and Desmond call "an X Club jamboree," with its priestly coterie of Huxley, Hooker, Lubbock, and of course Tyndall himself, all speechifying.[25] If indeed an assault was to be mounted by the new scientific priesthood on the old clerical guardians of revelation and respectability, scripture and social status, then what better venue could there be for a call to arms than the BA's meeting in Calvinist Belfast? Tyndall's pugnacious performance did not fall short of expectations. In an Ulster Hall garnished with accompanying orchestra, he delivered – with nothing short of evangelical fervor – a missionary call to "wrest from theology the entire domain of cosmological theory." His conclusion was that all "religious theories, schemes and systems which embrace notions of cosmogony ... must ... submit to the control of science, and relinquish all thought of controlling it."[26] The gauntlet had been thrown down.

Events moved quickly. On Sunday, 23 August, Tyndall's address was the subject of a truculent attack by Rev. Professor Robert Watts at Fisherwick Church in downtown Belfast. Watts, the Assembly's College Professor of Systematic Theology – and an erstwhile student of Princeton Seminary in the United States – had good reason for spitting blood. He had already submitted to the organizers of the Biology Section of the Belfast British Association meeting a paper congenially entitled "An Irenicum; or, A Plea for Peace and Co-operation between Science and Theology."[27] They flatly rejected it.[28] It must have seemed to Watts that the scientific fraternity was not interested in peace. As for the spurned lecture, which Watts – not prepared to waste good words already committed to paper – delivered at noon the following Monday in Elmwood Presbyterian Church, it revealed just how enthusiastic he could be about science.[29] But the BA's rebuff *had* stung, and chagrin over his expulsion from the program put him in a bad mood. Yet this was nothing compared to the anger that Tyndall's address aroused in him. So when on the Sunday following the infamous address he turned his big guns on Tyndall in a sermon preached to an overflowing evening congregation at Fisherwick Place Church in the center of Belfast, the irenic tone of the rejected paper was gone. Tyndall's mention of Epicurus was especially galling; that name had "become a synonym for sensualist," and Watts balked at the moral implications of adopting Epicurean values. To him it was a system that had "wrought the ruin of the communities and individuals who have acted out its principles in the past; and if the people of Belfast substitute it for the holy religion of the Son of God, and practise its degrading dogmas, the moral destiny of the metropolis of Ulster may easily be forecast."[30]

Watts, of course, was not a lone voice imprecating Tyndall-style science. On the very same Sunday that he was arraigning atomism before the downtown congregation of Fisherwick Place, the same message was buzzing through the ears of other congregations.[31] That evening none other than W. Robertson Smith found himself preaching from Killen's North Belfast pulpit. And while he did not speak out there and then on Tyndall's speech, he had certainly caught the prevailing mood, for he was moved within the week to write to the press challenging Tyndall's assertions about early religious history, castigating his "pragmatic sketch of the history of atomism," and complaining that he was "at least a century behind the present state of scholarship" concerning the Christian Middle Ages.[32] Small wonder that Tyndall reflected: "Every pulpit in Belfast thundered of me."[33] It was the BA event that set the agenda for Porter's opening speech at the Assembly's College that winter, and for the Belfast response to evolution for a generation.

In Princeton, things were different yet again. Of crucial importance, I think, was the fact that Hodge's powerful voice was not the only one to be heard. The judgments of James McCosh, president of the College of New Jersey since 1868, were also audible. In fact, as Gundlach has convincingly shown, there was little real difference of substance between Hodge and McCosh on the Darwinian question.[34] Both were determined to resist any removal of design from the natural order; teleology, not transmutation, was the key theological issue for both. But there *was* a difference in rhetorical style such as to secure McCosh's reputation as the foremost Protestant reconciler of theology and Darwinism in the New World. Thus at the 1873 New York meeting of the Evangelical Alliance, McCosh told his hearers that, instead of denouncing the theory of evolution, "religious philosophers might be more profitably employed in showing . . . the religious aspects of the doctrine of development; and some would be grateful to any who would help them to keep their new faith in science."[35] In later years he could produce such works as *The Religious Aspect of Evolution* – a title, I suspect, that Hodge would never have adopted.

McCosh, as I have implied, was every bit as unnerved by the naturalism of the Darwinian vision as Charles Hodge, and thus devoted considerable intellectual energy to the task of producing an evolutionary teleology. But by lapsing into a Lamarckian-style evolutionism that conceived of organic progress along predetermined paths, he was able to couch his responses in terms of a sympathetic rereading of the evolutionary story and thereby to do what he could to avert what one of his students, Rev. George Macloskie, called a potentially "disastrous war between science and faith."[36] It further enabled him, when reminiscing on twenty years as Princeton's president, to confess that much of his time had been devoted to "defending Evolution, but, in so doing, have given the proper account of it as the method of God's procedure, and find that when so understood it is in no way inconsistent with Scripture."[37]

It was back in Belfast, however, that McCosh had learned how to use words wisely – and on a rather different issue. During the momentous events of 1859 he found it possible to query the Ulster revival's "physiological accidents." But – and this is crucial – by casting his diagnosis in the language of defense he could insist that the "deep mental feeling" that induced somatic convulsions was itself "a work of God."[38] Thus he came to be seen as the key theological apologist for revival. Precisely the same was true of his evolutionary stance. Remote from the Tyndall episode, McCosh was free to maneuver his theology around science.[39] Like his Belfast colleagues, McCosh too had drunk deeply at the waters of Scottish common-sense philosophy. But whereas Tyndall's naturalism predisposed them to marshall its inductive principles in opposition to evolution, McCosh could mobilize William Hamilton's neo-Kantian reformulations to construct a rather more idealist response. By renegotiating natural theology in a less Paleyan way, he found resources to accommodate biological transformism.[40] So, whether one spoke of divine visitation or Darwinian vision, rhetorical nuance counted for much, cognitive content for little. For it was style of communication rather than substance of argument that secured history's judgment on McCosh as the defender of religious revival *and* evolutionary theory. By keeping the rhetorical space of Princeton open to the possibility of evolution theory – whatever his reservations about Darwinian naturalism – he did much to determine that the theory would at least be tolerated at American Presbyterianism's intellectual heartland.

Three stories

Given the different sets of circumstances prevailing in Edinburgh, Belfast, and Princeton, it is now understandable why the subsequent histories of conservative Christian responses to evolution were different in these Calvinist cities. In a nutshell, the theory of evolution was absorbed in Edinburgh, repudiated in Belfast, and tolerated in Princeton. This, I hope, can be illustrated by a brief survey of the judgments issued by a number of key individuals in these different places.

Henry Calderwood was a lifelong member of the United Presbyterian Church.[41] During the latter part of the nineteenth century, this Presbyterian body – which in 1900 joined with the Free Church to form the United Free Church – had increasingly adopted aspects of liberal theology which frequently issued in a relatively radical political outlook. According to S. D. Gill, "while other Presbyterian Churches in Scotland were being rocked by the issue of higher criticism the UPC remained comparatively unaffected."[42] Calderwood himself, however, remained substantially evangelical in his outlook, supporting the evangelistic activities of Moody and Sankey in Edinburgh, and retaining the modified Calvinism (which, according to some, demonstrated Arminian infiltration[43]) embodied in the United Presbyterian's Declaratory Act of 1879 with which he was closely associated.[44] Besides this, his staunch defense of Scottish common-sense philosophy enabled him to maintain philosophical continuity with the tradition of Scottish realism. This stance had involved him in a critique of his teacher William Hamilton's idealist inclinations. As Pringle-Pattison put it, "Calderwood may be said to have been the first to reassert ... the traditional doctrine of Scottish philosophy against the agnostic elements of Kantianism which

Hamilton had woven into his theory."[45] This early work,[46] which appeared in 1854, established his philosophical reputation and, after serving as a United Presbyterian minister in Glasgow from 1861 until 1868, he was appointed to the chair of moral philosophy in Edinburgh in 1868. Significantly, one of the first letters of congratulation was received from James McCosh at Princeton, who wrote enthusiastically of his appointment but also spoke of "a dark thread running through the white of my feeling ... I looked to you as my successor, and had laid plans to secure it. You would have had the support of the College and of my old pupils in the Presbyterian Church. As it is, I am now fairly at sea."[47] Clearly McCosh had hoped that Calderwood would follow in his own footsteps and cross the Irish sea to assume the professorship of logic and metaphysics at the Queen's College of Belfast.

In the case of Calderwood (and indeed McCosh), the commonly held view that common-sense philosophy was typically deployed against evolutionary theory certainly does not seem to hold good. In 1877 Calderwood took up the issue of ethical evolution in *The Contemporary Review*. Here he left to one side all questions of evolution as a strictly biological theory, and turned to its use as an account of moral development. For this is what Calderwood found objectionable – the attempt to read human values through the spectacles of natural selection and heredity. And so, using the example of human altruism, he argued for evolution's inability to account for self-denial without compromising the very principles of selection and competition intrinsic to the theory.[48] For all that, he was far from denying the evolutionary process and, like Rainy in the Free Church, certainly did not oppose it as a bona fide scientific theory. In 1881, for example, he observed:

> biological science must be credited with the new departure which has given fresh stimulus to thought, and has succeeded in gaining wide support to a theory of Evolution. To Mr. Darwin, as a naturalist, belongs the indisputable honour of having done service so great as to make its exact amount difficult of calculation, involving, as it does, vast increase of our knowledge of the relations and inter-dependence of various orders of animate existence. "Origin of species" may not be the phrase which exactly describes the sphere of elucidation, over the extent of which all students of Nature now acknowledge indebtedness to Mr. Darwin; but "development of species" does express an undoubted scientific conclusion, which has found a permanent place in biological science.[49]

That same year Calderwood delivered the Morse lectures at the Union Theological Seminary in New York on the subject "The Relation of Science and Religion." Here he insisted that even if "the theory of the

Development of Species by Natural Selection ... were accepted in the form in which it is at present propounded, not only would the rational basis for belief in the Divine existence and government not be affected by it, but the demand on a Sovereign Intelligence would be intensified." Again, he assured his hearers that Darwin's theory "is no more at variance with religious thought, than with ordinary notions of preceding times" and that "the fewer the primordial forms to which the multiplicity of existing species can be traced, the greater is the marvel which science presents, and the more convincing becomes the intellectual necessity by which we travel back to a Supernatural intelligence as the source of all."[50] And again:

> The scientific conception of the history of animal life is, that there has been a historical progression in the appearance of animals, in so far as lower orders took precedence of higher, while the higher have shown large power of adaptation to the circumstances in which they have been placed. In accordance with the whole principles regulating the relations of religion and science, religious men, scientific and non-scientific, will readily acquiesce in this modification of general belief, as largely favored by evidence which geology supplies, and supported by testimony drawn from the actually existing order of things; and they will do so with clear recognition that this view involves no conflict with scriptural statement, and in so far from containing in it any thing antagonistic to the fundamental conception of the supernatural origin of existence, that it harmonizes with it, even intensifying the demand upon a transcendent cause for the rational explanation of the admitted order of things.[51]

Not surprisingly, when Calderwood participated in the 1884 third General Council of the Alliance of Reformed Churches in Belfast, he steadfastly supported George Matheson's presentation on "The Religious Bearings of the Doctrine of Evolution" and urged that Darwin's theory afforded an even grander conception of the designer than had hitherto been glimpsed.[52] Such a perspective was not exactly shared by his Belfast colleagues present at the meeting!

Subsequently Calderwood delivered his mature reflections on the whole subject, first in a short article for the *Proceedings of the Royal Society of Edinburgh*,[53] and then at much greater length in *Evolution and Man's Place in Nature*. In two editions of this work – the second a more or less complete reworking – which came out in 1893 and 1896, respectively, he sought to exempt rational life from the reign of evolutionary law by arguing that "Animal Intelligence shows no effective preparation for Rational Intelligence."[54] Yet this did not prevent him from declaring that "Evolution stands before us as an impressive reality in the history of Nature,"[55] for he was convinced that the evidence for "descent with modifications" was "so abundant and varied, as to leave no longer any uncertainty around the conclusion that a steady advance in organic form and function has been achieved in our world's history."[56]

Now, to be sure, none of this can be construed as unqualified endorsement for the Darwinian scheme. Throughout, Calderwood was insistent on the need for creative activity in the *origin* of life and for constant providential superintendence. However, his rhetorical style was radically different from that of the Belfast Calvinists, who, as we shall presently see, maintained an attitude of "no compromise!" with the new biology.

I have been dwelling for some time on the work of Henry Calderwood, since my focus at this point is on Edinburgh, but it would be remiss to ignore altogether the contributions of men such as James Iverach and James Orr. By the time Iverach was appointed to a chair at the Free Church of Scotland's College in Aberdeen in 1887, he had already begun his assessment of Darwinism in *The Ethics of Evolution Examined* (1886); subsequently he would also publish *Evolution and Christianity* in 1894. His defense of Robertson Smith certainly condemned him in the eyes of more conservative elements in the Free Church. But his renowned intellectual stature and his strategic position in the Free Church, and its successor the United Free Church, meant that his acknowledgment of Darwin's theory of evolution by natural selection as a good working hypothesis – within limits – further contributed to obviating Scottish conflict between faith and science. As he put it, the conflict was "not between 'evolution' and what our friends call 'special creation.' It is between evolution under the guidance of intelligence and purpose, and evolution as a fortuitous result."[57]

Like Calderwood, James Orr's ecclesiastical home was the United Presbyterian Church. His major assessments of evolution appeared in print after he moved from Edinburgh to assume the chair of apologetics and systematic theology at the new United Free Church College in Glasgow in 1900.[58] Nevertheless, in the same year that he was appointed to the professorship of church history at the Divinity Hall in Edinburgh, he told the hearers of his Kerr Lectures in 1891 that "on the general hypothesis of evolution, as applied to the organic world, I have nothing to say, except that, within certain limits, it seems to me extremely probable, and supported by a large body of evidence."[59] Subsequently Orr confirmed this stance when he made it clear that there was no reason to dispute the "genetic derivation of one order or species from another,"[60] although, like Calderwood, he wanted to preserve a teleological metaphysics and an act of special creation in the origin of the human species.[61] Even in his contribution at the end of his life to the *Fundamentals*, published between 1910 and 1915, he tellingly quipped that "'Evolution,' in short, is coming to be recognized as but a new name for 'creation,' only that the creative power now works from *within*, instead of, as in the old conception, in an *external*, plastic fashion."[62]

Also of considerable importance at this late Victorian moment was Robert Flint, who moved from the chair of moral philosophy in St. Andrews to assume the professorship of divinity in Edinburgh University in 1876. He had grown up in the Free Church but as a student became disenchanted with what he took to be its intellectual narrowness and was ordained to the Church of Scotland. To be sure, his rationalistic defense of traditional theism, reminiscent of the eighteenth century, might well be out of spirit with aspects of Scotland's evangelical traditions, but his capacity to deliver a natural theology of evolution would have done much to confirm the stance taken by such men as Calderwood and Orr. For Flint, with his passionate concern to establish the rationality and objectivity of belief in God, was convinced that neither evolution in general nor Darwinism in particular had weakened the design argument; the "law of heredity," the "tendency to definite variation," the "law of overproduction," the "law of sexual selection," and even the "law, or so-called law, of natural selection" could all be seen as the expression of divine purpose. All in all he was convinced, as he put it in his first series of Baird lectures in 1876, that "the researches and speculations of the Darwinians have left unshaken the design argument. I might have gone further if time had permitted, and proved that they had greatly enriched the argument."[63]

Two other figures also bear mentioning as contributing significantly to the Edinburgh ethos on Presbyterianism and science – George Matheson and Henry Drummond. In Matheson's case, Spencer's "Unknowable" was ultimately nothing less than Almighty God, and so he could happily tell the Presbyterians gathered for the 1884 congress in Belfast that if we "agree to call [Spencer's] Force an inscrutable or unsearchable Will, we shall have already established a scientific basis, not only for the belief in a guiding Providence, but for the possibility of an effiacious prayer."[64] Undeterred by local opposition, Matheson elaborated these themes in extenso the following year. In *Can the Old Faith Live with the New*?, he not only welcomed an evolutionary account of human origins but insisted that evolution also shed light on such Christian concerns as the workings of Providence, the operations of the Spirit, and human immortality.[65] The self-same concern to place Christian theology on a solid scientific footing was the driving force behind Henry Drummond's life and work. A professor of natural science at the Free Church College in Glasgow who gave weekend classes to the Edinburgh divinity students, and a close personal friend of D. L. Moody, Drummond set forth his "Spencerization" of evangelicalism in *Natural Law in the Spiritual World* and *The Ascent of Man*.[66] It amounted to an attempt to find a law of continuity connecting the material and spiritual worlds, and it was widely interpreted as an underwriting of faith by the latest science. In the long run its opiating strategy contributed significantly to what James Moore calls the naturalization of the supernatural.

By retrieving these proevolutionary sentiments among Edinburgh Calvinists, I do not mean to give the impression that absolutely no opposition was expressed toward the new biology. Consider, for example, the stance advocated by John Duns, the successor of John Fleming as the New College professor of natural science since 1864.[67] Even while Rainy, the principal of New College, was flirting with evolutionary theses, Duns maintained a firm attitude of opposition. Thus in 1877 he found the speculative "hypothesis of organic evolution" distinctly harmful to theology, and those who were finding "scope for the principle in Scripture, or in history, [had] generally very indistinct notions as to its full scientific import."[68] Later in his investigation of the connection between natural selection and the theory of design, he repudiated the efforts of those such as Asa Gray who insisted on the possibility of an evolutionary teleology.[69] For all that, however, I think it is clear that Duns is the exception that proves the rule.

In Belfast, the Tyndall event cast terror on the hearts of solid Presbyterians for more than a generation. But there was still a sense that Tyndall's act of aggression was ultimately to be welcomed, for it displayed to the world the machinations of the materialist school. It was precisely because he had spoken so plainly that the editor of *The Witness* (the weekly Presbyterian newspaper) could observe that "We now know exactly the state of matters, and what is to be expected from Professor Tyndall and his school, and we shall be able to take our measures accordingly."[70] And measures the Presbyterian Church did take. Plans were hastily laid for a course of evening lectures to be given at the Rosemary Street Church during the winter months on the relationship between Science and Christianity. In time these addresses would be drawn together into a book, distributed on both sides of the Atlantic, under the title *Science and Revelation: A Series of Lectures in Reply to the Theories of Tyndall, Huxley, Darwin, Spencer, Etc.*

During that 1874–75 winter of discontent, eight Presbyterian theologians and one scientist joined together to stem, from the Rosemary Street pulpit, any materialist tide that Tyndall's rhetoric might trigger. Just as the villagers of medieval Europe and colonial New England annually beat the bounds – marked out the village boundaries – so did the Presbyterian hierarchy need to reestablish its theological borders. Indeed, it is precisely for this reason, I would contend, that the winter lecture series at the Rosemary Street Church included ministerial addresses – by A. C. Murphy on proofs of the Bible as divine revelation, by John McNaughtan on the reality of sin, by John Moran on the life of Christ, and by William Magill on the divine origin of scripture – which made no specific reference to Tyndall, Darwin, Spencer, or any version of evolutionary theory. That they were included in the series attests to the perceived need to reaffirm systematically the cardinal doctrines of the faith so as to ensure that Presbyterian theological territory remained intact.

Just over a year earlier the Catholic serial, *The Irish Ecclesiastical Record*, had presented an evaluation of Darwinism in which its correspondent castigated the "Moloch of natural selection" for its "ruthless extermination of . . . unsuccessful competitors," for its lack of evidence for transitional forms, for its failure to account for gaps in the fossil record, and for its distasteful moral implications.[71] As for the latest clash in Belfast, the Catholic Archbishops and Bishops of Ireland issued a pastoral letter in November 1874 in which they repudiated the "blasphemy upon this Catholic nation" that had recently been uttered by the "professors of Materialism . . . under the name of Science." This certainly was no new warfare, they reflected; but the Catholic hierarchy perceived that this most recent incarnation of materialism unveiled more clearly than ever before "the moral and social doctrines that lurked in the gloomy recesses of [science's] speculative theories." Quite simply it meant that moral responsibility had been erased, that virtue and vice had become but "expressions of the same mechanical force," and that sin and holiness likewise vanished chimeralike into oblivion. Everything in human life from "sensual love" to religious sentiment were "all equally results of the play between organism and environment through countless ages of the past." Such was the brutalizing materialism that now confronted the Irish people.[72]

The congruence between these evaluations and those of the Presbyterian commentators we have scrutinized is certainly marked. So why then should Watts, in a subsequent reprint of his pamphlet on "Atomism: Dr. Tyndall's Atomic Theory of the Universe Examined and Refuted," incorporate in an appendix "strictures on the late Manifesto of the Roman Catholic hierarchy of Ireland in reference to the sphere of science"? That he wished to distance himself from their proposals is clear: it was "painful," he noted, "to observe the position taken by the Roman Catholic hierarchy of Ireland in their answer to Professors Tyndall and Huxley." So, just what was the problem? Simply this: Watts felt that the Catholic willingness to compartmentalize science and religion would lead to "the secularisation of the physical sciences." Because Cardinal Cullen and his co-religionists seemed only too willing to follow Newman in restricting the physicist to the bare empirical, Watts insisted that since "the Word of God enjoins it upon men as a duty to infer the invisible things of the Creator from the things that are made," the Roman Catholic hierarchy of Ireland had, by their declaration, "taken up an attitude of antagonism to that Word, prohibiting scientists, as such, from rising above the law to the infinitely wise, Almighty Lawgiver."[73]

The sectarian traditions in Irish religion doubtless had a key role to play in these particular machinations. The furor surrounding the Tyndall event merely became yet another occasion for Ulster nonconformity to uncover its sense of siege. In his confrontation with evolutionary theory, Watts wanted to cultivate and tend to his own tradition's theological space, and not engage in extramural affiliations. And by seeking to cast secularization and Catholicism as subversive allies against the inductive truths of science and the revealed truths of scripture, he found it possible to conflate as a single object of opprobrium the old enemy, popery, and the new enemy, evolution. To Watts these were indeed the enemies of God. Writing to A. A. Hodge in 1881, he observed: "Communism is, at present rampant in Ireland. The landlords are greatly to blame for their tyranny, but the present movement of the Land League, headed by Parnell, is essentially communistic. He and his co-conspirators are now on trial, but with six Roman Catholics on the jury there is not much likelihood of a conviction."[74] Later in an 1890 letter to Warfield, on the eve of Gladstone's second Home Rule Bill, in which he castigated both the Free Church and the United Presbyterians in Scotland, he added: "Both these churches are so bent on Disestablishment that they are quite willing to sustain Mr. Gladstone's Irish policy & deliver their Protestant brethren into the hands of the Church of Rome, to be ruled by her through a band of unmitigated villains."[75]

For their part, the Catholic hierarchy did not miss the opportunity of firing its own broadsides at Protestantism. The bishops and archbishops felt it would "not be amiss, in connection with the Irish National system of education, to call attention to the fact that the Materialists of to-day are able to boast that the doctrines which have brought most odium upon their school have been openly taught by a high Protestant dignitary." It only confirmed them in their uncompromising stance on the Catholic educational system. Had it not been for their own vigilance, an unbelieving tide would have swept through the entire curriculum. Such circumstances justified "to the full the determination of Catholic Ireland not to allow her young men to frequent Universities and Colleges where Science is made the vehicle of Materialism." Accordingly, the bishops and archbishops rebuked "the indifference of those who may be tempted to grow slack in the struggle for a Catholic system of education."[76] Tyndall's speech, it seems, succeeded not only in fostering the opposition of both Protestants and Catholics in

Ireland to his own science, but in furthering their antagonism to each other.

As for Watts, he maintained his antagonistic attitude to evolution for the rest of his life. At that 1884 pan-Presbyterian meeting in Belfast, even while Matheson and Calderwood were endorsing an evolutionary viewpoint, the redoubtable Watts told the company that evolution was a "mere hypothesis" with "not one single particle of evidence" to support it. The need for embarking on any doctrinal reconstruction – including the exegesis of Genesis – did not exactly commend itself to him. Indeed, he had long been campaigning, both in Belfast and Scotland, against Spencer's system because to him it was explicitly designed "to overthrow the Scripture doctrine of special creations," as he had realized in *An Examination of Herbert Spencer's Biological Hypothesis* – his winter address of 1875. Again in 1887, he wrote: "[d]espite the efforts of evolutionists, the laws which protect and perpetuate specific distinctions remain unrepealed, and Science joins its testimony to that of Revelation in condemnation of the degrading hypothesis that man is the offspring of a brute."[77] Here we witness a much narrower biblicist rejection of evolution than was evident in either Edinburgh or Princeton.

Not surprisingly, Watts utterly repudiated the efforts of Henry Drummond to evolutionize theology and to import Spencer's evolutionary laws into the supernatural realm. To him Drummond's project was just a mess – theologically *and* scientifically. To be sure, his aim of removing the "the alleged antagonism" between science and religion was a worthy objective, but the Drummond strategy ultimately was "not an Irenicum between science and religion or between the laws of the empires of matter and of spirit" at all. Rather it only displayed the expansionist character of natural law. Watts found the whole project so bizarre that, as he worked up to his conclusion, he felt "inclined to apologise for attempting a formal refutation of a theory which, if it means anything intelligible, involves the denial of all that the Scriptures teach, and all that Christian experience reveals."[78]

By now, Watts had grown entirely disillusioned with the Edinburgh New College network and continued to cultivate links with Princeton. This is clearly apparent in extant correspondence from Watts to Warfield, though there is just the slightest whiff of suspicion that Watts was beginning to have concerns on the evolution front there too. Writing in 1889 to Warfield, he began: "It would seem as if Princeton is going to absorb Belfast. Here I am asked to introduce to you, I think, the fifth student, within the past few weeks. Well, over this I do not grieve but rather rejoice. I am glad that our young men are setting their faces towards your venerable and orthodox institution instead of turning their backs upon orthodoxy and seeking counsel at the feet of men who are trampling the verities of Revelation under foot."[79] In introducing yet another student to the Princeton campus, he wrote to Warfield in 1893: "I am greatly pleased to find that our young men have turned their eyes to Princeton instead of Edinburgh, and I owe you my warmest thanks for the great kindness you have shown them."[80] Indeed he had already reminded Warfield that he had been writing a series of articles on the changing attitudes toward inspiration in the Free Church, reporting at the same time on the current heresy hunting of A. B. Bruce and Marcus Dods. Predictably, Watts took a dim view of these teachers. "How has the fine gold of the disruption become dim!" he sighed. "The men of '43 would have made short work of these cases."[81] And later that same year, having sent Warfield two days earlier a copy of his new book *The New Apologetic; or, The Down-Grade in Criticism, Theology, and Science*,[82] he concluded yet another letter with the comment: "I dread the influence of the Scotch Theological Halls, as you may learn from my book."[83]

For all that, in a reply to Warfield in 1894 thanking him for sending a copy of William Brenton Greene's inaugural lecture, Watts felt constrained to comment:

> I admired Dr. Greene's Inaugural, but I am sorry that he passed such high eulogium upon Mr. Herbert Spencer. My revered father, Professor Wallace, author of an immortal work, entitled 'Representative Responsibility,' had, for Spencer, supreme contempt as a professed philosopher. Immediately after the meeting of the British Association in Belfast, in 1874, I reviewed Spencer's Biological Hypothesis, & proved that he was neither a philosopher, nor a scientist. Someone sent him a paper containing only one half of my review, &, assuming that what the paper gave was the whole of my address, he took occasion in an issue of his book, then passing through the Press, to charge me with misrepresentation. I called his attention to the mistake of his correspondent, but he had not the courtesy to withdraw the charge. Dr. McCosh's successor in the Queen's College here, trots out Spencer to our young men in their undergraduate course, and one of my duties, in my class work, is to pump out of them Mill & Spencer.[84]

Clearly even after twenty years the recollection of the Tyndall event remained unsullied and Watts neither would, nor could, release his grip on that bitter memory.

In 1916, reflecting on his undergraduate days at Princeton, B. B. Warfield recalled the coming of James McCosh to take up the presidency of the College of New Jersey. Yet in one sense Warfield made it clear that McCosh's coming had no influence on him ... and that was on the question of Darwin's theory of evolution. For Warfield already was, by his own admission, "a Darwinian of the purest water" before

McCosh's presence graced the Princeton campus. To be sure, Warfield would later depart from what he took to be McCosh's Darwinian "orthodoxy," but he would remain open to the possibility of evolutionary transformism through a range of mechanisms besides Darwinian natural selection.[85] We should recall that in 1916, when Warfield penned these recollections, Darwinism was in eclipse in the scientific world. In the interim, evolutionary theses flourished – neo-Lamarckism, orthogenesis, the mutation theory of Hugo de Vries and so on – though without assuming the gradualism or the central significance of natural selection, which Darwin's disciples had promulgated.[86]

Long before Warfield ever expressed himself in print on the evolution question, however, he had evidently thought seriously about the issue. This I believe stemmed from the long-standing interest both he and his father William had in short-horn cattle breeding. Indeed in 1873, prior to entering Princeton Seminary as a student for the ministry, he had been livestock editor for the *Farmer's Home Journal* of Lexington, Kentucky. He maintained this interest for the remainder of his life and carefully kept a scrapbook of clippings – mostly those of his father – on what he called "Short Horn Culture."[87] It is not surprising therefore to find the following words in the preface to William Warfield's *The Theory and Practice of Cattle-Breeding*: "I wish to take this opportunity to acknowledge the assistance my sons have given me in preparing all my work for the press. Without their aid much – even most – of it could never have been done. . . . Great credit is also due to my elder son, Prof. Benjamin B. Warfield, D. D., now of Princeton, N. J., whose energy and vigor of thought and pen gave me such essential aid in the earlier years of my connection with the press; nor has the pursuit of the more weighty things of theology destroyed his capacity for taking an occasional part in the active discussion of cattle matters. The papers which have appeared over my signature have thus to quite a large degree been of family origin."[88] This affirmation is, I think, particularly significant because in the course of his discussion of such key issues as variation, heredity, atavism, and reversion to type, William Warfield quoted liberally from Darwin's *Animals and Plants under Domestication*.[89] Moreover, he directly endorsed the theory of evolution by natural selection when he wrote:

> Nature's method seems to be a wide and general system of selection, in which the strong and vigorous are the winners and the weaker are crushed out. Among wild cattle the more lusty bulls have their choice of the cows in a way that under natural selection insures the best results to the race . . . . If the conditions of life should suddenly change the result on such wild cattle would be to deteriorate or to improve the average according as the change was for their advantage or disadvantage. It is

> quite apparent that no question of breeding intrudes itself here. Nature's selection, while always in favor of the maintenance of the animals in the best manner, yet is impartial, and under ordinary circumstances would maintain an average.[90]

Like Darwin himself, B. B. Warfield encountered natural selection through an intense study of breeding.[91] And this, I believe, is why he could, early in his career, describe himself as a pure Darwinian – he met the theory empirically, as it were, not theologically or philosophically.

While Warfield's espousal of Darwinism evidently predated the coming of McCosh to Princeton in 1868, some of his earliest published statements on evolutionary theory date from 1888 – significantly just the year after he returned from Western Theological Seminary near Pittsburgh to Princeton to take up the chair of didactic and polemic theology. By this stage Warfield's reservations about Darwinism as a comprehensive system had begun to manifest themselves. In his Lectures on Anthropology, begun that same year, he took up the subject "Evolution or Development" and made it clear that in his view Darwinism did not enjoy anything like the scientific status of the Newtonian system, for there were just too many unresolved geological, zoological, and palaeontological anomalies – anomalies he did not hesitate to itemize. For all that, he was prepared to leave the truth or falsity of the Darwinian theory an "open question." In sum, "the whole upshot of the matter is that there is no *necessary* antagonism of [Christianity] to evolution, *provided that* we do not hold to too extreme a form of evolution." If the constant supervision of divine providence and the "occasional supernatural interference" of God were retained then, he concluded, "we may hold to the modified theory of evolution and be [Christ]ians in the ordinary orthodox sense. I say we may do this. Whether we ought to accept it, even in this modified sense is another matter, & I leave it purposely an open question."[92] If Warfield's sense of equivocation is evident here, so too is the openness of his stance on the whole issue, and his willingness to test the theory by its empirical adequacy.

During the 1880s, of course, other issues were concerning Warfield and other conservative Presbyterian theologians – matters which, I suspect, rather cast the Darwinian threat in the shade. Of primary importance was the whole issue of biblical criticism, which dramatically manifested itself in the years 1881–1883 when the *Presbyterian Review* published an exchange of essays on the subject in the wake of the Robertson Smith affair in Scotland. Along with A. A. Hodge and Francis L. Patton, Warfield provided a conservative critique of the position advanced by Charles A. Briggs, Henry Preserved Smith, and Samuel I. Curtiss. Having said this, as Noll has pointed out, in the early 1880s the two sides personified in Briggs and Hodge enjoyed "a reasonably satisfactory relationship," and it was not until the following decade that the conservatives, increasingly worried about the critics' radicalism, took action that would lead to heresy charges against Briggs and his eventual suspension from the ministry. Because, as Noll put it, "ecclesiastical control seemed fully on the side of the conservatives," they retained a cultural hegemony in the last decade of the century that their Scottish counterparts did not enjoy.[93]

By the early years of the new century, Warfield's theological tentativeness over the evolution question progressively declined as he came to adopt a decidedly more partisan stance. The depth of evolutionary penetration into his thinking as the first decade of the twentieth century wore on is particularly evident on two issues of burning concern, namely, the origin and the unity of the human species. In 1906, a review of James Orr's *God's Image in Man* afforded him the opportunity to spell out in more detail his own thinking on the subject. In this work, Orr argued for the entirely supernatural origin of the first human, on the grounds that the complex physical organization of the brain – necessary he believed for *human* consciousness – made it impossible to postulate a special origin for the mind and not for the body. To Warfield, however, Orr's objection could effectively be undermined by the idea of evolution by mutation. Thereby Warfield proposed a kind of emergent evolutionary solution to the problem. As he put it, "if under the directing hand of God a human body is formed at a leap by propagation from brutish parents, it would be quite consonant with the fitness of things that it should be provided by His creative energy with a truly human soul."[94]

Warfield's self-perception as perpetuating orthodox Calvinism, even in this concession to the possibility of a human evolutionary history and a theologically permissible explanatory naturalism, is nowhere more dramatically displayed than in his 1915 exposition of "Calvin's Doctrine of the Creation" for the *Princeton Theological Review*. Here he made much of Calvin's insistence that the term "creation" should be strictly reserved for the initial creative act, for the subsequent creations were – technically speaking – *not* creations ex nihilo, but were, rather, modifications of the primeval "indigested mass" by "means of the interaction of its intrinsic forces." The human soul was the only exception, for Calvin held to the Creationist (as opposed to the Traducianist) theory that every human soul throughout the history of propagation was an *immediate* creation out of nothing. To Warfield, then, Calvin's doctrine opened the door to a controlled naturalistic explanation of natural history – including the human physical form – in terms of the operation of secondary causes. These, to be sure, were directed by the guiding hand of divine providence; but this conviction did not prevent him from asserting that Calvin's doctrine of creation actually turned out to be "a very pure evolutionary scheme." Besides all this, Warfield's architectonic defense of biblical inerrancy did not necessitate a literalistic interpretation of the early Genesis narratives. Thus while acknowledging that Calvin himself had understood the days of creation as six literal days, he still believed that Moses, "writing to meet the needs of men at large, accommodated himself to their grade of intellectual preparation, and confine[d] himself to what meets their eyes." Now in the twentieth century Warfield suggested that in order to perpetuate the spirit of Calvin's hermeneutic genius, "it was requisite that these six days should be lengthened out into six periods – six ages of the growth of the world. Had that been done Calvin would have been a precursor of the modern evolutionary theorists. As it is, he only forms a point of departure for them to this extent – that he teaches, as they teach, the modification of the original world-stuff into the varied forms which constitute the ordered world, by the instrumentality of second causes – or as a modern would put it, of its intrinsic forces."[95] Recalling that Warfield penned these words in 1915, around the time he reminisced on his student days with McCosh and observed that he later abandoned the Scotsman's Darwinian orthodoxy, certainly helps us to see that departing from classical Darwinism evidently did not mean abandoning evolutionary transformism.

Warfield, of course, was not alone in the evolutionary sentiments expressed at Princeton. Among figures such as A. A. Hodge, Francis L. Patton, and Casper Wistar Hodge, not to mention James McCosh and George Macloskie, similarly cautious attitudes may be discerned.[96] Of these the most considered treatments were forthcoming from the two arrivals from Belfast – McCosh and Macloskie. It seems as though their removal from Belfast to the New World facilitated a rhetorical liberation that allowed them to express their reservations about *Darwinism* in a muted way, and to find *evolution* theologically congenial. I have already commented on McCosh's essentially neo-Lamarckian proclivities; a few comments on Macloskie will suffice to complete my survey.

George Macloskie, a minister of the Presbyterian Church in Ireland, followed his teacher McCosh to Princeton University in 1875 to take up the position of professor of biology. From the time he arrived there, Macloskie vigorously promoted science among orthodox Presbyterians and consistently urged the value of a chastened Darwinian perspective, notwithstanding his personal friendship with Robert Watts.[97] Achieving this, however, required Macloskie to jettison the Scottish philosophy's inductive strictures on the scientific imagination. To him speculation was not only inevitable; it was a necessity.[98] Hypothesizing, even of the Darwinian variety, was thus to be welcomed,[99] and, when given a suitably Calvinist gloss, it might even be acceptable as an account of human origins:

> Evolution, if proven as to man, will be held by the biblicist to be a part, the naturalistic part, of the total work of his making, the other part being his endowment miraculously with a spiritual nature, so that he was created in the image of God. . . . As a member of the animal kingdom, man was created by God, probably in the same naturalistic fashion as the beasts that perish; but, unlike them, he has endowments which point to a higher, namely a supernaturalistic, order of creation.[100]

Not surprisingly, he applauded A. A. Hodge's words that "we have no sympathy with those who maintain that scientific theories of evolution are necessarily atheistic." More, he insisted that attempts to make evolution a heresy – he seems to have had the James Woodrow case in mind[101] – were disastrous for religion and science alike, and he applauded both McCosh and Asa Gray for their "great service" in demonstrating that evolution "is not dangerous."[102]

Managing theological space

In the latter decades of the nineteenth century, the intellectual leadership of the Presbyterian citadels of Edinburgh, Belfast, and Princeton were involved in the production and reproduction of theological space. In all three, the field of discourse these figures had done so much to manage set limits on the assertable – on what could be *said* about evolution, and on what could be *heard*. In Edinburgh the concerns with biblical criticism, idealist philosophy, confessional modification in the wake of ecclesiastical reunions, and the proevolution sentiments of key leaders all made rapprochment with evolutionary theory intellectually congenial. In Belfast, the infamous BA meeting made it exceptionally difficult to be sympathetic to those who might argue that evolution could be construed in terms other than those of Tyndall's supposed philosophical naturalism. Besides, even if this possibility had been grasped, it would have been exceedingly difficult to *express* it in the doctrinal locality they had labored so dilligently to reproduce. In Princeton the cautious words of Warfield and the strenuous efforts of McCosh to produce an evolutionary teleology, together with the conservative success in domestic Presbyterian debates over biblical criticism, permitted evolution, if not Darwinism, to be tolerated. It did not need to be absorbed or proscribed.

The portraits I have presented, of course, are not intended to be exhaustive surveys of the ways in which the intellectuals in these cities encountered Darwinian science. My focus has been resolutely upon the theological elites who occupied the key Calvinist spaces in Belfast,

Princeton, and Edinburgh. How Darwinism was more widely absorbed, repudiated, or modified within these urban worlds is a subject worthy of further examination. My task has been more circumscribed. But I believe my investigations have shown that local circumstances were crucial to understanding how Calvinists in these different places chose to negotiate their way around the issues that science seemed to be placing in one way or another on their agendas. By extension, to understand the encounter between science and religion, I submit, will require us to take with greater seriousness the situatedness of both scientific and religious discourses.

Notes

1 In the case studies which follow I make no claim that the various positions adopted were based solely on warranted rationality; my point here, rather, is that epistemology has recently displayed something of a spatial turn.

2 Clifford Geertz, *Local Knowledge: Further Essays in Interpretive Anthropology* (New York: Basic Books, 1983).

3 Erving Goffman, *The Presentation of Self in Everyday Life* (London: Allen Lane, 1969).

4 Anthony Giddens, *The Constitution of Society: Outline of the Theory of Structuration* (Oxford: Polity Press, 1984), p. 368.

5 I have surveyed some of these matters in "The Spaces of Knowledge: Contributions Towards a Historical Geography of Science," *Society and Space*, 13 (1995):5–35.

6 Patrick Carnegie Simpson, *The Life of Principal Rainy*, 2 vols. (London: Hodder and Stoughton, 1909), 1:285. A. C. Cheyne agrees, commenting that the proevolution sentiment of Rainy and Robert Flint "is a measure of the rapid intellectual transformation that had taken place in Scotland in less than half a century"; see Cheyne, *The Transforming of the Kirk: Victorian Scotland's Religious Revolution* (Edinburgh: St. Andrews Press, 1983), pp. 77–78.

7 Robert Rainy, *Evolution and Theology: Inaugural Address* (Edinburgh: Maclaren & Macniven, 1874), pp. 6, 9.

8 Ibid., p. 18.

9 I have discussed the Belfast Calvinist response to Darwin in "Darwinism and Calvinism: The Belfast-Princeton Connection," *Isis*, 83 (1992):408–28; and in "Darwin in Belfast," in *Nature in Ireland: A Scientific and Cultural History*, ed. John W. Foster and Helena Ross (Dublin: Lilliput Press, 1997), pp. 387–408.

10 J. L. Porter, *Theological Colleges: Their Place and Influence in the Church and in the World; with Special Reference to the Evil Tendencies of Recent Scientific Theories. Being the Opening Lecture of Assembly's College, Belfast, Session 1874–75* (Belfast: Mullan, 1875), p. 8.

11 Porter, *Science and Revelation: Their Distinctive Provinces. With a Review of the Theories of Tyndall, Huxley, Darwin, and Herbert Spencer* (Belfast: Mullan, 1874), pp. 3–4, 5, 20, 22.

12 Charles Hodge, *What Is Darwinism*? (New York: Scribner's, 1874), pp. 41, 48, 52, 108.

13 Ibid., p. 174.

14 K. R. Ross, "Rainy, Robert," in *Dictionary of Scottish Church History and Theology*, ed. Nigel M. de S. Cameron (Edinburgh: T&T Clark, 1993), q.v. For such reasons he remained suspect – as did James Orr – to figures such as John Macleod of the Free Church, who interpreted Scottish Theology to American Calvinists in his lectures at Westminster Theological Seminary. See John Mcleod, *Scottish Theology in Relation to Church History since the Reformation* (Edinburgh: Free Church of Scotland, 1943).

15 Andrew L. Drummond and James Bulloch, *The Church in Victorian Scotland, 1843–1874* (Edinburgh: St. Andrews Press, 1975), p. 234. It should be noted, however, that in his 1927 lectures on the Calvin Foundation at the Free University of Amsterdam, Donald Maclean, professor of Church History at the Free Church College in Edinburgh, did note that "in the scientific world the theory of Darwin profoundly affected theological thinking with resultant doctrinal accommodations that lessened the authority of the Word of God as a rule of faith and conduct." Maclean reported that Darwinism "has been severely challenged by the science of the twentieth century. All its outstanding principles have been the subject of the most adverse criticism. It has no explanation of the 'sudden apparition of species,' or of 'terminal forms,' or of the great 'gaps' in series of living forms, or of 'inversions' in the geological records. It has no concrete proof whatever of genetic continuity, which is fundamental to the theory." See Maclean, *Aspects of Scottish Church History* (Edinburgh: T&T Clark, 1927), pp. 105, 149. Nevertheless, Maclean's assessment does dwell at considerably greater length on the issue of biblical criticism.

16 Simpson, *Principal Rainy*, vol. 1, p. 406.

17 See Andrew L. Drummond and James Bulloch, *The Church in Late Victorian Scotland, 1874–1900* (Edinburgh: St. Andrews Press, 1978), Chap. 5.

18 Alasdair MacIntyre, *Three Rival Versions of Moral Enquiry: Encyclopaedia, Genealogy, Tradition* (London: Duckworth, 1990), p. 18.

19 See the discussions in Simpson, *Principal Rainy*, pp. 306–403; Cheyne, *Transforming of the Kirk*, pp. 44–52; and Richard Allen Riesen, *Criticism and Faith in Late Victorian Scotland: A. B. Davidson, William Robertson Smith and George Adam Smith* (Lantham: University Press of America, 1985), pp. 94–251.

20 Biographical details are available in J. S. Black and G. W. Chrystal, *The Life of William Robertson Smith* (London: Adam & Charles Black, 1912). The symbolic significance of this episode has been noted on numerous occasions. See, for instance, Don Cupitt, *The Sea of Faith* (London: BBC, 1995); and Peter Hinchliff, *God and History: Aspects of British Theology, 1875–1914* (Oxford: Clarendon, 1992).

21 George Elder Davie, "Scottish Philosophy and Robertson Smith," in *The*

Scottish Enlightenment and Other Essays (Edinburgh: Polygon, 1991), pp. 101–45, quotation on p. 131.

22 Davie, "Scottish Philosophy," p. 123.

23 Karl Barth pointed out, when he delivered the lectures in the late 1930s, that the entire enterprise of natural theology was based upon a radical theological error. See Barth, *The Knowledge of God and the Service of God* (London: Hodder and Stoughton, 1938).

24 *The Witness*, 19 August, 1874.

25 Adrian Desmond and James Moore, *Darwin* (London: Michael Joseph, 1991), p. 611.

26 John Tyndall, *Address Delivered Before the British Association Assembled at Belfast, with Addition* (London: Longman, 1874). See the discussion in Ruth Barton, "John Tyndall, Pantheist: A Rereading of the Belfast Address," *Osiris*, 2nd ser., 3 (1987):111–34.

27 Robert Watts, "An Irenicum; *or, A* Plea for Peace and Co-operation between Science and Theology," in *The Reign of Causality: A Vindication of the Scientific Principle of Telic Causal Efficiency* (Edinburgh: T&T Clark, 1888), pp. 2–27.

28 *The Witness*, 9 October, 1874.

29 *Northern Whig*, 25 August, 1874, p. 8. Also *The Witness*, 9 October, 1874.

30 Watts, "Atomism – An Examination of Professor Tyndall's Opening Address before the British Association, 1874," in *Reign of Causality*, pp. 27–43.

31 Rev. John MacNaughtan at the Rosemary Street Presbyterian Church, Rev. George Shaw at Fitzroy, and Rev. T. Y. Killen at Duncairn all took up the cudgels.

32 Robertson Smith, "Letter," *Northern Whig*, 27 August, 1874.

33 Cited in Barton, "John Tyndall, Pantheist," p. 116.

34 Bradley John Gundlach, "The Evolution Question at Princeton, 1845–1929," Ph.D. diss., University of Rochester, 1995.

35 James McCosh, "Religious Aspects of the Doctrine of Development," in *History, Orations, and Other Documents of the Sixth General Conference of the Evangelical Alliance*, ed. S. Irenaeus Prime (New York: Harper, 1874), pp. 264–71.

36 William Milligan Sloane, *The Life of James McCosh: A Record Chiefly Autobiographical* (Edinburgh: T&T Clark, 1896).

37 Ibid., p. 234. The standard biography of McCosh is J. David Hoeveler, *James McCosh and the Scottish Intellectual Tradition* (Princeton: Princeton University Press, 1981).

38 James McCosh, *The Ulster Revival and Its Physiological Accidents: A Paper Read Before the Evangelical Alliance, September 22, 1859* (Belfast: C. Aitchison, 1859).

39 This certainly does not mean that Princetonians were unaware of the Tyndall episode. For example, "An Open Letter to Professor Tyndall" from Rev. John Laing of Dundas, Ontario, appeared in *The Presbyterian Quarterly and Princeton Review*, 4 (1875):229–53.

40 I discuss this in *Darwin's Forgotten Defenders: The Encounter between Evangelical Theology and Evolutionary Thought* (Grand Rapids and Edinburgh: Eerdmans and Scottish Academic Press, 1987), pp. 107–08.

41 Biographical details are drawn from W. L. Calderwood and David Woodside, *The Life of Henry Calderwood* (London: Hodder and Stoughton, 1900).

42 S. D. Gill, "United Presbyterian Church," in Cameron, *Dictionary of Scottish Church History*.

43 See Mcleod, *Scottish Theology*, p. 307.

44 For all that, Calderwood, according to his biographer, "regarded with equanimity the new views promulgated by the late W. Robertson Smith and others in this country regarding the authorship of certain books of the Old Testament, and had a firm conviction that these did not touch the real question of inspiration or the abiding influence of the Word of God." Calderwood and Woodside, *Life of Calderwood*, p. 275. His concerns were much more focused on issues of moral philosophy and the creeping influence of German idealism.

45 A. Seth Pringle-Pattison, "The Philosophical Works," in *Life of Calderwood*, p. 423.

46 Henry Calderwood, *Philosophy of the Infinite* (London: Macmillan, 1854). Calderwood was also the author of the popular student text, *Handbook of Moral Philosophy* (London: Macmillan, 1888). By 1902 it had gone through fourteen editions.

47 Cited in Calderwood and Woodside, *Life of Calderwood*, p. 159.

48 Calderwood, "Ethical Aspects of the Theory of Development," *Contemporary Review*, 31 (1877):123–32.

49 Henry Calderwood, "Evolution, Physical and Dialectic," *Contemporary Review*, 40 (1881):865–76; 867–68.

50 Calderwood, *The Relations of Science and Religion* (New York: Ketcham, 1881), pp. 21, 134–35.

51 Ibid., pp. 153–54.

52 George D. Mathews, ed., *Alliance of the Reformed Churches Holding the Presbyterian System: Minutes and Proceedings of the Third General Council, Belfast 1884* (Belfast: Assembly's Offices, 1884), pp. 250–51.

53 Calderwood, "On Evolution and Man's Place in Nature," *Proceedings of the Royal Society of Edinburgh*, 17 (1890):71–79.

54 Calderwood, *Evolution and Man's Place in Nature* (London: Macmillan, 1893), p. 337. This aspect of the book drew praise from A. R. Wallace himself, who reviewed the book in *Nature*. See Calderwood and Woodside, *Life of Calderwood*, p. 190.

55 Calderwood, *Evolution and Man's Place*, p. 340.

56 Calderwood, *Evolution and Man's Place in Nature*, 2nd ed. (London: Macmillan, 1896), p. 33.

57 James Iverach, *Evolution and Christianity* (London: Hodder and Stoughton, 1894), p. 104.

58 See Alan P. F. Sell, *Defending and Declaring the Faith: Some Scottish Examples, 1860–1920* (Exeter: Paternoster Press, 1987), pp. 137–71.

59 James Orr, *The Christian View of God and the World* (Edinburgh: Elliott, 1893), p. 99.

60 Orr, *God's Image in Man and Its Defacement in the Light of Modern Denials* (London: Hodder and Stoughton, 1905), p. 88.
61 I discuss Orr's thinking on this matter in *Darwin's Forgotten Defenders*, pp. 140–44.
62 Orr, "Science and Christian Faith," in *The Fundamentals* (Chicago: Testimony Publishing [1910–1915]), vol. 4, p. 103.
63 Robert Flint, *Theism*, 3rd ed. (Edinburgh: Blackwood, 1880), pp. 201–06, 208.
64 George Matheson, "The Religious Bearings of the Doctrine of Evolution," in *Alliance of the Reformed Churches*, p. 86. See also Matheson's "Modern Science and the Religious Instinct," *Presbyterian Review*, 5 (1884): 608–21.
65 Matheson, *Can the Old Faith Live with the New? or, The Problem of Evolution and Revelation* (Edinburgh: Blackwood, 1885).
66 The best analysis of Drummond's efforts is James R. Moore, "Evangelicals and Evolution: Henry Drummond, Herbert Spencer, and the Naturalisation of the Spiritual World," *Scottish Journal of Theology*, 38 (1985):383–417.
67 Technically, Duns occupied the position as lecturer from 1864 to 1869 due to some doubts expressed by the General Assembly of 1864 about the propriety of retaining natural science in the theological curriculum. A bequest provided for the endowment of a professorial chair in 1869, and Duns occupied that position until his death during the General Assembly of 1903.
68 John Duns, *Creation According to the Book of Genesis and the Confession of Faith: Speculative Natural Science and Theology: Two Lectures* (Edinburgh: Maclaren & Macniven, 1877), p. 36.
69 Duns, "On the Theory of Natural Selection and the Theory of Design," *Transactions of the Victoria Institute*, 20 (1885):1–22. Duns also maintained this viewpoint in *Christianity and Science* (Edinburgh: William P. Kennedy, 1860), and *Science and Christian Thought* (London: Religious Tract Society, n.d.)
70 "The British Association," *The Witness*, 28 August, 1874. It should not be assumed that there were no less strident voices. Rev. George Macloskie, in a letter to the *Northern Whig* on 26 August, for example, insisted that there was no desire in Belfast "to stifle free scientific inquiry." *Northern Whig*, 27 August, 1874, p. 8.
71 J. G. C., "Darwinism," *Irish Ecclesiastical Record*, 9 (1873):337–61.
72 "Pastoral Address of the Archbishops and Bishops of Ireland," *Irish Ecclesiastical Record*, 11 (1874):49–70.
73 Robert Watts, *Atomism: Dr. Tyndall's Atomic Theory of the Universe Examined and Refuted. To which are added, Humanitarianism accepts, provisionally, Tyndall's impersonal atomic deity; and a letter to the presbytery of Belfast; containing a note from the Rev. Dr. Hodge, and a critique on Tyndall's recent Manchester recantation, together with strictures on the late Manifesto of the Roman Catholic hierarchy of Ireland in reference to the sphere of science* (Belfast: Mullan, 1875), pp. 34, 38, 39. The addition of Hodge's endorsement evidently meant a

great deal to Watts. He regarded Hodge as "the author of the greatest work on systematic theology which has ever issued from the pen of man." Later in a letter to A. A. Hodge in 1881 thanking him for sending a copy of the biography of his father, Watts wrote: "Precious indeed it is to me and mine. We feel towards it as if it were the history of one of our own kith and kin." Letter, Robert Watts to A. A. Hodge, 1 January, 1881, Archives, Speer Library, Princeton Theological Seminary. Earlier in 1879 he had supplied A. A. Hodge with copies of letters from Charles Hodge to Principal Cunningham. Letter, Robert Watts to A. A. Hodge, 2 December, 1879, Archives, Speer Library.

74 Letter, Watts to Hodge, 1 January, 1881, Archives, Speer Library.

75 Letter, Watts to B. B. Warfield, 18 June, 1890, Warfield Papers, Archives, Speer Library.

76 "Pastoral Address," op. cit.

77 Watts, "Evolution and Natural History," in *Reign of Causality*, pp. 316–17. This review originally appeared in the April 1887 issue of the *British and Foreign Evangelical Review*.

78 Watts, "Natural Law in the Spiritual World," *British and Foreign Evangelical Review* (1885), reprinted in *Reign of Causality*. See also Watts, *Professor Drummond's "Ascent of Man," and Principal Fairbairn's "Place of Christ in Modern Theology," Examined in the Light of Science and Revelation* (Edinburgh: R. W. Hunter, circa 1894).

79 Letter, Watts to Warfield, 23 September, 1889, Warfield Papers, Archives, Speer Library.

80 Letter, Watts to Warfield, 5 October, 1893, Warfield Papers, Archives, Speer Library.

81 Letter, Watts to Warfield, 18 June, 1890.

82 In a postcard to Warfield dated 11 February, 1891, he reported to Warfield that "Spurgeon's 'Sword and Trowel,' 'Expository Times,' 'United Pres. Mag.' & 'The Theological Monthly' have all given most flattering reviews of my 'New Apologetic &c.'"

83 Letter, Watts to Warfield, 13 October, 1890, Warfield Papers, Archives, Speer Library. These same sentiments were also expressed in a letter of 31 May, 1891. Of course he dreaded no less the inroads biblical criticism was making in the United States too, and he chose to provide a detailed critique of Briggs's biblical theology in his opening address to the Assembly's College in November 1891. See letter, Watts to Warfield, 22 July, 1891, Warfield Papers, Speer Library.

84 Letter, Watts to Warfield, 20 February, 1894, Warfield Papers, Archives, Speer Library.

85 Warfield, "Personal Reflections of Princeton Undergraduate Life: IV – The Coming of Dr. McCosh," *The Princeton Alumni Weekly*, 16 (1916):650–53, quotation on p. 652.

86 See the discussion in Peter J. Bowler, *The Eclipse of Darwinism: Anti-Darwinian Evolution Theories in the Decades around 1900* (Baltimore: Johns Hopkins University Press, 1983).

87 This scrapbook is extant in the Speer Library, Princeton Theological Seminary. Most of the clippings are from the *National Livestock Journal.* The Speer Library also retains a number of books on short-horn cattle from B. B. Warfield's personal library.
88 William Warfield, *The Theory and Practice of Cattle-Breeding* (Chicago: J. H. Sanders, 1889).
89 From the Darwin volumes in Warfield's personal library, it is clear that the second volume of this work was the first of Darwin's books to be purchased by B. B. Warfield. It is dated May 1st 1868. *The Origin of Species* and *The Descent of Man* are both dated 1871. In 1872, while in London, B. B. Warfield purchased *The Expression of the Emotions.*
90 W. Warfield, *Cattle-Breeding,* pp. 85–86. Warfield also enthusiastically commended a volume by J. H. Sanders, *Horse-Breeding: Being the General Principles of Heredity applied to the Business of Breeding Horses* ... (Chicago: J. H. Sanders, 1885), in which Darwin's evolutionary thinking was promulgated. Warfield particularly endorsed the long first chapter, which dealt with these general principles.
91 See James Secord, "Nature's Fancy: Charles Darwin and the Breeding of Pigeons," *Isis,* 72 (1981):163–86.
92 B. B. Warfield, "Lectures on Anthropology," December 1888, Speer Library.
93 Mark A. Noll, *Between Faith and Criticism: Evangelicals, Scholarship, and the Bible in America* (New York: Harper and Row, 1987), pp. 28–29. See also the introduction in Noll, *The Princeton Theology: Scripture, Science, and Theological Method from Archibald Alexander to Benjamin Warfield* (Grand Rapids: Baker, 1983).
94 B. B. Warfield, Review of *God's Image in Man,* by James Orr, *Princeton Theological Review,* 4 (1906):555–58.
95 B. B. Warfield, "Calvin's Doctrine of the Creation," *Princeton Theological Review,* 13 (1915):190–255.
96 See my discussion in *Darwin's Forgotten Defenders,* Chapter 4.
97 When Watts visited Princeton in 1880, he wrote to his wife: "After writing you at Professor Green's on Monday, Dr. M'Closkie who you remember lost his dredge in Belfast Lough when our little boys were out with him in a boat, called according to promise and took me over all the buildings of Dr. M'Cosh's College." Letter, Robert Watts to his wife, 17 September, 1880, in "Family Letters of Revd. Robert Watts, D.D., LL.D.," compiled by his wife (typescript). I am grateful to Dr. R. E. L. Rodgers for making this typescript available to me.
98 George Macloskie, "Scientific Speculation," *Presbyterian Review,* 8 (1887): 617–25.
99 See Macloskie, "Concessions to Science," *Presbyterian Review,* 10 (1889):220–28; idem, "Theistic Evolution," *Presbyterian and Reformed Review,* 9 (1898):1–22; idem, "The Outlook of Science and Faith," *Princeton Theological Review,* 1 (1903):597–615; idem, "Mosaism and Darwinism," *Princeton Theological Review,* 2 (1904):425–51. I discuss the contents of these articles in *Darwin's Forgotten Defenders,* pp. 92–96.

100 Macloskie, "The Origin of Species and of Man," *Bibliotheca Sacra*, 60 (1903): 261–75, quotation on p. 273.
101 Macloskie's papers at the Princeton University Library contain *The Examination of the Rev. James Woodrow D. D. by the Charleston Presbytery* (Charleston, S. C.: Lucas & Richardson, 1890); see Carton 2, Macloskie Papers, CO498, Princeton University Library.
102 Mss. notebook by Macloskie, "Drift of Modern Science," Carton 1, Macloskie Papers, CO498, Princeton University Library.

2

Darwin down under: science, religion, and evolution in Australia

BARRY W. BUTCHER

Darwinism came early to Australia. Charles Darwin's *Origin of Species* appeared on sale in Sydney only four months after its publication in Britain. Within a year, worried presidents of the local scientific societies were joining forces with church leaders to warn of the social and intellectual dangers of "the development hypothesis," and various Australian scientists were privately expressing fears about the adverse effects of Darwinism on traditional thinking. Yet within three years of the arrival of Darwin's book, an Australian scholar had pioneered, in the application of Darwinism to political economy, an achievement that elicited approval from no less a person than Herbert Spencer; and within fifteen years Darwin himself was praising an Australian botanist for his contributions to organic evolution. By the last decades of the century, Darwinism had become entrenched in the major teaching institutions of the country, and many churchmen were comfortably accommodating the new science of life to Christian theology.[1]

Assessing the impact of major ideas on any given society is both challenging and frustrating. The very term *society* slips through the fingers as soon as one attempts to grasp it. Who received new ideas: all the citizens or a select few? In a continent the size of Australia, which for most of the nineteenth century comprised little more than a handful of sparsely populated colonies clinging to the coastal fringe, the difficulties are compounded. Each of the colonies possesses a distinctive history. The British founded New South Wales and Tasmania, for example, as convict dumping grounds, while Victoria exploded into being after the midcentury discovery of gold. The Victorian gold rush attracted not just seekers after quick wealth but an influx of professionals, such as surveyors and doctors, who quickly organized themselves into efficient social elites.

Such a diversity of developmental histories makes a mockery of attempts to capture the spread and growth of science by means of models. No single paradigm accounts for the twists and turns associated with

Darwin's reception in Australia. The acceptance or rejection of Darwinism exhibited the full gamut of attitudes found elsewhere. In the 1860s small numbers of Darwinians were scattered throughout Australia, just as they were in Europe and North America. And into the 1880s *anti*-Darwinians remained active in Australia, just as they did in Europe and North America, though their influence had abated. Between 1875 and 1885 Robert David Fitzgerald, a surveyor in New South Wales, produced a Darwinian tour de force in the form of a survey of Australian orchids, which drew admiration from Darwin and the British botanist Joseph Hooker.[2] At the same time, most of the professional botanists in the colonies – those running the botanic gardens, for instance – were explicitly opposing evolutionary ideas.

In what now seems a naive attempt at model building, I once tried to account for this sort of complexity by suggesting that although there may have been dominant orthodoxies in science at the periphery of Western culture, there were also heterodoxies that thrived and laid the groundwork to some extent for the eventual triumph of Darwinism. In many respects, as Ann Moyal has claimed, Paleyan natural theology dominated Australian science until very late into the nineteenth century; nevertheless, many Australian scholars worked within a distinctively evolutionary framework from the 1860s on. There was no time lag between happenings at the center of Darwinian science in Britain and similar events at the Australian periphery; the same cornucopia of opinions, beliefs, philosophies, and religious allegiances influenced responses to Darwinism in both regions.[3]

This fact raises the issues of geographical isolation and distance: How much did the physical distance separating Australia from its cultural homeland, Britain, affect intellectual developments? Australia, a tiny fragment of European culture surrounded by a variety of Pacific and Asian cultures, sometimes out-Europeanized Europe. From the time of the earliest European settlements, colonists had attempted to recreate "home" in a landscape manifestly unsuitable for such yearnings. Small gatherings of British immigrants set up philosophical and scientific societies in outposts of "real culture," importing the trappings of British society life: theaters, museums, rabbits to feed the foxes that had been brought in to be hunted, journals from the London, Paris, and Edinburgh scientific societies, and, of course, books such as the *Origin of Species*. In a context in which the journey from Europe to Australia lasted about twelve weeks and the constant influx of immigrants brought all of the latest fashions, material and otherwise, with them, one has to question whether nineteenth-century Australians suffered much of an intellectual disadvantage. The historian James Secord once pointed out that a person lacking social and institutional support could be isolated even in the midst of London itself.[4] In the nineteenth century, distance was as much a function of social and professional class as of geography. There were too many Australians with good social contacts "back home" to make the alleged tyranny of distance much more than a minor inconvenience.

Among the culturally connected Australians were William Sharp Macleay and his scientific cronies in Sydney; George Bennett, a lifetime correspondent of the British anatomist Richard Owen; and Gerard Krefft, who exchanged letters with Darwin and other British naturalists. From the 1850s on, the cities of Melbourne in Victoria and Sydney in New South Wales possessed universities staffed by first-rank scientists personally familiar with the greatest names in British science. The small number of active scientists in Australia may have retarded the growth of science in the colonies, but their physical separation from Britain had little noticeable effect on their response to scientific ideas or on the kinds of debates that such ideas engendered.

The recipients of European cultural products such as Darwinism spanned the social spectrum, as the following story suggests. In 1876, when William Archer, a twenty-three-year-old Englishman (later a celebrated theater critic), visited Australia, he noted that the typical squatter's library contained inter alia "an encyclopaedia, Shakespeare, Macaulay's *England* and *Essays*, Mill's *Political Economy* [and] one or other of Darwin's works."[5] Though anecdotal, Archer's observation supports a common perception among antiquarian booksellers in Australia, who know only too well the extent to which scientific publications found their way into the libraries of isolated bush dwellers and (to the dealers' cost) the extent to which such works were fiercely and often knowledgeably annotated by readers.

Early responses

In 1863 George Britton Halford, newly appointed professor of medicine at the University of Melbourne, effectively initiated public debate over Darwinism in Australia when he challenged Thomas Henry Huxley's claim that the similarities between humans and apes suggested a common ancestor. In a series of public lectures over the next two years, Halford refused to accept Huxley's growing scientific authority and continued to assert that crucial anatomical differences between humans and apes could not be explained by evolutionary theory. What was at one level a technical argument about muscles and bones became a platform for debating the merits of evolutionary theory and the place of humans in nature. Politicians, churchmen, and colonial heads of state who attended Halford's lectures invariably favored his anti-Huxley stance. The local press, in contrast, split along lines that suggest a close relationship between political and social ideologies and attitudes toward Darwin's theory. The *Age*, edited by the economic protectionist David Syme, stoutly supported Halford, while the *Argus*, under the watchful eye of the liberal free trader Edward Wilson, backed Huxley. Syme later wrote two books critical of Darwinian natural selection; Wilson returned to England, became a neighbor of Darwin's in Kent, and supplied him with information about the Australian aborigines.[6]

Halford, a former student of Richard Owen's who owed his position in Melbourne partly to his mentor's influence, helped to entrench a non-Darwinian scientific attitude into the university, which remained in place for more than two decades. In combination with the often eccentric anti-Darwinian professor of geology, Frederick McCoy, a sometime collaborator of the prominent British naturalist and anti-Darwinist Adam Sedgwick, Halford succeeded in maintaining a type of providentialist science that rejected the "extreme" naturalism espoused by such agnostics as Huxley and the Irish physicist John Tyndall. Halford's success, however, came at a considerable price: local fame translated into international notoriety once the Huxley doctrine became dominant "back home." Halford's attempts to gain entry into the Royal Society came to naught, despite his appeals to Owen and, it must be said, his siding with Huxley in a later dispute over the nature of life. Thus while gaining the admiration of colonial governors, church leaders, and many of his scientific colleagues, Halford saw his wider scientific ambitions curtailed in large measure because of his outspoken rejection of Darwinism and the naturalistic science of which it was a part.[7]

Repercussions from the Melbourne disputes could be felt at both the local and international levels, affecting individual careers and exacerbating tensions among the intelligentsia of the city. The debates also brought into focus differing attitudes toward the aims and objectives of science. Although the crude "Paley versus Darwin" scenario, pitting natural theology against natural selection, makes a finely textured analysis difficult, it does illustrate the two ends of the spectrum of opinion prevailing among Australian scientists, churchmen, and lay persons on the general question of evolutionary theory and on the specific issue of what was called man's place in nature. The Melbourne debates over the merits of Darwinism took place in a context in which religious, philosophical, and social factors were constantly in play. Subsequent discussions occurred in a similar setting but tended to center on the claims of science to be an arbiter of knowledge in areas traditionally seen as the preserve of religion. They raised questions such as: Where should the line be drawn between theology and science? In what areas of knowledge should science remain silent? Who should be initiated into the ways of science?

Providence and Darwinian theory

If, as the historian Ann Moyal has argued, some form of providentialism dominated the thinking of the colonial intelligentsia about the natural world and the place of humanity in it, then it is not surprising to find many of these same people responding hostilely to the naturalistic explanatory base that underlay the Darwinian theory of evolution.[8] Well into the second half of the nineteenth century, most colonial scientists continued to believe in a divine Creator and a world that evidenced his activities. As a result, some of the most prominent men of science in Australia opposed the development hypothesis: outwardly through public and classroom lectures, privately through correspondence.

Their opposition took on extra meaning when some of the chief spokesmen for religious and political institutions joined with their scientific colleagues in opposing Darwinism. The topic of man's place in nature invariably provided the catalyst for debate, even when subjects as far removed from human–ape relations as "the physical basis of life" were being discussed. Philosophical and religious concerns remained central, as evidenced by the involvement of so many prominent churchmen in the discussions.

In 1860, when the Anglican bishop of Melbourne, Charles Perry, lectured publicly on the topic "Genesis and Geology," he set out to show how modern geological theories could be reconciled with the Mosaic account of creation. Although doing so necessitated reinterpreting some of the terminology of Genesis, especially with regard to the length of the days of creation, Perry saw no problem. Instead, he stressed the remarkable parallels between the sequence of creation as set out in the biblical account and recent geological discoveries that revealed the succession of life on earth. So scientifically accurate did the Genesis story appear to Perry, he concluded that only divine revelation could explain it. He did not require a detailed agreement between Scripture and science, only a correspondence along broad lines. The parameters he set out for grappling with the relationship between science and religion remained virtually unchanged through all of his later public utterances on the subject, when both Darwin and Huxley came in for considerable criticism.[9]

During the first "Darwinian decade," many Australian scientists privately expressed their opinions on the relationship between Darwinism and religion. Writing to a politician in 1860, the Sydney naturalist

Macleay reduced the issue to the question "What is man, a created being under the direct government of his creator or only an accidental sprout of some primordial type that was the common progenitor of both plants and animals?" Macleay possessed excellent scientific credentials. His own extremely complex quinarian system of classification was much discussed by some of the leading scientific minds of the day, including the Swiss-American naturalist Louis Agassiz. Darwin himself appears to have flirted with it in the earliest stages of his investigations into the species problem. When the question of man's place in nature attracted public discussion in 1863, Macleay let it be known that although he accepted the close physical relationship of men and apes, he denied the possibility of any intellectual evolution connecting the two groups. Rejecting Darwinism, Macleay opted instead for a divinely ordained universe where "even the black tuft of hair on the breast of the turkey cock" gave evidence of providential concern. In passing, Macleay railed against those who sought to "surrender everything to the many headed monster of democracy and the rush to embrace the brutality of the mob." This juxtaposition of political and scientific opinion illustrates the cogency of the historian Robert Young's claim that scientific content can never be isolated out of the debates over Darwinism, because social, religious, economic, and philosophical factors were ever present. When John Bleasdale, a Roman Catholic priest who served as president of the Colonial Royal Society, denounced Huxley's evolutionary reasoning as a "swindle ... fit only for the half educated intellect fashioned in mechanics institutes," he revealed the nexus of these factors.[10]

A similar situation occurred when eminent social and political figures spoke out on scientific issues. Support for Halford's attack on Huxley came from the highest possible sources, most notably the governor of Victoria, Henry Barkly. In 1860 and 1861 Barkly served as president of the Royal Society of Victoria, a position that by tradition required him to address the scientific elite of the colony on the general state of science and its recent advances. In 1860, before the Darwinian issue emerged into prominence, he pleaded for scientific freedom as a religious duty, arguing that scientists ought to be unfettered in their investigation of the natural world as a counter to religious skepticism. A year later, however, he warned against the "gross and subversive theory of progressive development." Politically, Barkly was grappling with the social turmoil arising from the aftermath of the gold rushes in Victoria; in his seven years as governor (1856–1863) he witnessed six changes of government and the encroachment of that "many headed monster" that so disturbed Macleay. Barkly publicly expressed concern about the possible breakdown of social relations between labor and capital and the ensuing collapse into anarchy that would surely follow. Against such a background, a scientific theory that seemed to threaten established religious values and institutions must have appeared doubly subversive, undermining both individual morality and social stability.[11]

Barkly's gubernatorial counterpart in New South Wales, William Denison, expressed less overt concern for the social consequences of the development hypothesis, but he clearly perceived in Darwinism the same dangers to the sociopolitical system. Like many anti-Darwinians, Denison probably never read Darwin; he based his criticism on a hostile review of the *Origin of Species* in the British *Quarterly Review* and on a series of lectures given by Simon Pittard, curator of the Australian Museum in Sydney and, like Halford, a protégé of Richard Owen's. Like many other colonial anti-Darwinians, Denison dismissed the *Origin of Species* as an unoriginal work, similar to the speculative writings of Jean Baptiste Lamarck and the anonymous author of *Vestiges of the Natural History of Creation* (1844), which mainstream scientists had rejected.[12]

Frederick McCoy, Pittard's opposite number at the National Museum of Victoria, used Adam Sedgwick's massive critique of *Vestiges* as the basis for his own attacks on the development hypothesis. A former colleague of Sedgwick's in Britain, McCoy used his position at the museum, along with his chair of natural science at the University of Melbourne, to oppose the Darwinian cause. In 1869 and 1870 he gave two long lectures on "The Order and Plan of Creation" to the Early Closing Association, formed by a coalition of prominent Melbourne citizens concerned about the influence that antireligious and free-thinking groups were having on the "uneducated" classes. By persuading local businesses to close for a short period each week so their workers could attend lectures by experts in scientific and political fields, the association hoped to counter "subversive" elements. In his lectures McCoy followed Agassiz in arguing for the special creation of the various life-forms found around the world. On other occasions he drew on the ostrich, rhea, and emu to show that similarity of form and habit did not necessarily indicate common ancestry, because "there could be no question of one of these forms having grown out of the other by a difference of surroundings for the ostrich has two toes, and the Australian and South American forms three – a change not required, and all three thriving when introduced into any one of the localities by man." An Anglican, McCoy, like Perry, happily allowed the Mosaic days of creation to be of indeterminate length. He assured the faithful that there was no disagreement between the scientific and biblical accounts of life on earth and no grounds for believing in a materialist interpretation of the origin and development of life.[13]

In later years McCoy claimed that he had attempted to organize the National Museum of Victoria along nonevolutionary lines. He wanted the museum to be more than a building stuffed with specimens; it should display the interrelationships of the animal and plant kingdoms, which a proper study of the geographical distribution of life forms uncovered. Continuing problems with funding ensured that his program of museum management failed to materialize.[14]

McCoy also opposed evolution in his classes at the University of Melbourne. When William Sutherland, Australia's first world-renowned physicist, attended McCoy's lectures as a student in the 1870s, he grew annoyed listening to him "thunder against the Darwinian doctrine." Examination papers for McCoy's courses invariably included questions aimed at forcing students to criticize evolution, and one biographer has claimed that "students who incorporated the results of independent reading on evolution in their examination papers did so at their own risk." During his tenure at the university McCoy tended to avoid theoretical biology – and anything else that was new. His lectures remained almost unchanged during his three decades of teaching, allowing notes from his lectures to be "handed down from father to son, with the jokes underlined in red ink." McCoy's suspicion of new ideas may have frustrated students eager to learn of the latest developments, but Sutherland surmised that McCoy probably turned out more believers in evolution that the "enthusiastic Darwinite" Ray Lankester, whose lectures he later attended on taking up a scholarship at University College in London in 1879. Whereas Lankester was "too slavish" in his "devotion to Huxley" and spent too much time going into the anatomical details "of a very limited number of animals," McCoy, after introducing students to the first chapter of Genesis, "had taken a broad survey of comparative anatomy, beginning with the lowest types and working up to man."[15]

Like the Anglican McCoy, Ferdinand Mueller, a devout Lutheran who served as the Victorian government botanist from 1853 to 1896 and as the director of the Melbourne Botanical Gardens from 1857 to 1873, rejected evolution in part for religious reasons. Arguably Australia's best known scientist before the twentieth century, Mueller spent much of his time collecting and classifying plant specimens and promoting the benefits of acclimatization and economic biology. His approach to science conformed to Susan Faye Cannon's description of Baconianism:

> something like the collection of facts, lots of facts, in all sorts of places, and on queer applied subjects; the absence of any analytical theory or of sophisticated mathematical tools; the belief that a hypothesis will emerge somehow from the accumulation of facts . . . it means an activity very suitable for a new country where collecting new rocks and species is easy.[16]

Like other Baconians, Mueller despised the tendency of the times to theorize on insufficient grounds.

In his presidential address to the Australasian Association for the Advancement of Science in 1890, Mueller coupled high praise for the achievements of modern science with warnings about accepting "a supposed evolutionary tendency of striving for higher development and further melioration wherever circumstances are favourable [which] arises from uncontrolled impulses, so that nothing is left in stationary distinctiveness." Yet by 1890 he had softened the hard-line rejection of evolution characteristic of his early years. Though still wary of Darwinism, he had come to accept evidence of biological change, such as the paleontological discoveries in the American West of the evolution of the horse. He now grudgingly expressed appreciation for the fact that the "momentous questions" associated with evolution had given Australians the chance to participate in the global debates over the mechanisms of biological development. Nevertheless, he warned his listeners that "the world would lose many of its charms to intellectual beholders if observers sink too much into materialistic explanations and speculative reasonings," adding that scientists should "deprecate extending theories beyond what is warranted by trustworthy observation." His continuing belief in the reality of species led him into some major controversies with George Bentham, the Darwin-leaning author of the *Flora Australiensis*.[17]

Although the Australian scientific community remained generally opposed to Darwinism until the late 1880s, some educated laymen adopted a more positive attitude. Henry Gyles Turner, a Unitarian and founding member of the Melbourne Eclectic Association, a free-thought society, converted to Darwinism after reading the *Origin of Species* in 1862. While admitting to a flimsy grasp of the scientific details, he found the logical structure of the book impressive. He applauded the way in which Darwin had pulled together a number of disparate facts in natural history and approvingly quoted an unnamed friend "with scientific credentials" who had described natural selection as "the imperfect working of natural law . . . the means are cumbersome, cruel and but shiftily adapted to the end, and you are scarcely able to bring your mind to regard it as the work of omniscience." When Turner later read *Monism: The Confession of Faith of a Man of Science*, by the German Darwinian Ernst Haeckel, he described it as "a terrible book in its relentless logic . . . [but] as far as I can follow the argument the conclusions appear irresistible. . . . It would be interesting to know how the world takes this outspoken deliverance, which knocks away all the props that support the hope of a personal immortality."[18]

Upon hearing Perry attack Darwin in 1862, Turner took exception to the way in which the bishop

> in about an hour demolished to his hearts content Mr Darwin and his theory [and] turned to Sir Henry Barkly who was in the chair, and said

> that being but an indifferent naturalist himself he had felt some delicacy in lecturing on such a subject before one whose attainments in that branch of science ranked so high! Not a word about his delicacy in putting the opinion of an "indifferent" naturalist against unquestionably one of, if not the first naturalist of the age. . . . The presumptous impertinence of the remark is only surpassed by its profound snobbishness.[19]

Perry's lecture left a lasting effect on Turner. When he came to write his two-volume *History of the Colony of Victoria* (1904), he painted the bishop as a man out of touch with modern thought, one who "had so little conception of the trend of scientific investigation as to be satisfied that he had demolished Darwin and all his theories in the course of an hour's lecture." Perry's disdain was all the more lamentable in view of his longstanding interest in scientific matters. For fifteen years prior to his arrival in Australia, the Anglican divine had worked alongside William Whewell at Cambridge. In 1833 he had attended the third meeting of the British Association for the Advancement of Science, where he made the acquaintance of many prominent scientists and clerics, including Thomas Chalmers, author of one of the *Bridgewater Treatises*. Like other evangelicals, Perry did not reject the scientific enterprise per se but only the attempt to formulate and structure it in ways that made it inimical to religious faith. His critique of Darwinism rested on the sort of inductivism that underlay the providentialist approach to science so popular among nineteenth-century Australians.[20]

Protoplasm and human antiquity

The late 1860s witnessed two public arguments over the nature of science in general and Darwinism in particular, which, like the earlier Halford–Huxley controversy, echoed developments in Britain. The first of these contests centered on the meaning of Huxley's work on protoplasm as the physical basis of life; the second focused on the antiquity of the human race.

In 1869 Huxley published an article in the British *Fortnightly Review* titled "Protoplasm: The Physical Basis of Life."[21] Although there was little original in the piece, controversy flared because of two factors. First, Huxley had originally presented his paper to an audience of working men, raising in some minds the specter of antireligious materialism becoming rife among an ill-educated social class. Second, though the article contained no overt reference to Darwinism, non-Darwinians commonly linked it to the development hypothesis, because they viewed both Darwinism and protoplasm as essentially related materialistic doctrines.

Within months of appearing in Britain, the article became widely available in Melbourne in pamphlet form. The resulting furor prompted the British journal *Nature* to run a piece headlined "The Protoplasm Excitement at the Antipodes," which pilloried the intellectual condition of Australia and decried the ignorant manner in which clergymen were attacking scientific ideas that they barely understood and restauranteurs were haranguing their customers over lunch. Although the editors of *Nature* noted encouraging signs of Australia taking on its own "national character," they remained unconvinced that "scientific zeal will be one of its chief features." The University of Melbourne's refusal to allow a lecture supportive of Huxley, because it treated a "religious subject," further fueled the dispute – while simultaneously uniting some members of both sides of the Darwinian debate on the issue of censorship. Thus what might have remained a minor squabble at the edges of the wider debate about evolution turned into a major debate over scientific freedom.[22]

The protoplasm controversy focused attention on the nature of science and the correct procedures to be used in undertaking scientific investigation. Partisans of traditional providentialist science envisioned science and revelation as complementary forms of truth and embraced a simple Baconian philosophy that subordinated theory to fact (and sometimes eschewed it altogether). Such critics of Darwin and Huxley objected to what they saw as an overeagerness to speculate first and seek corroborating factual data later. The Anglican divine R. B. Higginson, for example, who accused "modern physiologists" of offering bold theories "with little or no proof," relegated the ideas of Darwin and Huxley to the same level of scientific credibility as mesmerism and spiritualism.[23]

Supporters of Darwinian evolution and the Huxleyan view of protoplasm tended to reject the notion that scientific theories could ever be proven in some ultimate sense. Like the theory of the ether in the physical sciences, wrote one newspaper reader, Darwinism explained many facts "in a wonderfully clear manner" and encountered "no facts . . . diametrically opposed to it." Besides, those wishing to explain the origin and development of life on earth had two choices: Darwinism or supernaturalism in the form of miracles. The latter was plainly unscientific because it violated the tenet that "the scientific mind" should "not accept as cause to effect those agencies which belong to the marvellous when causes less improbable can be rationally conjectured." Fortunately for theists, Darwinism did not invalidate religious belief but simply provided a mechanism for understanding "the manner in which the Creator produced the forms known to have existed."[24]

Protoplasm as a topic of public discussion had quickly become entwined in the debate over Darwinism, but talk of protoplasm virtually ceased when a more directly Darwinian issue arrived on the scene: the "antiquity of man." As one anonymous correspondent in the *Argus* put it, "Man, pre-historic or otherwise, is a vastly more interesting subject than protoplasm."[25]

The debate over the antiquity of humans began with the publication of Charles Lyell's *The Antiquity of Man* (1863), which argued that humans had been around for far longer than 6,000 years. As was often the case in Australia, a clergyman first took up the issue; in this instance, the individual in question, Julian Tenison Woods, was not only a Catholic priest but one of the best qualified geologists in Australia. Over the years he overcame sectarian prejudice to earn the respect of his scientific peers, eventually becoming president of the Linnaean Society of New South Wales in 1879. In a pamphlet facetiously titled *Not Quite as Old as the Hills* (1864), based on a lecture given the previous year in a small country town in South Australia, Woods attacked both the alleged antiquity of humans and the uniformitarian doctrine in geology, of which Lyell was the best known champion. Woods argued that there was no certain way of dating the distant past or of obtaining an objective measure of the time humans had been on earth, which he estimated to be about 6,000 years.[26]

Although Wood's pamphlet provoked much discussion about human antiquity, the Reverend John Bromby's 1869 lecture on "Prehistoric Man," given to the Early Closing Association in Melbourne, generated even greater public interest. Politically, Bromby, headmaster of the prestigious Melbourne Church of England Grammar School, advocated Christian socialism; theologically, he embraced "conditionalism," an anti-Calvinist doctrine that held that humans possessed the option of choosing eternal life by accepting the saving grace of Christ, that "God created man mortal but with the capacity of immortality." After the publication of the *Origin of Species*, conditionalists often used the language of Darwinism to describe their position, for example, calling persons who would win eternal life the "fittest to survive," because they had made the right choices on earth. The use of such analogies often allowed conditionalists to accept some of the more controversial aspects of scientific investigation. Bromby himself welcomed new scientific truths because he firmly believed that they could do no real damage to correctly interpreted Christian beliefs.[27]

Though Bromby stopped short of fully embracing Darwinism, he sympathetically used contemporary scientific literature favorable to evolutionary theory in arguing that geological and archaeological evidence made nonsense of the widely accepted 6,000-year dating of human history. Bromby, for instance, extracted much of the empirical content of his lecture on prehistoric man from Lyell's book, a work contemporaries commonly regarded as supportive of Darwin's claims. To harmonize scientific findings about prehistoric beings with the biblical account of Adam and Eve, he invoked a pre-Adamite world populated by socially developing humanlike creatures that might have resulted from biological evolution. When God singled out Adam for special blessings, he created a new creature, a fully human being, who possessed body, soul, and spirit.[28]

Among the critics of Bromby's lecture was his own bishop, Charles Perry, who presided over the meeting and found Bromby's presentation "able and interesting" though unconvincing. In a lecture entitled "Science and the Bible," Perry ranged widely across the spectrum of controversial scientific theories, including those associated with the *Vestiges*, Darwin's *Origin of Species*, and Huxley's essay on protoplasm. As in his lecture on "Genesis and Geology" nine years earlier, Perry stressed at the outset that he was not pitting the Bible against science or seeking to define the boundaries of each in ways that divided them into separate intellectual compartments. However, as one who believed in the accuracy of "every statement" contained in the Bible, he repudiated Bromby's endorsement of pre-Adamites and the social evolution of humans. For him, the coexistence of the Australian aborigines' stone-age culture with that of modern Europeans illustrated the inadequacy of claims that culture must develop progressively. It seemed more likely, and more biblical, that the aborigines and other primitive peoples had degenerated from perfectly created human beings.[29]

Perry's address set the high-water mark of the debate over human antiquity. An impressive array of Melbourne's social elite turned out to hear his message, indicating the extent to which he was responding to establishment fears about growing liberalism in science and religion. The audience included the governor's wife and daughters, a number of well-known judges, prominent politicians and members of parliament, the head of the Presbyterian church in Victoria, and Frederick McCoy. At the conclusion of Perry's impassioned talk, the audience broke into "long and repeated applause," and McCoy moved a vote of thanks that was carried by acclamation. Perry had upheld the uneasy alliance between true science and revealed religion – and with it the authority of those invested with the task of maintaining social order in the face of "the brutality of the mob."[30]

Darwinism triumphant?

In 1883 the socially prestigious Scots Presbyterian Church in Melbourne sacked its pastor, Charles Strong, for allowing George Higginbotham to lecture on science and religion from the pulpit. Higginbotham, the greatest legal figure in Australia at the time, had shocked the congregation by telling them that they were living through the death throes of Christianity as they knew it because science had shown it to be false. The furor generated by "the Strong Case," as it came to be called, symbolized the extent to which debates over the relation of science and religion in general, and Darwinism in particular, extended into the very heart of Australian cultural life. Despite Sir Henry Barkly's fear of social ruin coming on the heels of the Darwinian theory, that theory itself quickly became simply another social resource that even some within the establishment could profitably use. Not only could a respected judge expound on the subject, but a prominent minister could use it as inspiration for new religious insights. Strong turned his back on Presbyterianism to found the Australian Church, a curious theological amalgam of science, as represented by Darwin, Huxley, and Herbert Spencer, and religion, as represented by liberal Christianity and spiritualism.[31]

The early debates over Darwinism in Australia, including the Strong case, belie the claim that science and theology were engaged in an ongoing "warfare." The surveyor Fitzgerald, introduced at the beginning of this chapter, found time to produce a massive testament to Darwinism with his work on Australian orchids, findings that Darwin incorporated into the second edition of his *Fertilization of Orchids* (1877). Nevertheless, Fitzgerald remained on friendly terms with several of the leading anti-Darwinians in the country: the botanist Ferdinand von Mueller, the internationally respected cleric-scientist William Woolls, and Charles Moore, the director of the Sydney Botanic Gardens. Fitzgerald, Woolls, and Moore made countless botanizing trips around New South Wales, while apparently never shying away from ridiculing one another's beliefs on evolution. When Fitzgerald died, Woolls wrote a moving poetic tribute to his friend, honoring him as the foremost authority on Australian orchids and referring sympathetically to his belief in "development."[32]

The concept of divine design in nature may have underlain the debates about human origins and protoplasm, but behind that notion stood a greater challenge: identifying the true nature of biological science and its role in arbitrating knowledge claims. What would replace the older teleological view of biology? Huxley explicitly aimed to separate biology from theology and thus make it as scientific as physics or astronomy. To men such as Perry, McCoy, Mueller, and Macleay, this was tantamount to amputating one of the major limbs from the body of science. Worse still, as both sides recognized, the naturalization of biology was the thin end of a wedge that would sever the realms of science and religion, secularize public policy, and undermine traditional mores and social institutions.

Participants in the Australian Darwinian debates did not divide clearly along political or denominational lines, though a mixture of social, religious, and philosophical commitments often influenced which side an individual or group would take. The implications of Darwinism for such commitments, rather than Darwin's theory itself, inflamed passions the most. The clerics Perry, Bromby, and Higginson and the scientists McCoy and Halford were all Anglicans; yet they responded to Darwinism in at least three different ways. Bromby tried to accommodate evolutionary theory to religious belief; Perry, Higginson, and probably McCoy wanted no truck with evolution under any circumstances; and Halford wished to adopt Huxley's secularist biology and to keep the clergy out of science while continuing to embrace a providentialist epistemology. The Roman Catholic Woods was prepared to examine the possibility of evolution, while his fellow priest Bleasdale believed that the reasoning behind the theory of evolution was fundamentally a "swindle."

Major shifts in the scientific and religious communities occurred in Australia in the 1880s and 1890s. The winds of change blew through the universities in Melbourne and Sydney, bringing in new men trained in what Roy MacLeod has termed the Huxleyan education program. William Haswell, a European-trained biologist who had studied under Huxley himself, arrived in Australia in 1878. After serving for a period as a demonstrator and lecturer at the University of Sydney, in 1889 he became Challis Professor of Biology at that institution, where he fought mightily for science "against the entrenched forces of Arts." With T. J. Parker of the University of Otago in New Zealand he wrote *A Text-book of Zoology* (1897), an international success that went through four editions by 1928. The hand of Huxley is clearly evident in the authors' desire to place the life sciences on a "sound" scientific footing, that is, within a an evolutionary framework.[33]

In Melbourne the science curriculum received a heavy shakeup when Walter Baldwin Spencer and Arthur Dendy arrived at the university in 1887. Spencer, who trained under Arthur Milnes Marshall, the "fiery apostle of evolution" at Victoria University, Manchester, put evolution squarely before his students in the form of Darwin's and Alfred Russel Wallace's books and in talks to the university science club. In the 1890s he collaborated with a South Australian postmaster, Frank Gillen, to document the cultural life of the aboriginal population of central and northern Australia, influencing the work of James Frazer and Emile Durkheim among others. Dendy, also a former student of Marshall's, joined Spencer as demonstrator and quickly threw himself into studying the local fauna, especially terrestrial invertebrates, and promoting evolution. Other colonial Darwinists, such as the plant pathologist Daniel McAlpine of Victoria, worked for the burgeoning government bureaucracies of pre-Federation Australia, while William Farrer, Australia's most famous wheat breeder, worked independently in the Darwinian tradition.[34]

In 1888 Australian scientists from the various colonies came together to form the Australasian Association for the Advancement of Science (AAAS), modeled on its British and American counterparts. This organization provided a forum where professional scientists could meet with educated amateurs once a year to hear the latest in scientific knowledge. Revered elder statesmen such as Mueller might call for caution in too readily accepting new theories, but the great majority of scientists who presented papers to the biology section of the AAAS expressed no qualms over adopting the theory of evolution. In his 1890 presidential address to the section, the New Zealand biologist A. P. Thomas lamented that the "*Origin of Species* had never entered the doors of a museum." The following year Haswell surveyed the various evolutionary theories associated with Carl Nageli, Thomas Eimer, Hugo de Vries, and others, while the president of the association, Sir James Hector, a former pupil of Huxley's, praised the "generalisation" of Darwin and Wallace.[35]

The evidence strongly suggests that by the last decade of the century, Darwinism in its broadest sense had become firmly entrenched within the teaching and research institutions of Australia, and Australian scientists were gaining the self-confidence to participate in the theoretical debates over biological evolution. The late nineteenth century also saw a range of visitors carrying their various views on evolution to Australia. In 1883 the American secularist Moncure Conway, a one-time student of Agassiz's, drew large crowds in Melbourne, Sydney, Hobart, and Brisbane, lecturing on such topics as "Darwin," "The Evolution of Woman," and "The Pre-Darwinite and Post-Darwinite World." By this time the tensions over Darwinism were beginning to subside. In Melbourne Conway dined with both the sacked cleric Charles Strong and James Moorhouse, a theological modernist who had succeeded Perry as Anglican bishop. The governor of Victoria invited Conway to dinner, and the premier of New South Wales attended his lectures in Sydney.[36]

Conway's free-thinking attitudes, reflected in his views on ethical principles and popular science, especially Darwinism, clearly appealed to many Australians. His disdain for organized religion, however, repelled some Christians. Their needs were better met by another visitor, the Scottish theologian Henry Drummond, who arrived in 1890, spreading the gospel of evolutionary Christianity that four years later he would publish under the title *The Ascent of Man*. Drummond's talks to selected audiences were closed to the press, but by his own account they stressed the compatibility of Christianity and evolution. One of his Melbourne meetings drew 140 doctors, all but four of them non-churchgoers. Drummond's openness to evolution helped him convert large numbers of Australian unbelievers to Christianity and strengthened others racked by the "Victorian crisis of faith."[37]

Conclusions

In the decades following the first arrival of Darwinism in Australia, the social and intellectual life of the continent underwent dramatic change. By the end of the century, four universities were offering instruction in science, government bureaucracies were employing an increasing number of professional scientists trained under the Huxleyan program, and the AAAS was promoting the production of local scientific research. The "new men" at the helm of Australian science still came largely on the recommendation of committees "back home" in Britain, but they brought with them a science stripped of theological and providential aspirations. Spencer, Haswell, McAlpine, Farrer, and a host of lesser minds represented a breed altogether different from such predecessors as McCoy, Mueller, Macleay, and Halford. The former were not only confident of their own abilities and excited by their prospect of contributing to the advancement of universal science, but sure that science could satisfy the material wants of humanity.

In the religious sphere, at least in the mainstream churches, the thoroughgoing biblical faith of men such as Perry was conceding to a more cautious, even doubting, faith, where questions rather than answers seemed to be the order of the day. Moorhouse, Perry's successor, had learned his theology during the heyday of critical biblical studies and schemes for harmonizing evolutionary theory with Christian belief. In his later years he praised Bergson's work, noting that it was "stimulating to find the conclusions of science called in aid of the speculations of philosophy: to find the haughty a priori methods of metaphysics forced to stoop their proud heads to look prosaic facts in the face." Such a Huxley-like phrase could never have been penned by Perry, for whom science and religion "rightly understood" could never come into conflict.[38]

As free thought and spiritualism prospered in Australia in the 1890s, the fledgling theosophical movement found a local footing, albeit largely temporarily. Combining elements of traditional belief with scientific language, the prophetess Annie Besant told her Australian audiences during her 1908 speaking tour that materialism was untenable, that science was discovering the spiritual side of human beings. "The man of genius," she declared, "shows you what the human race shall be. He is the prophecy of the future. He is not the product of degeneracy. He shows us what all men shall become at a stage of higher evolution."[39]

The dark forebodings of Barkly and Macleay regarding the likely social effects of Darwinism on "the mob" had come to nothing in the intervening years. The Australian colonies peacefully formed a federation and in the main prospered. Popular journals such as the *Bulletin* looked forward to the day when a new and superior variety of the Anglo-Saxon race would appear in the favorable climate of Australia, a day when the "uncivilized" indigenous population would disappear, swept aside by the inevitable process of biological and social evolution.[40] Australia's relation to "the development hypothesis" had changed immeasurably since the first arguments over gorilla feet and protoplasm. Darwinism may not have been triumphant at the end of the century, but it certainly was triumphalist.

Notes

1 Advertisement, *Sydney Morning Herald*, 18 April, 1860; William Edward Hearn, *Plutology; or, The Theory of the Efforts to Satisfy Human Wants* (Melbourne: George Robertson, 1863). Leslie Stephen, in "Review of *Plutology*, by William Edward Hearn," *Reader*, 19 March, 1864, p. 142, noted that Hearn's was the first attempt to apply evolutionary theory to economics. The following essay is based in large part on Barry W. Butcher, "Darwinism and Australia, 1836–1914," Ph.D. diss., University of Melbourne, 1992.

2 Butcher, "Darwin's Australian Correspondents: Deference and Collaboration in Colonial Science," in *Nature in Its Greatest Extent*, ed. Philip F. Rehbock and Roy MacLeod (Honolulu: University of Hawaii Press, 1988), pp. 131–58.

3 Ann Moyal, "Evolution and the Climate of Opinion in Australia, 1840–76," *Victorian Studies*, 10 (1965):411–30. See also Butcher, "Darwin's Australian Correspondents."

4 James A. Secord, "John W. Salter: The Rise and Fall of a Victorian Palaeontological Career," in *From Linnaeus to Darwin*, ed. A. Wheeler and J. H. Price (London: Society for the History of Natural History, 1985), pp. 61–75.

5 William Archer, *Tourist to the Antipodes*, ed. Raymond Stanley (St. Lucia: University of Queensland Press, 1977), p. 33.

6 Butcher, "Gorilla Warfare in Melbourne: Halford, Huxley and 'Man's Place in Nature,'" in *Australian Science in the Making*, ed. R. W. Home (Cambridge: Cambridge University Press, 1988), pp. 153–69.

7 Ibid., p. xx.

8 Moyal, "Evolution and the Climate of Opinion."

9 Perry's lecture was covered extensively in the *Age*, 28 September, 1860.

10 William Sharp Macleay to Robert Lowe, May, 1860, quoted in Moyal, *Scientists in Nineteenth-Century Australia: A Documentary History* (Melbourne: Cassell Australia, 1976), pp. 190–92; Louis Agassiz, *Essay on Classification* (London: Longman, 1859), pp. 344–45; Dov Ospovat, *The Development of Darwin's Theory* (Cambridge: Cambridge University Press, 1981), pp. 101–13;

Macleay to W. B. Clarke, 27 June, 1863, Clarke Papers, Mitchell Library MSS 139/11, State Library of New South Wales, Sydney; Robert M. Young, "Darwinism *Is* Social," in *The Darwinian Heritage*, ed. David Kohn (Princeton: Princeton University Press, 1985), pp. 9–38; and Young, *Darwin's Metaphor: Nature's Place in Victorian Culture* (Cambridge: Cambridge University Press, 1985), pp. 126–63. Bleasdale's comments, which came at the conclusion of Halford's lecture on the gorilla at the Royal Society of Victoria, appeared in *Argus*, 25 July, 1865.

11 Henry Barkly, "Presidential Address to the Royal Society of Victoria," *Transactions of the Royal Society of Victoria*, 5 (1860):12; Barkly, "Presidential Address to the Royal Society of Victoria," *Transactions*, 6 (1860):xix-xxxiv; and Geoffrey Serle, *The Golden Age: A History of Victoria, 1851–61* (Melbourne: Melbourne University Press, 1963), p. 245.

12 William Denison to Charlotte Denison, 5 November, 1860, quoted in William Denison, *Varieties of Vice-Regal Life* (London: Longman, 1870), p. 495.

13 Frederick McCoy, "The Order and Plan of Creation: The Substance of Two Lectures Delivered before the Early Closing Association, 1869–70," in *Lectures Delivered before the Early Closing Association, Melbourne, 1869–70* (Melbourne: Mullen, 1870). For the views of Agassiz and Sedgwick, see Adrian Desmond, *Archetypes and Ancestors: Paleontology in Victorian London, 1850–75* (Chicago: University of Chicago Press, 1982), especially pp. 102–07.

14 McCoy Papers, National Museum of Victoria. This reference was pointed out to me by Sally Gregory Kohlstedt.

15 W. A. Osborne, *William Sutherland: A Biography* (Melbourne: Lothian, 1920), pp. 23, 30–31; and Ernest W. Skeats, "Sir Frederick McCoy," *Melbourne University Magazine*, 9 October, 1928, pp. 18–19.

16 Susan Faye Cannon, *Science in Culture: The Early Victorian Period* (New York: Dawson/Science History, 1978), p. 73. For Mueller's life and work, see Margaret Willis, *By Their Fruits: A Life of Ferdinand von Mueller* (Sydney: Angus & Robertson, 1949).

17 Ferdinand von Mueller, "Presidential Address," *Report of the Second Meeting of the Australasian Association for the Advancement of Science* (Melbourne: The Australasian Association, 1890), pp. 8–9.

18 Henry Gyles Turner, "Notes on Books Read Subsequently to June 1862, Resumed 1895," Turner Papers, LaTrobe Library, Victoria, M1670, Box 464/4. For Turner's life and work, see Iain McCalman, "Henry Gyles Turner," in *Australian Dictionary of Biography*, 16 vols. (Melbourne: Melbourne University Press, 1966), vol. 6, pp. 311–13.

19 Turner Papers, LaTrobe Library.

20 Ibid.; Turner, *A History of the Colony of Victoria*, 2 vols. (London: Longman, 1904), vol. 1, p. 269; and A. de Q. Robin, *Charles Perry, Bishop of Melbourne* (Nedland: University of Western Australia Press, 1967), p. 15. Perry's attendance at the BAAS meeting at Oxford is documented in Jack Morrell and Arnold Thackray, *Gentlemen of Science: Early Years of the British Association for the Advancement of Science* (Oxford: Clarendon Press, 1981), p. 173. On evangelicals and science, see George M. Marsden, "Evangelicals and the

Scientific Culture: An Overview," in *Religion and Twentieth-Century American Intellectual Life*, ed. Michael J. Lacey (Cambridge: Cambridge University Press, 1989), pp. 23–48.

21 Thomas Henry Huxley, "On the Physical Basis of Life," *Fortnightly Review*, new ser., 5 (1869):129–45. On the protoplasm affair in Britain, see Gerald L. Geison, "The Protoplasmic Theory of Life and the Vitalist-Mechanist Debate," *Isis*, 60 (1969):273–92.

22 Huxley, *Protoplasm: The Physical Basis of Life* (Melbourne: Peter C. Alcock, 1869), pp. 15–16; "The Protoplasm Excitement at the Antipodes," *Nature*, 4 November, 1869, p. 13; *Argus*, 1, 2, and 31 July, 1869; *Australian Medical Journal*, August (1869):259. On the Australian debate, see also *Argus*, 23 July and 20 and 27 August, 1869; McCoy, "The Order and Plan of Creation," pp. 12–16; John Bromby, "Creation versus Development: A Lecture," in *Lectures Delivered before the Early Closing Association*, pp. 7–16; Archer to Huxley, 5 October, 1869, Huxley Papers, Imperial College, London. For Higginson's views, see *Argus*, 7 and 11 June, 1869.

23 *Argus*, 1 July, 1869. See also Charles Perry, *Science and the Bible* (Melbourne: Mullen, 1869), pp. 3–4.

24 *Argus*, 2 July, 1869.

25 *Argus*, 24 August, 1869.

26 D. H. Borchardt, "Julian Tenison Woods," in *Australian Dictionary of Biography*, vol. 6; pp. 254–55; Julian Tenison Woods, *Not Quite as Old as the Hills: A Lecture on the Evidences of Man's Antiquity* (Melbourne: Dolman Dwight, 1864). In later years Woods seems to have softened his stand on Darwinism; see Woods, "Presidential Address," *Proceedings of the Linnaean Society of New South Wales*, 4 (1879):474–76.

27 *Argus*, 10 August, 1869, which reported extensively on John Bromby's lecture on "Prehistoric Man," given the day before to the Early Closing Association; C. M. H. Clark, "John Bromby," in *Australian Dictionary of Biography*, vol. 3, pp. 242–43. For a discussion of conditionalism, see Geoffrey Rowell, *Hell and the Victorians* (Oxford: Clarendon Press, 1974), Chap. 9.

28 *Argus*, 10 August, 1869.

29 Ibid., which mentions Perry's positive comments about Bromby's lecture; Perry, *Science and the Bible*, pp. 3–4, 16–23. For additional criticism of Bromby, see *Argus*, 12, 19, 21, 23, and 24 August, 1869. See also Bromby's replies to his critics in *Argus*, 23 and 31 August, 1869; and Bromby, "Creation versus Development," pp. 16–23.

30 *Argus*, 21 September, 1869.

31 For a rather antagonistic review of the Higginbotham incident and the Strong case, see Aeneas MacDonald, *One Hundred Years of Presbyterianism in Victoria* (Melbourne: Robertson & Mullens, 1937), pp. 125–40. Strong's theological position after his dismissal from the Presbyterian church can be found in his collection of sermons, *Christianity Re-Interpreted and Other Sermons* (Melbourne: George Robertson, 1894).

32 Woolls's poem appears in MS 581.995/W, Mitchell Library, Sydney.

33 Patricia Morison, "William Aitcheson Haswell," *Australian Dictionary of Biography*, vol. 8, pp. 226–27.

34 For Spencer's life and work, see D. J. Mulvaney and J. H. Calaby, *"So Much That Is New": Baldwin Spencer, 1860–1929* (Melbourne: Melbourne University Press, 1985). On Dendy, see Brian J. Smith, "Arthur Dendy," *Australian Dictionary of Biography*, vol. 8, pp. 279–80; and Peter J. Bowler, *The Eclipse of Darwinism: Anti-Darwinian Evolution Theories in the Decades around 1900* (Baltimore: Johns Hopkins University Press, 1983), pp. 79, 170, 172. McAlpine's work is discussed briefly in J. H. Willis, "Botanical Pioneers in Victoria," *Victorian Naturalist*, 66 (1949): 107–08. For Farrer's life and work, see Archer Russell, *William James Farrer* (Melbourne: F. W. Cheshire, 1949).

35 F. von Mueller, "Presidential Address," *Report of the Second Meeting of the Australasian Association for the Advancement of Science* (Melbourne: The Association, 1890), pp. 8–9; A. P. Thomas, "Address to Section D," ibid., pp. 103–09; William Haswell, "Address to Section D," *Report of the Third Meeting of the Australasian Association for the Advancement of Science* (Wellington, NZ: The Association, 1891), pp. 174–86; James Hector, "Presidential Address," ibid., pp. 15–16. For criticisms of Wallace, see F. W. Hutton, "Address to Section D,"*Report of the Fourth Meeting of the Australasian Association for the Advancement of Science* (Hobart: The Association, 1892), pp. 365–66; and C. H. Hedley, "Address to Section D," *Report of the Fifth Meeting of the Australasian Association for the Advancement of Science* (Adelaide: The Association, 1893), pp. 444–46.

36 On Conway's Australian visit, see his book, *My Pilgrimage to the Wise Men of the East* (London: Archibold Constable, 1906), pp. 70–104.

37 For a description of Drummond's Australian tour, see George Adam Smith, *The Life of Henry Drummond* (London: Hodder & Stoughton, 1899), pp. 358–72.

38 On Moorhouse, see Edith C. Rickards, *Bishop Moorehouse of Melbourne and Manchester* (London: John Murray, 1920).

39 Annie Besant, *Australian Lectures: 1908* (Sydney: George Robertson, 1908), p. 157.

40 For a discussion of the *Bulletin*'s attitude toward the future of the "Australian race," see Beverley Kingston, *Glad, Confident Morning*, vol. 3 of *The Oxford History of Australia* (Melbourne: Oxford University Press, 1988), pp 106–07.

3

Darwinism in New Zealand, 1859–1900

JOHN STENHOUSE

In 1860 Samuel Butler, a young Englishman later famous as a novelist, sailed into Lyttelton harbor with the Bible and *The Origin of Species* in his baggage, leaving far behind an angry clergyman father and the Anglican Establishment. Within two years, Butler, now a high-country farmer, began enlightening New Zealanders on the merits of Darwin's theory. His first article, published in *The Press* of Christchurch, presented a dialogue between free thinker "F," an ardent Darwinian speaking for Butler himself, and "C," a devout and simple-minded Christian, who found evolution "horrid" and "utterly subversive of Christianity." Free thinker attempted to enlighten Christian by pointing out that illustrations of evolution could be observed everywhere in New Zealand. For example, the competition within a population of wild cats on sheep stations such as Butler's own Mesopotamia illustrated Darwin's struggle for life. Competing with one another for a diminishing supply of quail, only the fittest cats survived.[1]

The Reverend C. J. Abraham, the Anglican bishop of Wellington, writing under a pseudonym, picked up the gauntlet that Butler had thrown down. As far as he was concerned, the *Origin of Species*, which he had recently read, simply rehashed the speculations of earlier writers such as Erasmus Darwin. "Were it not for their supposed effect upon religion," Abraham declared, "no one would waste his time in reading about the possibility of polar bears swimming about and catching flies so long that they at last get the fins they wish for." These disparaging remarks angered Butler, who identified the author by consulting the editor of *The Press*. For "nicknames" to misrepresent a theory that "the British Association is discussing with great care in England," he fulminated, simply demonstrated how low some people would go in opposing modern science. Butler and Bishop Abraham continued to argue about evolution for some months, with Butler publishing two further articles in *The Press*, including "Darwin among the Machines," which foreshadowed his novel *Erewhon*.[2]

Though fascinating, this stereotypical controversy between Butler and the bishop inadequately represents New Zealand's early response to Darwinism. Few New Zealanders took the extreme positions of either Butler or Abraham. Julius Haast, for example, the only professional scientist to respond to the dispute, took a middle-of-the-road stance. Haast arrived in New Zealand from Austria in 1858, became Canterbury provincial geologist in 1861, founder of the Philosophical Institute of Canterbury in 1862, and a leading figure in the scientific and cultural life of the country until his death in 1887. In a letter to *The Press* written after the first exchange between Butler and Abraham, Haast suggested that "Darwin's great theory" could satisfactorily explain the origins of every living creature except humans. "We happen to know that man was created in a certain image," Haast argued, "and that he is specially distinguished above all animals."[3] On this point colonial scientist and Anglican bishop agreed.

A growing number of people in the major Protestant denominations followed Haast in accepting the biological evolution of plants and animals. Debate in New Zealand reached a peak in the decade 1875–1885, during which some colonial institutions, especially in Dunedin, were rocked by controversy. But by 1885 the country's main scientific, educational, and religious institutions had integrated evolution into their views of the world. Opposition to evolution existed, but on the margins of mainstream culture. In fact, New Zealanders responded to Darwin with more ease, greater enthusiasm, and less social division than the inhabitants of any other region in the English-speaking world.

To understand why, we must appreciate the youth of the cultural institutions of mid-nineteenth-century New Zealand when compared with those of the Old World and the English-speaking parts of the New, such as the United States, Canada, and Australia. The youngest of Britain's white settler colonies, and the first to be established after the Enlightenment, New Zealand remained largely untouched by Pakeha (that is, whites in Maori) until organized settlements were established from 1840 in the North Island, 1848 in Otago, and 1850 in Canterbury. The total population grew to almost a million by the end of the century, around 50,000 of whom were Maori. Slightly over half the Pakeha population came from England, hardly any from Wales, about a quarter from Scotland, and just over a fifth from Ireland.[4] A slight English majority faced large Scottish and Irish minorities. Culture, ethnicity, and history differentiated the white population.

Old World class divisions failed to flourish, however. Few upper-class aristocrats settled permanently in New Zealand. Most settlers came from lower-middle-class or respectable working-class backgrounds, disliked class distinctions, and hoped to build a freer and more egalitarian society in the Antipodes.

Most New Zealanders, able to read and write, could familiarize themselves with Darwin's writings if they so chose. The 1877 Education Act, which established a free, secular, and compulsory system of state schools, consolidated New Zealand's reputation as one of the most literate societies in the world. Four towns grew large enough to facilitate the development of intellectual culture: Auckland in the far north, Wellington on the southern end of the North Island, Christchurch in the middle of the South Island, and Dunedin in the far south. The cultural institutions established in these cities – churches, universities, schools, public libraries, scientific and literary societies, and so on – brought a diverse Pakeha population together, sometimes creating sparks.

Around eighty percent of the population professed to be Protestants in 1881, compared with about fourteen percent who identified themselves as Roman Catholics. New Zealand possessed a larger proportion of Catholics than England or Scotland, but proportionally fewer than Ireland, the United States, Canada, or Australia. Almost all New Zealand Catholics came from Ireland.

No denomination possessed a majority of even nominal adherents. The Church of England (almost entirely English, but including some Anglo-Irish) came closest at slightly over two fifths of the population; Presbyterians (largely Scottish) comprised about a quarter; Methodists (mostly English), almost a tenth. The first waves of settlers came to New Zealand a decade after Britain began lifting legal restrictions on dissenters and Catholics. This liberal tide flowed over to New Zealand.

No state church took root in New Zealand soil. Determined non-Anglican opposition destroyed the few early attempts made to transplant the English Establishment. Contemporaries remarked on the religious freedom of the colony. Charles Southwell, for example, a lecturer, publisher, and notorious English atheist, had been imprisoned for blasphemy in the Bristol jail in the early 1840s. But, shortly after arriving in Auckland in 1856, he praised his adopted land for opening "to all alike the paths of political fame and social power" without regard to religious belief. Southwell saw little need to advance atheism aggressively in New Zealand. Compared with the *Oracle of Reason* that he had published in England, Southwell's *Auckland Examiner and People's Journal*, which he brought out between 1856 and 1860, appeared almost pious.[5]

The religious temperature of the colony seldom rose above lukewarm. The revival fires that scorched North America and parts of Britain during the nineteenth century found few parallels in nineteenth-century New Zealand. There, all the denominations, including the Catholics, sponsored touring revivalists and missioners, but compared with the mercurial Methodists and Bible-thumping Baptists blazing a gospel trail across North America, revivalists made comparatively little impact on the major churches. The most effective Protestant revivalists, such as those associated with the Plymouth Brethren and the Salvation Army, represented only a tiny minority of the population. The affable Anglicans who made up almost half the colony's population generally disliked religious enthusiasm. Thus conservative Protestantism, though popular, never entrenched itself as solidly in New Zealand as it did in the United States and Canada.

Comparatively few New Zealanders attended church regularly. The 1851 census in Britain revealed that over a third of the population regularly attended services. In New Zealand, in contrast, less than one quarter of the population attended church in 1874, and the proportion of regular attenders never rose above 30 percent during the entire nineteenth century.[6] Most New Zealanders had emigrated to get on in life rather than to build the new Jerusalem.

In this raw, young colony, many churchmen found themselves too busy establishing churches and schools and ministering to widely dispersed flocks to concern themselves greatly about science. Bishop Abraham's early dismissal of Darwinism must not lead us astray on this point. No Anglican discussed evolution in general synod or Auckland diocesan synod sermons between 1859 and 1875. Pastoral concern to avoid disturbing the faith of recently transplanted and intellectually unsophisticated flocks probably accounts for some of this neglect. The Reverend H. W. Harper, for example, Archdeacon at Hokitika on the west coast of the South Island, believed it "unwise" to discuss evolution with "our local Coast people," because the materialistic views of scientists such as Darwin and Huxley "might disturb minds utterly incompetent to discuss them." Colonial Anglicans appear also to have looked to English authorities to provide a lead. In 1873 the New Zealand *Church Gazette* offered the theistic evolutionary views of leading English scientists, such as Dr. William Carpenter, and churchmen, such as Canon H. P. Liddon, to enlighten New Zealanders on the subject of Darwinism.[7]

Race relations rather than science preoccupied Anglican churchmen during the interracial warfare of the 1860s and 1870s. Bishop George Selwyn, Abraham, and Archdeacon Octavius Hadfield devoted enormous energy, time, and ink to racial matters, criticizing government policy, settler racial attitudes, and Maori behavior. Occasionally they linked racism and biology, attacking in particular the idea, popular among many settlers, that the Maori were doomed to disappear before the incoming white man. This issue will be discussed in more detail below.

Evangelical churchmen (that is to say, non-Anglican Protestants) began publicly accepting evolution from about 1870, slightly earlier perhaps than their American brethren.[8] In Dunedin in 1870, Mr. Smythies, a local barrister, assured a small audience in Trinity Wesleyan Church that evolution should be seen as compatible with Christianity, providing that the Genesis stories of Creation and Fall were interpreted allegorically rather than literally. In Christchurch in 1872 the Reverend William Habens, a young Congregationalist minister, told his audience that Christians had nothing to fear from Darwin's theory. "Far from leading to infidelity," Habens argued, "natural evolution was even a more wonderful thing than the creation itself."[9]

Protestant evolutionists were generally well-educated men determined to keep the churches abreast of the latest science and scholarship. The Reverend Charles Fraser, for example, the leading Presbyterian minister in Anglican-dominated Christchurch in the 1870s, had long been active in the cultural and intellectual life of the city. He helped to found the Philosophical Institute of Canterbury in 1862 and counted the geologist and explorer Julius Haast among his friends. Fraser, an able mathematician and self-taught naturalist, was elected a fellow of the Geological Society of London in 1867. As early as 1873 he expounded the merits of evolution to a small audience at his St. Andrews church, clearly explaining the argument of *The Origin*, even distinguishing between natural selection and other mechanisms of evolution.[10] Fraser's espousal of theistic evolution does not seem to have aroused much controversy.

Otago, the province immediately to the south of Canterbury, saw debate break out in earnest three years later. It had systematically been settled in 1848 by Scottish Presbyterians five years after the Reverend Dr. Thomas Chalmers had led the evangelical party out of the established Church of Scotland to form the Free Church. The first leaders of the Otago settlement, Captain William Cargill and the Reverend Thomas Burns, ardent Free Churchmen, hoped to shape the province along Calvinist lines. Yet Otago never became an exclusively Free Church settlement, and Presbyterians, though more numerous in the far south than in the rest of the country, soon lost majority status even in their own province. Conflict between Scottish Presbyterians and a powerful English and Anglican minority had occurred from first settlement. The latter group, prosperous, urbane, often appointees of the Anglican governor, refused to allow Dunedin to be turned into the Edinburgh of the south, much to the annoyance of the Scots.[11] Dunedin boomed during the Otago gold rush of the 1860s, becoming the largest, richest, most vibrant city in the country.

In 1876 the Dunedin Young Men's Christian Association (YMCA) blew apart over evolution. This incident is worth examining in detail for three reasons. First, it represents the most explosive controversy involving evolution to occur within New Zealand religious institutions. Second, it illustrates how powerfully factors such as sectarianism, ethnicity, politics, and personality informed the evolution debates, and how incomplete scientific and theological ideas alone are in explaining such conflicts. Third, it highlights the significance of religious geography. Like most explosions over Darwinism in New Zealand, this incident occurred in Dunedin. That the far south almost monopolized evolutionary controversy suggests that New Zealand, though small, contained sufficient regional diversity to render problematic nationwide generalizations about science and religion.

Early in 1876 the Reverend Alfred Robertson Fitchett, the liberal-minded minister of the city's leading Methodist church, advocated evolution in sermons to his Trinity Wesleyan Congregation and published a pamphlet reconciling evolution and Christianity. He dismissed the traditional understanding of the Genesis account of creation as "the popular craving for creation by flat." As he saw it, the accounts of Creation and Fall in Genesis constituted allegories. These passages taught profound moral and spiritual truths not in precise scientific terminology but in those of "a child's first lesson book." Humans had evolved, body and mind, from apelike ancestors.[12]

Shortly after announcing these evolutionary views, Fitchett applied to join the Dunedin YMCA. Five out of eleven members of the board of management blackballed his application, and a general meeting of the association was called to consider the case. The YMCA had originally been set up as an interdenominational organization in which evangelical Protestants could co-operate in evangelizing young men. Membership was open to all who accepted Christ as their "only Savior" and acknowledged the "Divine inspiration and authority of Scripture." But, as this debate over Darwinism would reveal, even this minimal credo could not contain the growing social, intellectual, and religious tensions within the Dunedin Protestant community.[13]

For the Fitchett hearing, the YMCA hall was packed with 120 worthy Dunedin citizens: Scottish Presbyterians, English Anglicans, Methodists, Congregationalists, and Baptists. Robert Borrows, a physician, a member of Fitchett's Trinity Wesleyan Congregation, and a close personal friend of his, spoke first. Evidently annoyed at the insult to Methodism that the blackballing represented, he began by criticizing the YMCA's board of management for rejecting a trained and respected theologian who had been ordained by the Wesleyan Methodist Conference. Fitchett believed in the inspiration and authority of Scripture, Borrows declared, and the board had no right to insist that members hold a particular view of the opening chapters of Genesis. Up to this point Borrows stood on solid Protestant ground.

His next suggestion, however, helped seal Fitchett's fate. Borrows suggested that because the YMCA had adopted a broad evangelical platform it should be prepared to consider enrolling "enlightened Roman Catholics . . . even if they call the Virgin Mary Queen of Heaven and say their prayers before an image, so long as they acknowledge her Son as King of Glory." Anti-Catholicism flourished in Dunedin, exacerbated by the local Catholic Bishop, Patrick Moran, a fiery, outspoken Irishman. Outraged Protestants greeted Borrows' optimistic ecumenical suggestion with "shouts of 'No' " and "prolonged hisses." Borrows' retort to his detractors inflamed the meeting still further: "Some of you require a little more Evolution. My ancestors have given up that language ages ago." With champions such as Borrows, Fitchett scarcely needed enemies.

The chairman eventually restored order. Fitchett's opponents, represented by the Reverend James Copland, a fierce Calvinist, then articulated their disagreements. For Copland, the Fall represented the main theological problem with evolution. On Fitchett's view that man had evolved from moral innocence to moral consciousness, Copland argued, man could not originally have possessed a moral sense. If man had no moral sense before the Fall, then he could not have been responsible for sin. Sin and the Fall thus became God's fault – and Fitchett's views became heretical. Applause greeted Copland's argument, and his motion to uphold the board's decision to blackball Fitchett carried by fifty votes to thirty-one. As soon as the result was announced, the Reverend Thomas Roseby, the minister of Moray Place Congregational Church, rose to declare that he would leave the YMCA.

A little over a week later some forty-three members withdrew from the YMCA in protest at Fitchett's treatment. Non-Presbyterian Protestants of English background led the exodus, notably the Reverend Thomas Roseby, the minister of the city's leading Congregationalist church, and the Reverend John Upton Davis, the minister of Hanover Street Baptist Church. A number of Methodists, including the Reverend L. M. Isitt and Dr. Borrows, joined Roseby and Davis, as did the strong-minded Anglican laymen Chetham Strode, an English justice of the peace known to forget to tell Scottish J. P.'s about joint meetings. In leaving the YMCA, this group expressed its refusal to be dictated to by "one or two self-made popes" such as the Reverend James Copland. The *Otago Daily Times*, which had long detested Copland, condemned him and his supporters as "extreme bigots."[14]

The Copland-led conservatives consisted mostly of Scottish Presbyterians but included the ultraevangelical Anglican Lorenzo Moore. This group, strongly Calvinist in outlook, believed that the Fall had plunged humans down into such radical moral and intellectual depravity that they could not save themselves. Denying the Fall, as Fitchett had done, implicitly denied the supernatural work of Christ as fallen humanity's only hope for salvation. These latter-day Tertullians looked sceptically upon the wisdom of this world, especially upon such modern science and scholarship as contradicted their reading of Scripture. J. Aitken Connell, a surveyor and lay Presbyterian, ascribed Fitchett's errors to his being "an intellectual man," a victim of pride, the besetting sin of intellectuals.[15] Moore spoke for many when he announced that he preferred to trust "the Word of God" over Fitchett's speculations.[16]

Liberal Protestants such as Fitchett, by contrast, placed more faith in human reason, softpedalled miracles and the supernatural, and refused to read Genesis literally. By such means they hoped to sustain the pre-Darwinian alliance between science and Christianity. On an intellectual level, then, the Fitchett affair concerned biblical theology, hermeneutics, and the ultimate ground of religious authority.[17] Biblical, theological, and metaphysical issues became vastly more important for most Christian thinkers than details of evolutionary biology.

Still, Otago Protestants did not divide into two neatly polarized camps of Calvinist antievolutionists versus liberal Protestant evolutionists. Some conservative Calvinists, such as the Reverend James MacGregor of Oamaru, the most powerful theologian in nineteenth-century New Zealand, saw no theological problem in biological evolution – though as late as 1882 he still had scientific reservations. And some of those who walked out of the YMCA remained sufficiently conservative to nurse suspicions about Fitchett's theistic evolutionary views. As one of them put it, not only the "Evolutionists" but also the "nonEvolutionists" who left the YMCA objected to having "the whole of the Scriptures strained through the brains of this so-called Christian Association."[18] These nonevolutionists, upholding the traditional Protestant right to interpret the meaning of Scripture for themselves, simply disliked dictatorial Calvinists even more than they disliked theological liberals. The Fitchett affair thus represented a power struggle between local Calvinists and Christians who were neither Scottish nor Presbyterian nor prepared to let the former dominate the cultural and religious life of Dunedin. Otago's ethnic and religious divisions constituted a fault line that brought the YMCA tumbling down.

This incident represented the major controversy over evolution to occur within the New Zealand religious community during the nineteenth century. Compared with those that wracked Christian institutions in the United States, such as the furor within Presbyterianism that resulted in James Woodrow's dismissal from Columbia Theological

Seminary, it did not amount to very much. No New Zealand university teacher ever lost his job for teaching evolution.

The Fitchett case also illustrates the impact powerful personalities could have in a small-scale society. James Copland constituted the leading Protestant antievolutionist in nineteenth-century New Zealand. Born and raised in Scotland, Copland studied arts and theology in Edinburgh and Berlin, obtained a Ph.D. from Heidelberg University in 1858, and an M.D. from the University of Aberdeen in 1864. Ordained to the Presbyterian ministry shortly after his arrival in New Zealand in 1865, he accepted a call to the North East Valley Church in Dunedin in 1871.

Copland illustrates how powerfully Christian reactions to biblical criticism informed subsequent reactions to Darwinism. Continental biblical critics such as F. C. Baur, D. F. Strauss, and E. Renan, whose work Copland had read, historicized Scripture and humanized Jesus. They represented the Bible, for example, as a fallible human document, containing the religious perceptions of particular peoples, whose status as a divine revelation was uncertain. Copland, who perceived such critics as denying the supernatural character and authority of Scripture, reacted by divinizing it. The Bible contained "only the expressions of God's mind unmingled with anything of merely human authority," he declared in 1873, some two years before the Fitchett affair.[19] Such a high view of Scripture made it difficult for Copland to interpret Genesis in the kind of nonliteral manner that enabled more moderate Protestants to come to terms with evolution.

Copland commenced his crusade against Darwinism at the beginning of the decade. Speaking at an evening tea to celebrate the opening of the University of Otago in 1871, he rejected the idea of human evolution. "Every tyro in comparative anatomy" knew that humans and higher animals resembled each other, Copland declared. That simply showed that all animals had been constructed on the same basic plan, not that they had common evolutionary origins. Darwin got it wrong in *The Descent of Man*; Moses told us all we needed to know about human origins.[20]

Copland's suspicions focused on Dr. Duncan MacGregor, inaugural professor of mental and moral philosophy at the university and holder of a chair endowed by the Presbyterian Synod. MacGregor taught ethics and political economy to students, some of whom were training for the Presbyterian ministry, and in 1876 he published an article advocating advanced social evolutionary views and fiercely criticizing the churches. Copland saw MacGregor as a secular intellectual out to overthrow ministerial authority and undermine the credibility of Christianity. He attempted to mobilize support within the Presbyterian Synod of Otago and Southland to withdraw theological students from MacGregor's lectures. Rumors began to circulate throughout the country that the Synod was about to sack its own appointee. In 1878 Robert Stout, a Dunedin lawyer, a member of the House of Representatives, and an admiring exstudent of MacGregor's, introduced a bill into parliament to give the university council the sole right to appoint and dismiss professors. This would relieve the university of "the intolerable incubus of ecclesiastical censorship," fumed Stout, a militant free thinker.[21] Although the bill failed, this secularist offensive spurred Copland to suggest that the synod divert its endowment to a new chair in the Presbyterian Theological College. Moderates in the synod, led by the Reverend Dr. D. M. Stuart, vice chancellor of the university, opposed this idea as sectarian and illegal, and MacGregor stayed put until he left Dunedin for Wellington in 1885.

MacGregor's departure in that year prompted Copland to suggest that he himself should fill the vacant chair. To support his candidacy, he published an attack on evolution entitled *The Origin and Spiritual Nature of Man*. Quoting extensively from Darwin's *Origin of Species* and *Descent of Man*, Copland argued that evolution denied divine design in nature and entailed thoroughgoing materialism. Genesis, which contained the true account of human origins, could not be "impugned without destroying the divine authority of the Bible." He found no form of evolution acceptable at all. As new Zealand's foremost clerical anti-Darwinian, Copland played an analogous role to that played in the United States by the Presbyterian theologian Charles Hodge of Princeton, though the Dunedin doctor's theology seems to have been narrower, drier, and more scholastic than that of Hodge. Yet by the 1880s Copland had become somewhat isolated; most educated New Zealand Protestants adopted more conciliatory positions on evolution.[22]

Thus Copland's attempt to obtain the Otago chair by getting up an antievolution crusade failed, even in the heartland of New Zealand Presbyterianism. The position went to a liberal rival, the Reverend William Salmond, a theistic evolutionist who had previously been professor of divinity at Knox College, a local Presbyterian school. Not long afterward Copland shook off the dust of Dunedin, moving south to Gore, where he died in 1902.

Though Copland grew out of touch with elite opinion, he articulated a position that remained popular enough among lay Protestants into the twentieth century. Some followed Copland in rejecting evolution on biblical grounds, as Salmond, professor of mental and moral philosophy, discovered when he addressed the Otago University Christian Union on Christianity and evolution in 1902. Speaking in a lecture hall overflowing with hearers, Salmond declared that evolution represented the method by which "God made man from the dust of the earth." Far from representing a historic individual, Adam symbolized generic humanity and the unity of the human race. Salmond called on his audience to cling no longer "to the beggarly letter" of Scripture but instead to "march on with all the life of reason and science, knowing that they, too, are divine."[23]

Such science-exalting, Scripture-abasing theology failed to convince a number of local Christians. A correspondent to the *Otago Daily Times* rejected the argument that Adam symbolized generic humanity, citing 1 Timothy 2:13, in which Paul assumed the truth of the Mosaic account of creation. According to this critic, Salmond had closed "the Old Book" and manufactured "a Christianity to suit the circumstances." Another writer, "Veritas," argued that because Christ in Matthew 19:5 quoted Genesis 2:24 on the creation of Eve out of Adam's rib, and Paul did the same in Ephesians 5:31, then the matter was settled.[24] By this time the gap between university and pew had widened further than the Presbyterian founders of the University of Otago, inspired by the Scottish tradition of the "democratic intellect," had ever envisaged.

In the opening decades of the twentieth century, another Presbyterian minister, the Reverend P. B. Fraser, succeeded Copland as the main Protestant opponent of evolution. Fraser never attended a theological college but had taught himself what theology he knew. Influenced by American fundamentalism, and disturbed by new guidelines from the Ministry of Education encouraging the teaching of evolution in schools, he entitled an antievolutionist tract in 1929 *Evolution and Ape-man-ism in Schools* and adopted a populist approach that scholarly nineteenth-century conservatives such as Copland had eschewed. Fraser attracted 400 Christchurch citizens to a public meeting against the teaching of evolution in schools in 1929, a substantial number by local standards. Yet he never mobilized sufficient support to mount a large-scale antievolution crusade comparable to the American movement of the 1920s.[25]

Darwinism and the scientific community

The appearance of scientific institutions in New Zealand coincided with the beginning of the Darwinian debates. Short-lived attempts had been made to found local scientific and literary societies in the 1840s and 1850s, but colonial science acquired a solid institutional infrastructure only in the 1860s. That decade witnessed the founding of the Philosophical Institute of Canterbury, the Auckland Institute, the Otago Institute, and the Wellington Philosophical Society, as well as their governing body, the New Zealand Institute. Though other leading provincial towns established institutes or philosophical societies, most of the colony's scientific research and discussion took place in these four main institutions.

The timing of their foundation is of crucial importance in explaining why evolution won rapid acceptance in New Zealand. No powerful scientific establishment with a substantial committment to pre-Darwinian modes of thought existed in 1859. In New Zealand it so happened that all of the most eminent natural scientists of the 1860s through the 1880s – Julius Haast, James Hector, F. W. Hutton, William Travers, Walter Buller, T. J. Parker of Otago University, and A. P. W. Thomas of Auckland University College – embraced evolution as the best scientific explanation available of the origin and development of organic beings.

Some of these scientists had become apostles of evolution before arriving in New Zealand. F. W. Hutton, for example, had written one of the first reviews of *The Origin of Species* to be published in Britain, in which he savaged Adam Sedgwick's direct creationism as "a mere assertion, an evasion of the question, a cloak for ignorance." In Hutton's view, supernatural causes erected insuperable barriers to scientific explanation. Darwinism, in contrast, offered an opportunity to explain a host of biological problems in terms of scientific naturalism. Impressed by the percipience and originality of Hutton's review, Darwin congratulated its twenty-four-year-old author. Arriving in New Zealand in 1865, Hutton did more than any other single person to plant evolution in New Zealand soil, becoming the New Zealand equivalent of the Harvard botanist Asa Gray, Darwin's chief champion in the United States – except that Hutton, a liberal Anglican, enjoyed comparatively more influence in New Zealand than Gray did in America.[26]

James Hector, director of the government Geological Survey and Colonial Museum, and governor of the New Zealand Institute, became the patriarch of New Zealand science in the nineteenth century. A Scotsman educated in medicine at the University of Edinburgh, he had read *The Origin of Species* in North America and converted to evolution before his arrival in the colony in 1861.[27] He encouraged local scientists to put evolution to work to explain the natural history of New Zealand, and published their findings in the *Transactions and Proceedings of the New Zealand Institute*.

Julius Haast, one of the country's leading geologists, compared Darwin with Galileo, Kepler, and Newton in his inaugural presidential address to the Philosophical Institute of Canterbury in 1862, lauding *The Origin of Species* as "the great work of the age." Evolution would enable local naturalists to explain the origins and development of New Zealand's unique bird species such as the Moa and the Kiwi, he declared. Haast sent a copy of his lecture to Darwin, who returned warm congratulations.[28]

William Travers, a lawyer, politician, and self-taught botanist and ornithologist, introduced evolution to the citizens of Wellington in a series of public lectures delivered at the Colonial Museum in 1868–69. The direct supernatural creation of species made a scientific explanation of their origins and development impossible, Travers charged. Evolution by natural selection, in contrast, made scientific explanation possible by relying on natural causes. Like Hutton, Hector, and Haast, Travers became an enthusiastic Darwinian not because he believed natural selection to be an adequate explanatory mechanism, but because the new theory eschewed the supernatural. Invoking natural causes alone, Darwinism enabled biology to develop into a respectable, modern, professional science. As Haast put it, Darwin had approached the problem of the origin of species in a "truly philosophical spirit."[29]

The major debate in a scientific society occurred in the Otago Institute in Dunedin in 1876, just prior to the YMCA debate. When the council of the institute decided to hold a series of popular lectures on scientific subjects, it selected the incumbent president, Robert Gillies, an evangelical Presbyterian and a fellow of the Linnaean Society, to give the first lecture, a review of Ernst Haeckel's *The History of Creation* (1868). Gillies accepted Haeckel's general outline of human evolution, but condemned the "purely materialistic portions" of the book. He also expressed reservations about Haeckel's human phylogeny of twenty-two stages, beginning with protoplasm and moving up through sea squirts and the apelike progenitors of humans to the emergence of modern men and women. In Gillies's opinion, Haeckel's schema left the beginning of life unexplained, failed to close the gap between vertebrates and invertebrates, and inadequately accounted for the origin of amniote animals (reptiles, birds, and mammals), the sudden appearance of placental mammals, and the beginning of articulate speech. Too many holes existed in Haeckel's explanation of human evolution for it to be convincing, Gillies concluded.[30]

Hutton, lecturer in geology and natural science at the university, worried that some members of the institute might throw out Darwin's evolutionary baby with Haeckel's materialist bathwater. He rose at once to declare that, although Haeckel had speculated too brashly, no theory of the origin of species but evolution had the slightest scientific evidence in its favor. Darwin's theory enjoyed "irresistible" empirical support. Humans had unquestionably descended from apelike ancestors, Hutton insisted. Samuel Nevill, the Anglican bishop of Dunedin, disagreed, arguing that evolution did not explain the growing complexity and organization in the history of organic life. In fact, the evidence suggested that the Creator had proceeded according to a definite plan, "superadding" to a species new faculties or structures that it did not previously possess. The Reverend Salmond, newly arrived Presbyterian professor of divinity, agreed with Nevill. For Hutton to represent evolution as proven struck him as "the height of folly."[31]

Three weeks later Hutton gave the second talk in the institute's series, on "The Inductive Method as Applied to the Theory of Descent." The Dunedin public was evidently captivated, for the University Hall was filled to capacity. Hutton began with the "most disagreeable" task of disposing of special creation by showing that it implied that the Creator was an incompetent bungler. Hutton compared the competing hypotheses of evolution and special creation by using J. S. Mill's "method of difference," designed to choose between competing explanations. Europe possessed a large number of terrestrial mammals. New Zealand had none, though the success of introduced animals showed that its environment favored them. Europe also had fossil mammals, while New Zealand had none. Evolution explained this difference by assuming that the fossil mammals of Europe were the progenitors of the presently existing species. Special creation did not. Why did the Creator fail to populate New Zealand's eminently hospitable environment? Furthermore, reasoned Hutton, small oceanic islands contained large numbers of distinct species closely allied to mainland species. Here again evolution explained the phenomenon. The progenitors of the island species had migrated from the mainland, and the intense struggle for existence on the confined islands had led to more rapid speciation than on the mainland. Direct creation implied that God had been busier on the islands, placing more species from the same genus there than on nearby continents. This seemed arbitrary, if not absurd, to Hutton, who concluded his defense of Darwinism to warm applause.[32]

Bishop Nevill gave the final lecture in the series. He summarized the failings of Darwin's theory on October 17 in a crowded University Hall. No new organic forms had evolved during recorded history, and fossil evidence of the intermediate forms that evolution demanded was lacking. Besides, natural selection could not explain the marvellous adaptation of structure to function. The highly developed teeth of the beaver, for instance, so well adapted for felling trees, could only be the work of an Intelligent Author, Nevill argued, not of the blind forces of nature. On purely scientific grounds, evolution offered no advantages over special creation. The audience politely applauded Nevill's exposure of the scientific weaknesses of Darwinism, and his idealist and teleological view of nature.[33]

Yet Nevill's dissection of Darwinism had convinced few. The editor of the *Otago Daily Times* observed that Hutton "ruthlessly despatched" Nevill's arguments in discussion following the lecture. Shortly after the Otago Institute series the Reverend William Salmond, who had initially been Hutton's fiercest opponent, announced his conversion, declaring himself "an Evolutionist of the Hutton [i.e., theistic] kind." He continued to believe that some evolutionists were "atheists, " but in the future, he confessed, he would reserve that term for "certain extreme men ... chiefly of the German school." The credit for Salmond's conversion belonged entirely to Hutton, observed the editor of the *Times*. He had never heard a "more charming speaker" able to bring "great knowledge out in a perfectly lucid and clear way, and convincing those who desire to pick holes almost against their will." The editor surmised that most persons attending the series had concluded that evolution rather than creation had been scientifically proven.[34]

The Hutton–Nevill debate represents the closest New Zealand came to the famous Huxley–Wilberforce debate at the Oxford meeting of the British Association for the Advancement of Science in 1860. The differences between the two encounters are significant. In New Zealand both sides sought peace rather than conflict. Robert Gillies, reviewing the lecture series in his retiring presidential address, observed that "the most kindly feeling" and "courteous consideration" for others' opinions had been shown by everyone, with the exception, initially, of Salmond. For this "happy result" the institute was indebted to Captain Hutton and Bishop Nevill; "high tone and gentlemanly feeling" had characterized their discussions.[35]

Hutton's attitude toward Anglican critics of evolution in New Zealand in 1876 contrasted markedly with his attitude in England in the early 1860s. As a young man of twenty-four, he had denounced clerical revilers of Darwin such as Adam Sedgwick as "wilful pervertors [sic]" of evolution, spouting "mere verbosities, baseless assertions," and "gross ironical misrepresentations."[36] He never employed such fierce language in New Zealand, apparently because of the different social and religious conditions he found there. In Britain the rising breed of younger scientists, led by Huxley and Tyndall and including Hutton in 1860–61, turned science against religion in order to break the power of the clerical establishment and to assert their own professional autonomy and hegemony. They employed scientific naturalism and religious agnosticism as weapons to wrest control of the nation's scientific and educational institutions from the conservative, Anglican establishment.[37] In New Zealand, by contrast, no established church existed. Clergymen could hardly dominate science for they were too busy establishing churches and ministering to widely scattered flocks. Bishop Nevill, one of the most scientifically literate and active ministers in the country, illustrated this when he apologized to the Otago Institute in 1877 for being unable to review the latest scientific discoveries as the retiring president normally did:

> One who is debarred by the ceaseless pressure of other duties from conducting a course of independent enquiry, and who can do little more

> than skim the pages of a scientific journal amid the inconveniences of a coach journey or the difficulties of a lively railway carriage . . . can hardly on that account venture on so ambitious a flight.[38]

The absence of a pre-Darwinian scientific establishment also proved significant. Scientists such as Hutton, Hector, and Haast, all of whom arrived in the 1860s, never encountered an entrenched elite dominating science and demanding obeisance to orthodoxy. New Zealand's religious freedom had a positive, liberating effect on science. James Hector remarked on this, telling the Australasian Association for the Advancement of Science in 1891 that biology had forged ahead in New Zealand thanks to the fact that "energetic teachers" such as Hutton, T. J. Parker of Otago University, and A. P. W. Thomas of Auckland University College never felt constrained in expressing evolutionary views by fear of "religious controversy."[39] New Zealand scientists could dominate scientific institutions, establish professional autonomy, and gain prestige and recognition with comparative ease. They found little need to wield antireligious ideologies such as agnosticism, positivism, and materialism against interfering clerics.

Personality must also be borne in mind in assessing the relative calm of the Otago Institute debates. Hutton, who played such a crucial role in smoothing the way for Darwin in New Zealand, remained a devout Broad Church Anglican to the end of his life. From his review of *The Origin* in 1860 and his debate with Bishop Nevill in Dunedin in 1876 to the writing of his last book, *The Lesson of Evolution*, in 1907, Hutton argued that evolution should be interpreted theistically. Though he had no time for the direct creationism of fellow Anglicans such as Sedgwick and Nevill, he had equally little use for materialists or pantheists such as Ernst Haeckel. In his presidential address to the Australasian Association for the Advancement of Science in 1902, Hutton also gently disavowed the agnosticism fashionable in the late Victorian period. Hutton's skill in debate, his combination of firmness and courtesy, and his ability to christianize evolution helped clerical opponents such as Salmond and Nevill to come to terms with the theory. More than any other single figure, he smoothed the way for Darwinism in New Zealand's loosely defined scientific community, and outside it as well.[40]

The New Zealand scientific establishment, such as it was, favored Darwin from beginning to end. No anti-Darwinian of the scientific stature of Richard Owen in Britain, J. W. Dawson in Canada, or Louis Agassiz in the United States, around whom antievolutionists could rally, existed in New Zealand. In the Australian colonies, too, Darwinism seems to have made headway in the scientific community more slowly than in New Zealand, thanks to the presence there of eminent men of science holding antievolutionary views.[41]

William Miles Maskell, a devout Roman Catholic, constituted the only anti-Darwinian in New Zealand with any scientific status. Even he regarded the evolution of plants and animal as "of absolutely no importance whatever" for Christianity. But he drew the line at human beings. For evolutionists to assert that "all the body and soul of man, proceeded simply and solely" by "gradual development from the brute" seemed to him utterly materialistic, and subversive of Christianity. Furthermore, it angered him that "the most eminent professors in the colony," paid to teach science, were indoctrinating their students, many of them trainee teachers, in philosophical materialism. These teachers in turn indoctrinated their pupils in "antiChristian" science, which was thus beginning to dominate "all the public educational institutions in New Zealand." Though Maskell exaggerated – evolution was not in fact taught in New Zealand schools until the 1920s – his fears possessed some foundation in reality. Free thinkers and agnostics indeed occupied chairs at Canterbury College and the University of Otago. A. W. Bickerton, professor of chemistry at Canterbury, had trained in London under militant agnostics T. H. Huxley and John Tyndall, and tirelessly popularized scientific materialism in Christchurch, to the annoyance of local Christians besides Maskell.[42]

By the 1880s, as a result of innovative taxonomical work with the microscope, Maskell had established an international reputation as the world authority on the Coccidae (scale insects). His new-found eminence may have encouraged him to launch an antievolutionary crusade. His career as New Zealand's foremost scientific anti-Darwinian is worth examining, because it reveals how isolated he was. In May of 1881, with F. W. Hutton in the president's chair, the Philosophical Institute of Canterbury declared "its high appreciation of the great services that have been rendered to science by the late Dr. Charles Darwin and its deep sense of the loss that science has sustained through his death." Maskell alone voiced dissent. Two years later he called members' attention to Darwin's heterodox religious views as revealed by Ernst Haeckel. In 1882 the German biologist had attempted to enlist Darwin against Christianity by quoting a particularly agnostic passage from a letter Darwin had written to a student at Jena. Most English-speaking journals, including *Nature*, had suppressed the passage, preferring to present to the public a pious rather than an agnostic Darwin. Early in 1883 Maskell, addressing the Philosophical Institute of Canterbury, condemned the "disingenous omission by the Editor of *Nature* of this letter showing Darwin's want of religious beliefs." Before Maskell had finished his attack, Hutton simply ended the meeting, declaring that the Philosophical Institute of Canterbury forbade religious controversy. Members had no choice but to trudge home. Maskell had been summarily silenced.

Yet the scientific community appears to have taken care not to ostracize the crusading Catholic in their midst. Maskell was elected treasurer of the Philosophical Institute of Canterbury the following year and a governor of the New Zealand Institute in 1885. He eventually tired of crying alone in the wilderness, admitting in 1890 that, in the context of scientific discussions, religious belief might perhaps remain "a matter of private opinion." By the end of the nineteenth century most New Zealand scientists had, like Maskell, largely privatized their religious beliefs.[43]

From the little evidence available, it seems that New Zealand Catholics resisted scientific naturalism in general, and evolution in particular, more systematically than the dominant Protestant and secular community. In 1884 R. H. Bakewell, for example, a Catholic physician, perhaps encouraged by Maskell's recent forays, told the Canterbury Institute that "life" represented a distinct force or form of energy which, when directed by a "Supreme Intelligence," led to growth, differentiation, and reproduction. Bakewell's argument for theistic vitalism appeared in the *Transactions* of the New Zealand Institute, which went out to scientific societies all over the world. The thought of distinguished foreign scientists reading such stuff annoyed the hardheaded Presbyterian botanist G. M. Thomson, who declared that Hector, as editor of the *Transactions*, should have consigned Bakewell's paper to "oblivion" as soon as it reached his desk.[44] Although Maskell and Bakewell made themselves heard, they hardly constituted a significant anti-Darwinian bloc within the scientific community. Because Catholics constituted only about fourteen percent of the total population, these two had relatively few co-religionists to mobilize, far fewer than if they had lived in Ireland, Canada, the United States, or Australia.

On one major issue, however, Maskell and Bakewell stood in the mainstream. Like these two Catholics, the vast majority of the colony's scientists continued to believe in God. Hutton, Haast, Hector, Travers, Thomson, the eminent ornithologist Walter Buller, and A. P. W. Thomas, liberal Anglican professor of natural science at Auckland University, all believed that, as Thomson put it, "Darwin's view of life and of the Creator of life was a far grander one than any which it subverted." Hector and Travers articulated religious views that were at least deistic, while the others interpreted evolution theistically.[45]

If Maskell and Bakewell should be placed on the conservative end of the religious spectrum, with most men of science somewhere in the middle, a couple of professional scientists should be assigned to the opposite pole. Bickerton, foundation professor of chemistry at Canterbury College, spread Huxley's gospel of scientific naturalism and religious agnosticism after arriving in New Zealand in 1874. His often spectacular public science lectures drew audiences of up to 400 people, and he numbered among his students the young Ernest Rutherford. Bickerton argued that new heavenly bodies were formed not by a Divine Creator but by stars colliding at a tangent. This "partial impact theory" proved to Bickerton's satisfaction that "the principle of evolution" constituted "a universal law controlling the cosmos." In 1879 he told the Canterbury Institute that organic life, including human beings, had material rather than divine origins. A scientific observer of life's beginnings, he speculated, might have watched "viscous silicon building itself into complex protoplasmic molecular skeletons, developing organ after organ, and breathing forth its halogen breath." Speculative materialism verging on science fiction did not alarm Christchurch citizens as much as Bickerton's social and moral radicalism. Denouncing marriage as an institution, he experimented with communal living at his Wainoni estate just outside Christchurch. Running conflicts with colleagues and university administrators led to his sacking in 1902.[46]

Thomas Jeffery Parker had also studied under Huxley at the Royal School of Mines. He arrived in Dunedin in 1881 to succeed Hutton as professor of natural science at the University of Otago, and managed to alienate a number of local Christians with his inaugural lecture, which, in a manner worthy of Huxley himself, damned direct creation and endorsed evolution. By that time, however, Dunedin people had been discussing evolution and religion for years. Hutton, Nevill, Salmond, and others had done their work so well that Parker faced little opposition. Besides, though he called himself a "disciple" of the "Master" Huxley, his own religious views lay somewhere between Huxley's militant agnosticism and the Christianity of his father, W. K. Parker, an eminent English anatomist. T. J. Parker assured the audience at his inaugural lecture that it would be wrong to think that "that power which could form worlds out of a nebula" could not also "evolve a horse from an hipparion or even a speck of living protoplasm from the elements of the primaeval sea." Parker, perhaps hedging his bets, remained a parishioner and seatholder at St. Matthew's Anglican church until his premature death in 1897.[47]

Free thinkers and evolution

Those scientists professing theistic or deistic evolutionary views were probably not merely making ritual obeisance to tradition. The social pressures for purely nominal religious conformity appear to have been fairly weak in New Zealand. Just how little Christian belief mattered in New Zealand's public domain is perhaps best illustrated by the career of Robert Stout. He was born in the Shetland Islands, and acquired a dislike of dogma and sectarianism before arriving in Dunedin in 1864. There he trained as a lawyer, studied mental and moral science and political economy under Duncan MacGregor at the University of Otago, and emerged a militant agnostic and disciple of Darwin and Herbert Spencer. By the early 1870s Stout had become the leading free thinker in the country, editing the rationalist paper *The Echo* for many years, and fighting to reduce church influence at all levels of the education system.

Stout seized on evolution as a weapon with which to attack Christian belief and the power of the churches. In Christchurch in 1881, for example, he delivered a lecture entitled *Evolution and Theism*, in which he argued that because evolution was true, theism must be false. Intelligent modern people must choose between the Bible, which Stout derided in Huxleian rhetoric as "the Baal of authority," and modern science, which he defined as "truth – obtained after careful research." The time was coming, he declared in Dunedin, when intelligent people would regard religious beliefs, miracles, and creeds as geologists now regard fossils: out of them "something better has been developed." Unlike free thinkers in England, Stout suffered hardly at all for his outspoken agnosticism. He not only won a seat in parliament but became premier of the country – in the same year (1884) that he was elected president of the New Zealand Freethought Federal Union. To Christian voters in Dunedin, Stout's impeccable personal morality and political liberalism outweighed his rejection of orthodox belief. By 1890, New Zealanders had elected not just Stout but also John Ballance, another avowed free thinker, to the highest political office in the land. Many others, such as Edward Tregear, Duncan MacGregor, William Pember Reeves, William Collins, and George Hogben, prospered in New Zealand politics at century's end.[48]

Even at its height, however, the free-thought movement never attracted more than a few thousand adherents. It found the temperate colonial religious climate somewhat inclement. Little in the way of an oppressive religious establishment existed to provide the free thinkers' reason for being. They succeeded in arousing antagonism only from the small, sectarian Protestant groups they most nearly resembled in style. Eventually, free thought got tolerated to death. The movement dwindled in the late 1880s, and by century's end, riven by internecine disputes, virtually died out. Most New Zealanders seem to have found the free thinkers' scientific atheology little more riveting than Christian theology. Workingmen preferred beer.[49]

Still, though free thought as an organized movement declined, free thinkers had already begun to exercise cultural influence out of proportion to their numbers. As the cases of MacGregor, Stout, Bickerton, and Parker suggest, free thinkers were playing significant roles in politics, law, and higher education in late-nineteenth-century New Zealand. In the twentieth century Stout, for example, became Chief Justice of the Supreme Court and Chancellor of the University of New Zealand. Quietly penetrating parliament, the judiciary, and the universities, centers of cultural authority in the modern world, free thinkers did so without provoking massive Christian protest. Mainstream Protestantism, willing to accept these unbelievers, would seem in this post-Enlightenment colony to have evolved in highly tolerant, liberal directions.

Darwinism and the Maori, 1860–1900

Finally, New Zealanders embraced Darwinism for racist purposes. Before examining the views of these colonial social Darwinists, however, we must acknowledge that the biological determinisms they deployed did not burst out of the blue in the 1860s. The idea that Nature (and/or Nature's God) had doomed the Maori race to disappear before the Pakeha considerably predated the *Origin of Species*. In his *Journal of Researches*, published in 1839, for example, gentle Darwin himself observed that "the varieties of man seem to act on each other in the same way as different species of animals – the stronger always extirpating the weaker." The "fine energetic" Maori, the young naturalist noted sadly, themselves acknowledged that their "land was doomed to pass from their children." Many of the land-hungry settlers trickling into the colony during the 1840s eagerly embraced the dying Maori and, unlike Darwin, saw in the prospect of Maori extinction little to regret. William Fox, chief agent for the New Zealand Company and a prominent colonial politician, announced in 1851 that "statistics prove the fact of the progressive extinction of the natives," and gave the race fifty years at the outside. According to Edward Shortland, a physician, anthropologist, and moderate Anglican, by mid-century "many" Pakeha were anticipating the "ultimate extinction of the Aboriginal race" as "a matter of certainty."[50]

Christian humanitarians, or "philo-Maoris" as they were called at the time, determined that the Maori should not suffer the same fate as the American Indian and the Australian Aborigine, attacked the dying out of Maori on both moral and scientific grounds. According to Bishop Selwyn, for example, those Britons who invoked "mysterious dispensations" such as the "inevitable law of nature" or "decree of Providence" that "coloured races" simply "melted away" before "civilization" were trying to "veil" their "damnable crimes" of "violence" and "neglect" in decorous scientific and religious verbiage. Shortland agreed, contending that Pakeha colonizing bodies such as the New Zealand Company were spreading the dying Maori myth in order to "mislead the intending colonist" into believing that Maori were "merely sojourning for a time on the earth" and would provide no permanent obstacle to colonization. The real "duty" of the colonists, said Shortland, was to treat Maori as "highly intellectual human beings" destined not to die out but to "take their place side by side with the white man, as equals in civilization." Shortland used his anthropological expertise to attack the dying out of Maori on scientific grounds as well, arguing that the theory's proponents had misinterpreted the relevant demographic evidence, which by no means proved the race was doomed.[51]

Some settlers, undaunted by Anglican criticism, pressed Darwinism into service after 1859, giving established racial ideologies new life and a new vocabulary. Before examining post-Darwinian racism, however, we must sketch in the relevant context. The most sustained period of racial warfare in New Zealand history began in the Taranaki province in 1860, following a land dispute within the Ati Awa tribe. Te Teira Manuka, determined to settle an old score, offered to sell a portion of tribal land near the Waitara river. The paramount chief of the tribe, Wiremu Kingi Te Rangitake, forbade the sale. Local settlers had long coveted this lovely, flat, and fertile area. They put pressure on Governor Thomas Gore Browne, who in 1860 buckled and purchased the Waitara block, disregarding Te Rangitake's protests. When Te Ati Awa peacefully resisted attempts to survey and occupy the area, and built a pa (fortified village) to assert their ownership, government soldiers attempted to capture the pa. Fighting soon spread to other parts of the Taranaki province, and in 1863 to the Waikato area, engulfing much of the central North Island.

The Anglican hierarchy, led by Selwyn and Archdeacon Hadfield, denounced Governor Browne's "illegal conduct" and "premeditated wickedness" in buying disputed land, plunging the country into war. According to Hadfield, Te Rangitake and his people constituted "peaceable and loyal" subjects, upon whom Governor Browne had "illegally and unjustifiably made a hostile attack" in order "forcibly" to take "property guaranteed them by the Treaty of Waitangi." Such "illegal proceedings" had driven "good and loyal" Maori into defending their lands by "force." Yet defending what was rightfully theirs constituted neither "rebellion" nor "a denial of the Queen's sovereignty," insisted the missionary. The Maori were simply doing what any red-blooded Englishmen would have done in similar circumstances.[52]

Angry colonists took a different view. In Auckland, the *Southern Monthly Magazine*, dedicated to disseminating science, literature, and politics to the reading public, and reflecting the views of Auckland businessmen and land speculators eager to get their hands on Maori lands in the Waikato to the south of Auckland, declared it entirely natural that Maori and Pakeha had begun killing each other in Taranaki. "Whatever may be thought of Mr. Darwin's views concerning natural selection and the origin of species," wrote the editor, "no one will be disposed to deny the existence of that struggle for life which he describes, or that a weak and ill-furnished race will necessarily have to give way before one which is strong and highly endowed." Adopting "the fitness of things" as "the basis of all rights and obligations" toward other races, the editor condemned the "fallacious philanthropy" of Christian humanitarians, ridiculing in particular their "false theory" that Maori owned the whole of New Zealand. There could be "no fitness," he observed, in allowing Maori, like the proverbial "dog in the manger," to "retain in a barren state the land which is the gift of providence" to the entire "human race." The colonists should from the beginning of settlement have "taken and held by force" any land not actually cultivated by Maori "without violating any law of justice or humanity." Far from having a moral duty to "save and elevate" the Maori race, the settlers ought to "admire" the "natural laws" effecting "the decline and extinction of the inferior race." Such a view, applying "the scientific curiosity of zoologists" to "human society," might appear somewhat "frigid," the editor admitted. But "enlightened benevolence" and "sound reason" permitted no other course of action. To "squander money, and waive our just rights in the attempt to save from extinction a declining race" was pointless. "Our war is a war of colonization, and justifiable on that ground."[53]

Meanwhile, in the Taranaki province, Arthur Samuel Atkinson, a settler prominent in local politics, was reading Darwin in similar fashion. The Richmond–Atkinson "mob," intermarried English families to which Atkinson belonged, settled in Taranaki in the early 1850s, where they found themselves confined to unrewarding, labor-intensive farming in rugged bush areas, while much of the best land remained in Maori hands. Atkinson devoured the *Origin of Species* in 1862–63, and soon after joined the local settler militia, the Taranaki Volunteers. Though he had been friendly with local Maori before the war, now he dismissed saving and preserving the race as "not possible." Rather, he lay in wait to do his "scientific duty" by shooting Maori.[54]

Others interpreted the *Origin* in more decorous, but no less imperialist, ways. The editor of the *Lyttelton Times* of Christchurch, for example, in a piece entitled "Mr Darwin at the Antipodes," drew attention to the importance of evolutionary theory for New Zealand by quoting a common Maori saying: "As the white man's rat has driven away the native rat, as the European fly drives away our own, as the clover kills our fern, so will the Maori disappear before the white man himself." The fact that European species were displacing indigenous ones so ubiquitously demonstrated to the editor's satisfaction that the "hardier and subtler vitality of 'selected' – i.e., civilized, – nature" quickly "beats the luxuriance of wild nature" in the struggle for life.[55] The *Times* commended its observations to readers as "entirely in the spirit of Mr Darwin's great work."

Some of the colony's most enthusiastic evolutionary naturalists endorsed such views in the colony's fledgling scientific societies. William Travers, an Anglo-Irish lawyer and politician, for example, linked scientific and political communities by serving as chief spokesperson for science in the colonial parliament during the second half of the nineteenth century. He discussed the impact of the "civilized races" of Europe on New Zealand natural history in a series of public lectures delivered at the Colonial Museum in Wellington between 1869 and 1870. The large audience listening to the first talk included the governor of the colony and many of Travers' fellow politicians. Indigenous plants and animals "even on their own ground" were proving "unable to cope" with Old World "intruders," Travers declared. The evidence appeared overwhelming. The European rat had displaced the Maori rat. The European honey bee, proliferating in New Zealand forests, was taking the food of indigenous honey-eating birds, which were rapidly decreasing in numbers. Introduced animals such as pigs, cattle, and goats, their populations exploding, were dominating vast areas of bush and high country. The displacement of indigenous by "civilized" forms seemed to Travers to constitute a law of nature that applied as surely to human races as to plant and animal species. Wherever a "white race comes into contact with an indigenous dark race on ground suitable to the former," he observed, then "the latter must disappear in a few generations." Travers invited "even the most sensitive philanthropist" in his Wellington audience to accept Maori extinction with "resignation, if not with complacency." Only one other option remained: the rapid assimilation of Maori into the dominant Pakeha culture.[56]

Other naturalists used the concept of natural selection to gloss Maori depopulation. William Colenso, for example, a botanist, ethnologist, and exmissionary, noted in an influential 1865 essay "On the Maori Races of New Zealand" that Maori women often had ten to twelve children. Yet disease killed off so many that "only the strongest lived." This struck Colenso as "a very good kind of natural selection, no doubt highly beneficial to the race." Colenso's enthusiasm for natural selection must not mislead us. He remained ambivalent about evolution on religious grounds until the end of his life.[57]

By the 1880s, as the colonial economy entered a long depression, some Pakeha found more infuriating than ever the fact that Maori continued to own large areas of land and showed little interest in either selling or developing it. With the Maori population continuing to decline, some scientific commentators abandoned even the hope of racial miscegenation, insisting that nothing whatever could save the Maori. Alfred K. Newman, a Wellington physician, politician, and scientific materialist, told the Wellington Philosophical Society that "the two races will never mingle." Half-castes possessed such weak and shallow chests that they "all die young." The children of half-caste parents, "a very feeble race," disappeared even faster. Newman found the "utter effacement" of the Maori "scarcely subject for much regret. They are dying out in a quick, easy way, and being supplanted by a superior race." Walter Buller, an ornithologist and self-styled "disciple of Darwin," argued in 1884 that the "inscrutable laws of Nature" decreed that aboriginal races must give way before "more civilized" ones. Racial amalgamation could not help, because when half-castes married Maori partners, their offspring, lacking "stamina," seldom "reached maturity." Buller, whose name had been proposed by Darwin for membership in the Royal Society of London, encouraged the "good, compassionate" colonists in his audience to "smooth down" the Maori's "dying pillow."[58]

As these ideas illustrate, evolutionary racism certainly existed in Victorian New Zealand. Yet we must be careful not to exaggerate its political significance. Newman, for example, depicted by many historians as the quintessential Victorian scientific racist, advocated humane racial policies during an almost forty-year career in the House of Representatives. Far from suggesting that the Pakeha should step aside and let Nature take its course, Newman called fellow politicians to do their "duty" by doing whatever they could to "stop the decay." Two years later, after condemning the fact that Maori in the Otaki area were "dying out in a shockingly rapid manner," he called parliament to employ a doctor to visit the area once or twice every week. Newman's scientific naturalism, repugnant to modern sensibilities, does not seem to have underpinned a ruthless ethical naturalism.[59]

Maori evidence indirectly confirms this point that social Darwinism played a modest role in colonial politics. A variety of Maori leaders, some of them well-educated in Western science, attacked the dying Maori theory in parliament and in the scientific societies from the 1890s onward. None, to my knowledge, identified any strong connection between the dying Maori and Darwinism. Some leaders, like Pakeha humanitarians before them, argued that Pakeha sinfulness, not science, lay behind the theory. Te Rangi Hiroa (Peter Buck), a physician and eminent anthropologist, for example, condemned Newman's dying Maori as a Pakeha "excuse" for their arrogance and rapacity.[60]

Conclusions

As these applications of Darwinism to Maori affairs suggest, the cultural leaders of New Zealand embraced evolution for a wide variety of reasons. Evolution interpreted theistically as God's method of creation enabled thinking Christians to support their faith with the authority of modern science. Darwinism enabled colonial scientists to explain a host of scientific problems naturalistically, and it helped them to turn natural history into a respected and useful activity. Most men of science adopted evolution without abandoning their religious tenets. Because they faced neither an entrenched religious establishment nor significant opposition from fellow scientists, they arguably enjoyed greater intellectual freedom than their colleagues in the rest of the English-speaking world. Free thinkers deployed Darwinism as a weapon in their crusade against the churches, and suffered little as a result. A number of scientist-politicians and settlers turned evolution into a weapon to justify racial warfare and colonial conquest. For all these diverse reasons, New Zealand's cultural mainstream absorbed Darwinism with both equanimity and enthusiasm.

Notes

1 *The Press*, 20 December, 1862.

2 Ibid., 17 January, 1863; Samuel Butler, *A First Year in Canterbury Settlement with Other Early Essays*, ed. R. A. Streatfield (London: A. C. Fifield, 1914), pp. 154–55; *The Press*, 21 February, 1863.

3 *The Press*, 2 April, 1863.

4 These and the following statistics are taken from the *New Zealand Census 1881*.

5 *Auckland Examiner and People's Journal*, 11 December, 1856; for Southwell's religious views, see "Religion in Relation to Education," ibid., 8 January, 1857.

6 *New Zealand Census 1921*. For a fuller discussion, see H. Jackson, "Churchgoing in Nineteenth Century New Zealand," *New Zealand Journal of History*, 17 (1983):43–59.

7 H. W. Harper to C. W. Richmond, 16 August, 1869, in *The Richmond-Atkinson Papers*, ed. G. H. Scholefield, 2 vols. (Wellington: Government Printer, 1960), vol. 2, p. 291; *New Zealand Presbyterian*, 1 (1866):7; *Church Gazette*, 1 March, 1873, p. 29; ibid., 1 April, 1873, p. 57; and ibid., 1 September, 1879, p. 89.

8 Jon Roberts has argued that in the United States few Protestant thinkers regarded evolution as scientifically plausible until about 1875, when the conversion of most scientists to evolution encouraged a change of mind. See Jon H. Roberts, *Darwinism and the Divine in America: Protestant Intellectuals and Organic Evolution, 1859–1900* (Madison: University of Wisconsin Press, 1988).
9 *Otago Daily Times*, 30 July, 1870, p. 4; *The Press*, 27 August, 1872.
10 [Charles] Fraser, "Darwinism," *Canterbury Presbyterian and Record of Church News*, 1 (1873):125–34.
11 See further Erik Olssen, *A History of Otago* (Dunedin: John McIndoe, 1984), pp. 39–41.
12 A. R. Fitchett, *The Ethics of Evolution* (Dunedin: Mills, Dick, 1876), pp. 9, 11–17.
13 The report on which the following account is based appears in the *Otago Daily Times*, 24 October, 1876, pp. 2–3. Quotations are from this source unless otherwise acknowledged.
14 *Otago Daily Times*, 17 January, 1876, p. 2; and ibid., 25 October, 1876, p. 4.
15 Ibid., 25 October, 1876, p. 4.
16 Ibid., 24 October, 1876, p. 3.
17 Roberts argues thus in this volume.
18 James MacGregor, *Regarding Evolution: The Previous Question of Science Considered* (Dunedin: J. Horsburgh and A. Tracer; n.d. [1885?]), and *Otago Daily Times*, 4 November, 1876, p. 3 (italics in original).
19 James Copland, "The Divine Authority and Inspiration of the Holy Scriptures," *The Evangelist*, 5 (1873):76.
20 *The Evangelist*, 3 (1871):242–43.
21 *Otago Daily Times*, 3 October, 1878, p. 2.
22 Copland, *The Origin and Spiritual Nature of Man* (Dunedin: James Horsburgh, 1885), passim. For a fuller discussion of Copland's life and thought, see John Stenhouse, "The Rev. Dr James Copland and the Mind of New Zealand 'Fundamentalism,'" *Journal of Religious History*, 17 (1993):475–97.
23 *Otago Daily Times*, 4 September, 1902, p. 6; and ibid., 6 September, 1902, p. 2.
24 ibid., 6 September, 1902, p. 4; and ibid., 11 September, 1902, p. 7. See also ibid., 15 September, 1902, p. 7.
25 Allan K. Davidson, "A Protesting Presbyterian: The Reverend P. B. Fraser and New Zealand Presbyterianism, 1892–1940," *Journal of Religious History*, 9 (1986):193–217; P. B. Fraser, *Evolution and Ape-manism in Schools: New Syllabus challenged*, (Dunedin: Otago Daily Times, 1929); and *Lyttelton Times*, 25 June, 1929; for a useful survey of evolution in school textbooks and curricula in the late nineteenth and twentieth centuries, see Colin McGeorge, "Evolution and the New Zealand Primary School Curriculum, 1900–1950," *History of Education*, 21 (1992):205–18.
26 On Hutton, see Stenhouse, "Darwin's Captain: F. W. Hutton and the Nineteenth-Century Darwinian Debates," *Journal of the History of Biology*, 23 (1990):411–42.
27 *The Press*, 16 January, 1891, p. 6.
28 *Canterbury Standard*, 9 October, 1862; letter, Charles Darwin to Julius Haast, 22 January, 1863, Haast MS 35, Folder 51, Alexander Turnbull Library, Wellington.

29 W. T. L. Travers, "On the Changes Effected in the Natural Features of a New Country by the Introduction of Civilized Races," *Transactions and Proceedings of the New Zealand Institute* (hereafter *TPNZI*), 1 (1868):31–33; and *Canterbury Standard*, 9 October, 1862.

30 *Otago Witness*, 2 September, 1876, p. 3; Ibid., 9 September, 1876, p. 3; and *Otago Daily Times*, 23 August, 1876, p. 3.

31 Ibid., 23 August, 1876, p. 3.

32 *Otago Witness*, 30 September, 1876, p. 3; and *Otago Daily Times*, 20 September, 1876, p. 3.

33 *Otago Daily Times*, 18 October, 1876, pp. 2–3.

34 Ibid., 19 October, 1876, p. 2.

35 R. Gilles, "Presidential Address," *TPNZI*, 9 (1876):655–64.

36 F. W. Hutton, "On the Origin of Species by Means of Natural Selection; or, The Preservation of Favoured Species in the Struggle for Life," *Geologist*, 3 (1860):464–72, on p. 464.

37 Frank Miller Turner, "The Victorian Conflict Between Science and Religion: A Professional Dimension," *Isis*, 69 (1978):356–76; and Adrian Desmond, *Archetypes and Ancestors: Palaeontology in Victorian London, 1850–75* (London: Blond and Briggs, 1982).

38 S. T. Nevill, "Presidential Address," *TPNZI*, 10 (1877):562–66.

39 James Hector, *Inaugural Address to the Australasian Association for the Advancement of Science, 15 January 1891* (Christchurch: Australasian Association for the Advancement of Science, 1891), pp. 15–16.

40 For a fuller discussion see Stenhouse, "Darwin's Captain."

41 Anne Mozley, "Evolution and the Climate of Opinion in Australia, 1840–1876," *Victorian Studies*, 10 (1967):411–30; and Barry W. Butcher, "Gorilla Warfare in Melbourne," in *Australian Science in the Making*, ed. R. W. Home (Cambridge: Cambridge University Press, 1988), pp. 153–69; see also Butcher's paper in this volume.

42 W. Miles Maskell, *Christianity, Modern 'Science,' and Evolution* (Christchurch: A. Simpson, 1881), pp. 3–4, 9.

43 *New Zealand Journal of Science*, 1 (1882): 188; *The Press*, 4 May, 1882, p. 2; *TPNZI*, 18 (1885):xiii; 23 (1890):598.

44 Robert Hall Bakewell, "Is Life a District Force?" *TPNZI*, 17 (1884):410–17; see also *New Zealand Journal of Science*, 2 (1885):518.

45 For Buller's views, see *TPNZI*, 27 (1894):103; the Thomson quotation appears in *New Zealand Journal of Science*, 2 (1885):371; for Thomas on evolution and religion, see *New Zealand Herald*, 7 June, 1904, p. 6.

46 Alexander William Bickerton, "On the Genesis of Worlds and Systems." *TPNZI*, 12 (1879):197; and R. M. Burdon, *Scholar Errant: A Biography of Professor A. W. Bickerton* (Christchurch: Pegasus Press, 1956).

47 T. J. Parker, *Professor Huxley from the Point of View of a Disciple* (London: Rait, Henderson, and Co., 1896); Parker, *Inaugural Lecture Delivered in the University Library, May 2, 1881* (Dunedin: Otago Daily Times, 1881), pp. 1–2; letter, Canon Curzon-Siggers to T. M. Hocken on the memorial service for T. J. Parker, 13 November, 1897, New Zealand Pamphlet Collection, 11 (1868–1870), Hocken Library, Dunedin.

48 Robert Stout, *Evolution and Theism* (Christchurch: 1881), pp. 1, 2, 7; and Stout, *The Resurrection of Christ: A Reply to Messrs Brunton and Forlong* (Dunedin: Joseph Braithwaite, 1881), pp. 17–18.
49 Peter J. Lineham, "Freethinkers in Nineteenth Century New Zealand," *New Zealand Journal of History*, 19 (1985):61–80; and Lineham, "Christian Reaction to Freethought and Rationalism in New Zealand," *Journal of Religious History*, 15 (1988):236–50.
50 Quoted in Alfred W. Crosby, *Ecological Imperialism: The Biological Expansion of Europe, 900–1900* (Cambridge: Cambridge University Press, 1986), p. 217; William Fox, *The Six Colonies of New Zealand* (London: Parker, 1851), p. 54; and Edward Shortland, *The Southern Districts of New Zealand: A Journal, with Passing Notices of the Customs of the Aborigines* (London: Longman, 1851), p. 40.
51 G. A. Selwyn to W. E. Gladstone, 15 September, 1846, microfilm MS 426, Hocken Library; Shortland, *The Southern Districts*, pp. 40–41.
52 Octavius Hadfield, *A Sequel to 'One Of England's Little Wars': Being An Account Of The Real Origin Of The War in New Zealand: Its Present State, and The Future Prospects of The Colony* (London: Williams and Norgate, 1861), pp. 14–15.
53 Joseph Giles, ed., "Waitara and the Native Question," *Southern Monthly Magazine*, 1 (1863):209–16, on p. 215; "Our Colonization and its Ethics," Ibid., 548–52, passim.
54 A. S. Atkinson, personal journal, 5 June, 1863, Alexander Turnbull Library, Wellington.
55 *Lyttelton Times*, 8 July, 1867, p. 3.
56 Travers, "On the Changes," pp. 308, 312; and *TPNZI*, 1 (1868):443.
57 William Colenso, "On the Maori Races of New Zealand," *TPNZI*, 2nd ed. 1 (1875):339–424, quotation on p. 342, originally published in 1868.
58 Alfred K. Newman, "A Study of the Causes Leading to the Extinction of the Maori," *TPNZI*, 14 (1881):459–77, quotations on pp. 475–76; Walter Buller, "Presidential Address," *TPNZI*, 17 (1884):443–45, quotation on p. 444; Ibid., p. 446.
59 *New Zealand Parliamentary Debates*, 38 (1884):231–32; Ibid., 56 (1886):105; for a fuller discussion of Newman's life and thought, see Stenhouse, " 'A Disappearing Race before We Came Here': Doctor Alfred Kingcome Newman, the Dying Maori, and Victorian Scientific Racism," *New Zealand Journal of History*, 30 (1996):124–40.
60 Te Rangi Hiroa, "The Passing of the Maori," *TPNZI*, 57 (1924):362–75, on p. 364.

4

Environment, culture, and the reception of Darwin in Canada, 1859–1909

SUZANNE ZELLER

Historians attempting to assess the responses of naturalists in British North America[1] to Charles Darwin's theory of *The Origin of Species by Means of Natural Selection* soon encounter two significant problems. First, the country's leading Victorian scientists, Sir John William Dawson (1820–99) and Sir William Edmond Logan (1798–1875), were both geologists. As such, they were positioned to address the theory's weaknesses more than its strengths, in the fossil record that appeared to preserve not a single example of one species evolving into another.

Yet their responses differed considerably, as neither Dawson nor Logan reacted primarily out of scientific considerations. Dawson, the principal of McGill College, Montreal (from 1855), longtime president of the Natural History Society of Montreal, and founding president of the Royal Society of Canada (1883), became known internationally for his unremitting resistance to Darwin. In 1860, if not earlier (and, Darwin suspected, before having read *The Origin*), he determined to defend the traditional place of natural theology in the cosmology of Design. Dawson's long-awaited scientific biography, Susan Sheets Pyenson's *John William Dawson: Faith, Hope, and Science* (1996), recently confirmed earlier claims that it was "a calling to natural theology that Dawson heard as a calling to science." Yet while Pyenson also documented the detrimental effects of this decision upon Dawson's subsequent scientific reputation, historians have yet to evaluate definitively not only the impact, but even the nature and extent, of his sustained critique. Only then will we know its actual place in Dawson's long and complex career as well as in the larger Canadian context.[2]

The author thanks Ron Numbers, John Stenhouse, Pam Henson, Keith Benson, George Urbaniak, Ken Dewar, Cynthia Comacchio, and Graeme Wynn for their valued comments on earlier versions of this paper; and gratefully acknowledges financial support for this research from a grant funded by Wilfrid Laurier University Operating Funds and the Social Sciences and Humanities Research Council of Canada Institutional Grant to WLU.

Logan, the founding director of the Geological Survey of Canada (1842–69), represented a public institution whose continuance depended upon annual parliamentary grants. He therefore felt most keenly and personally his Survey's vulnerability to public opinion. In scrupulously cultivating friends across the Province of Canada's volatile political spectrum, he ordered his staff explicitly to eschew every hint of public controversy, including any discussion that touched on religion.[3] This contrast between the outspoken Dawson and the close-mouthed Logan underscores the importance of circumstance in shaping responses to Darwin.

It also offers a microcosm of the second problem on the larger Canadian scene. For unlike the situation in New Zealand and elsewhere, naturalists in British North America generally remained unusually taciturn on the subject of Darwin. No strong public defender of the theory emerged there during the half-century after 1859. As a result, Canadian historians have felt challenged, first and foremost, to explain this "puzzling" reticence[4] among scattered communities of British North American naturalists.

Historical analyses have, moreover, rested overwhelmingly upon the outlooks of geologists. What of the views of other naturalists? How did circumstances – scientific, historical, geographical, personal – influence *their* responses to Darwin? The broader evidence suggests that Darwin's theory imposed no abrupt discontinuity upon the experiences of most naturalists in Canada, except insofar as it lent new purpose and direction – often indirectly – to their scientific activities. As was the case elsewhere, it penetrated British North American society in successive phases, sustaining new myths within a cultural and environmental context in which older ones were losing ground.

The historical literature

Perhaps not surprisingly, the apparent reluctance of British North American naturalists to enter the Darwinian fray has been interpreted by historians in largely negative terms, to exemplify a conservative consensus in British North American society during the nineteenth century. In a 1976 conference paper, Patricia Roome contrasted the Canadian reception of Darwin's theory with the galvanized debate that transformed approaches to science, social science, and religion in Britain and the United States. A smaller, less secure educated elite in Canada, explained Roome, was instead "preoccupied with building an academic structure and intellectual tradition in a young country," a difficult task in the face of both "the immature state of scientific research and advanced education" and "the predominance, strength, and influence of evangelical

Protestantism." In Roome's analysis, these limitations demanded intellectual conformity; with an elite "united in ... opposition to Darwin's wild speculations," she concluded, "there was no necessity for debate."[5]

Roome's cultural consensus held firm in the institutional framework explored by A. B. McKillop, in *A Disciplined Intelligence: Critical Inquiry and Canadian Thought in the Victorian Era* (1979). McKillop exposed the intellectual underpinnings of a powerful "moral imperative" in English-Canadian culture that derived from both Scottish Common Sense philosophy and natural theology. He agreed that critical reviews of *The Origin of Species* published in 1860 by both J. W. Dawson and Daniel Wilson, a Scottish archaeologist who taught history and literature at the University of Toronto, "reflected to a very high degree" the reception of Darwin's evolutionary theory more generally in Canada.[6]

Dawson's review, published in the *Canadian Naturalist and Geologist*, the journal of the Natural History Society of Montreal, precluded questions of origin from the domains of both natural law and Baconian (inductive) inquiry. Concerned more with religious implications than with scientific validity, Dawson dismissed the importance of Darwin's theory in what one subsequent observer termed "a posture of abnegation unique in the history of science": "We may well ask," Dawson chastened, "what is gained by such a result [as Darwin's], even if established." Wilson (1816–92), McKillop noted, shared Dawson's commitment to the Baconian ideal. His review in the *Canadian Journal of Industry, Science, and Art*, the organ of Toronto's Canadian Institute, also emphasized the lack of fossil evidence "for any such gradations of form as even to suggest a process of transmutation." The fact that, on the one hand, Dawson's natural theology was shaped by "the Hebraic legacy of his Calvinism," while, on the other, Wilson's science drew more upon his own poetic sensibility, highlighted for McKillop the existence of a broader cultural consensus in Canadian society.[7]

During the 1980s two more general contributions to the history of Canadian science nuanced our understanding of this cultural consensus, along the fault lines of the country's fundamental dualism. In a 1983 lecture series on the natural-history tradition in English Canada, Carl Berger revealed the power of Dawson's own growing isolation to illuminate his vehemence against Darwin. Personal disappointments, including his failure (twice) to accede to coveted chairs at the University of Edinburgh (1855, 1868) and the unprecedented refusal of the Royal Society of London to publish his Bakerian Lecture (1870), scarred Dawson with a deepening sense of alienation from British centers of science. Time and distance only hardened his conviction that, as the metropolis lost sight of "the true Baconian mission" of science (i.e., by accepting Darwin's hypothetico-deductive method), the torch had "somehow" to be taken up in Canada. Dawson's siege mentality was reinforced as he felt Montreal's anglophone community surrounded by an increasingly aggressive new ideology, ultramontanism, that dominated Quebec's largely Catholic society after 1840. Ultramontanism, a term derived from the French view of Rome "on the other side of the mountains" from France, denoted a form of Catholicism centered largely on the leadership of the papacy. In Quebec the church's traditionally monolithic position allowed it to proclaim its own power to hold sway in civil society, including politics. Ultramontanism there identified the French language with the survival of the French-Canadian nation in Quebec. The self-appointed champion of a "double-handed fight" over the very future of Canada, Dawson rallied naturalists to preserve their privileged vision of a larger British North American nation: "We hold its natural features," he explained in 1858, "as fixing its future destiny, and regard its local subdivisions as arbitrary and artificial."[8]

Dawson's dichotomized perception of Canadian society as a struggle between Anglo-Protestant modernity and francophone ultramontane "popery" overlooked some potential anti-Darwinian allies in his own province. In *Histoire des Sciences au Québec* (1987), Luc Chartrand, Raymond Duchesne, and Yves Gingras described a surprisingly broad range of French-Canadian responses to Darwin, from more liberal to more ultramontane outlooks culminating in the fundamentalist *Cercle catholique*. In particular, their work suggests some interesting parallels between Dawson's scientific objections to Darwin and those of the Abbés, T.-E. Hamel (1830–1913) and J.-C.-K. Laflamme (1849–1910), both geologists at Laval University, Quebec. Hamel and Laflamme resisted ultramontane strictures generally accepted by their scientific colleagues, especially in Montreal, even after the publication of Pope Pius IX's antiliberal *Syllabus of Errors* in 1864. Much like Dawson, the Laval geologists rejected Darwinian evolution for falling short of inductive standards of science. Over time, all three geologists gradually reconciled themselves to a form of evolution that nevertheless continued to exclude human origins.[9]

The studies by Berger and Chartrand, Duchesne, and Gingras offered new insights into the dearth of Canadian responses to Darwin. Berger's more rigid reading of the conservative cultural consensus speculated that Darwin's method grew "muddled" in subsequent editions of *The Origin*, so deviating from accepted inductive tradition that Canadians may quite understandably have recoiled from such "alien" theoretical discussion. Moreover, Berger surmised, since the theory of evolution by natural selection "did not impinge directly upon the collecting and classifying activities in which most of them were engaged," it was quite possible for Canadian naturalists simply to continue collecting specimens without addressing theoretical issues at all.[10] And while the Quebec historians discerned important nuances in the responses to Darwin even among clerics in an increasingly conformist ultramontane culture, they also underlined the isolation and dispersion of French-Canadian as well as English-Canadian scientific communities as major hindrances to sustained public debate of *The Origin* in Quebec, at least until the 1870s.[11]

Meanwhile, Roome also recognized the other side of her case for a conservative consensus. While it remained prudent for members of fledgling public institutions, dependent upon public financial support and the goodwill of the churches, to emphasize the harmony of "science and religion, revelation and reason," she noted, it does not necessarily follow that individual private convictions toed this party line. Roome referred specifically to Canadian universities, but her judgment can certainly be extended to the deafening silence observed by Sir William Logan and his staff on the Geological Survey.[12]

Indeed, scientific biography has since demonstrated the accuracy of Roome's implication that interesting exceptions might be found flourishing beneath the formal consensus in British North American society. In *David Boyle: From Artisan to Archaeologist* (1983), Gerald Killan documented the fascinating intellectual journey of a Scottish blacksmith who became Canada's foremost archaeologist at the Royal Ontario Museum, Toronto. A schoolteacher in Elora, Ontario, during the 1870s, Boyle had developed a passionate interest in Darwinian biology. He urged its inclusion in the elementary-school curriculum as part of a local "intellectual awakening," which he promoted almost single-handedly.[13]

A. B. McKillop and others have also shown that by the 1870s, "new magazines of informed opinion" on a national scale reached growing numbers of middle-class Canadians, whose more open discussions breached the "cloistered confines of Canadian educational institutions and the pages of religious newspapers." These intensified deliberations, especially after the publication of *The Descent of Man* (1871), often focused more generally on evolution and religion than specifically on Darwin and his science.[14]

This broader focus of public concern had important roots in the historical experiences of various religious denominations. A recent revival of scholarly interest in Canadian religious history has pursued two divergent interpretive approaches to the problem of responses to Darwin. In the more inclusive view, Protestant culture in Canada had long been predisposed to accept evolution through the idea of progress, which helped to bridge the gap between secular and sacred in the modern world. Anglicans, Methodists, Presbyterians, and Baptists were thus able by the mid-1870s to meet many of the religious challenges of Darwinism without alarm. For example, this spirit of accommodation declawed the human implications of the mutability of species, reassigning to the evolutionary process an overseeing deity in place of a self-regulating natural law.[15]

In the more exclusive view, Canadian evangelical responses to evolution, forged by the Great Awakening after 1780, differed fundamentally from those of Enlightenment-based Anglicans. In *The Evangelical Century* (1991), Michael Gauvreau argued that not only creed but college as well in English Canada combined biblical sources, Baconian method, and historical perspectives to render natural theology superfluous to the evangelical outlook. For Gauvreau, the creative edge of the Baconian method of acquiring knowledge had been blunted in Canada by the 1850s, leaving only "a rhetorical structure promoting caution and reverence." The resulting conservatism helps to explain why Canadian evangelical churches offered no "single major contributor to the transatlantic discussion" of Darwin's theory or, indeed, of more general issues raised by science.[16]

Excluded from Gauvreau's interpretation was the Presbyterian J. W. Dawson, on the questionable grounds that Dawson's critique "stemmed less from his religious or theological beliefs than from his scientific methodology"; following Berger's interpretation, Gauvreau judged Dawson's position as "even more atypical of evangelical attitudes toward modern thought than of the scientific reaction to Darwinism."[17] Yet this restriction may be unnecessary. The explanatory power of Gauvreau's insight into evangelical appropriations of the Baconian method is, if anything, enhanced when Dawson is included. These important religious connotations go a considerable way to contextualize Dawson's dogged critical fixation upon Darwin's methodology.

Consensus in context

While Berger estimated that among Canadian naturalists in general, a tide turned in favor of evolution only shortly after 1900,[18] there is evidence that positive responses eluded the formal ideology of consensus considerably earlier, apparently with impunity, in three phases. The first, during the 1860s, confirmed the extent of the consensus that an insecure colonial society could actually muster; yet almost immediately Canadian botanists took up Darwinian themes indirectly, quietly adapting their scientific researches accordingly. The second, after the publication of *The Descent of Man*, marked the rise of public debate and the eventual accommodation of evolution within a larger conservative context of philosophical idealism; yet with it arrived a new generation of Canadian scientists trained in Darwinian and other modern approaches to research. The third, from about the mid-1880s, saw a broadening acceptance of Darwinian science, as Canadian entomologists joined their botanical colleagues in this enterprise. During this phase, idealist and imperialist ideologies incorporated evolutionary interpretations of Canadian history and culture, often with Darwinian, neo-Lamarckian, and Spencerian overtones of development through struggle in a harsh environment. By the 1890s, university curricula also turned to evolutionary biology, as an older generation of faculty left the scene. By the 1890s, industrial expansion and material progress in Canada formed a backdrop for the open acceptance of evolutionary themes and approaches, with broad Darwinian applications not only in science but also in applied and social sciences.

This stepwise acceptance of Darwinian evolution after 1860 lay rooted in British North Americans' prior experience of both history as an evolutionary process and geography as a source of struggle in a harsh environment. Unlike the situation in the United States, a longstanding heritage of constitutional development based the colonies' political existence in North American upon a Loyalist myth that rejected revolution in favor of maintaining the British connection. In everyday life, this process had proven slow and often arduous. British North America's diverse and disparate fringes of population along the American border remained preponderantly rural, conservative, and regionally isolated. A plethora of largely Protestant denominational churches competed for scarce resources among the majority of anglophone inhabitants, while Roman Catholic minorities of francophones and others added complex dualities. The result was a vulnerable and relatively unstable culture that was both British and yet increasingly North American in character. Unlike the experience with a friendlier environment in New Zealand, pioneer immigrants to the British North American backwoods found a relatively unsympathetic land in which the irresistible force of agricultural expansion soon met an immovable object in the Precambrian Shield that limited severely the possibility of settlement farther north.[19]

After midcentury the country's "greatest single fact" was still nature, "and a most unWordsworthian nature" at that. A longstanding literary response to its harshness, "cold, empty and vast," included a sense of being overwhelmed by the corresponding "vast unconsciousness of nature" that is said to have confronted Canadian settlers as late as the 1880s.[20] In a prize-winning essay for the Paris Exposition in 1855, John Sheridan Hogan detailed the experience of the pioneer settler in Canada in strikingly bleak Malthusian (and perhaps even Spencerian) terms that would soon find scientific echoes in the "moral silence" that permeated Darwin's view of nature:

> In the middle of a dense forest, and with a "patch of clearing" scarcely large enough to let the sun shine in upon him, he looks not unlike a person *struggling for existence* on a single plank in the middle of an

> ocean. . . . The same still, and wild, and boundless forest every morning rises up to his view; and his only hope against its shutting him in for life rests in the axe upon his shoulder. A few blades of corn, peeping up between stumps whose very roots interlace, they are so close together, are his sole safe-guards against want; whilst the few potatoe [sic] plants, in little far-between "hills," and which *struggle for existence* against the briar bush and luxuriant underwood, are to form the seeds of his future plenty. Tall pine trees, girdled and blackened by the fires, stand out as grim monuments of the prevailing loneliness, whilst the forest itself, like an immense wall round a fortress, seems to say to the settler, – "how can poverty ever expect to escape from such a prison house."[21]

Nor did it seem self-evident to everyone that God intended human populations to follow, as the fur trader George Barnston urged, the northward diffusion of edible plants from the cabbage family into "those dismal regions where ice holds almost eternal empire."[22]

Yet earlier pioneer settlers had found solace in the belief that clearing and cultivation would eventually moderate British North America's climate to resemble that of Europe in similar latitudes. Such optimism harked back to long-standing climatic theories absorbed by Enlightenment culture and lent authority by Edward Gibbon's famous *History of the Decline and Fall of the Roman Empire* (1776). Relying on earlier writers of the 18th century, Gibbon designated "Canada, at this day, [a]s an exact picture of ancient Germany. Although situated in the same parallel with the finest provinces of France and England, that country experiences the most rigorous cold." Virtually every educated person in the English-speaking world was familiar with Gibbon's *History*, its claims vulgarized in popular expectations that Canada's climate would one day be ameliorated by settlement and cultivation. The end of the pioneer era in Canada, when settlement reached the Shield, forced reluctant admissions by the late 1850s that no improvement at all could be discerned.[23]

It may not seem surprising, then, that some British North Americans concerned themselves with the long-term implications of the country's apparently permanent climatic condition. There remained, for example, the Comte de Buffon's infamous theory of physical degeneration among European immigrants to the North American environment. In 1854 the Clear Grit (frontier liberal) politician William McDougall (1822–1905), an agricultural reformer and editor of the *Canadian Agriculturist* in Toronto, offered a more positive alternative derived from Jean Baptiste de Lamarck's theory of the inheritance of acquired characteristics. McDougall suggested that North American's historical time scale was still too short to interpret conclusively the significance of physical modifications outside of the vegetable kingdom. Recognizing that second- and third-generation North Americans already differed visibly from their British forebears, he postulated instead of degeneration a slow process of adaptation in which the human body required more time to acclimatize, to "acquire and transmit to its offspring the new and requisite physiological peculiarities" that would ultimately enable it to thrive in its North American environment.[24]

Such hopes revitalized in Canada another long-standing climatic myth that had been popularized and disseminated, once again, through Gibbon's *History*. This one idealized the north as a source of liberty, physical strength, and hardiness of spirit, and had found frequent expression in pioneer settlers' accounts of their experiences. In 1858 Alexander Morris, a young Conservative politician from Montreal, added an imperial dimension when he predicted that, in the "Great Britannic Empire of the North," descendants of immigrant northern peoples would be tempered over time to thrive in the frozen crucible of their new homeland, the vanguard of an incipient transcontinental nation. The Canadian people, Morris believed, inherited an "energetic character" fostered by "the fusion of races and the conquering from the forest of new territories"; the result would be a "harmonious whole – rendered the more vigorous by our northern position."[25]

Just as Darwin was publishing his theory, then, popular expectations regarding Canada's potential were reaching an important crossroad. Realities had imposed certain limitations upon earlier optimism about the country's future. Yet as older dreams of climatic progress faded during the late 1850s, new alternatives rose to replace them: in the northern environment that now formed an apparently permanent condition of Canadians' struggle for existence, Darwin offered a scientific framework within which to comprehend this predicament more fully and, for some, strengthened the resolve to overcome its negative implications.[26]

First responses

In a colonial society in which the University of Toronto rejected both Thomas Huxley and John Tyndall as prospective additions to its faculty during the 1850s, it is perhaps surprising that Canadian naturalists responded to Darwin as quickly as they did. For the publication of Darwin's theory* in 1859 introduced into cosmology a factor of randomness that could be discomfiting on several levels. It undermined not only traditional "artificial" taxonomic structures for comprehending nature, but also the integrity of the very intellectual basis upon which those

* That is, that the tendency toward variation in nature, combined with the relative scarcity of the necessities of life, resulted in a struggle for existence for all species; that changes in the conditions of life gave advantages to varieties of species able to survive them, extinguishing those that could not; and that over time new species thus developed from older forms through this process of natural selection.

structures had been built. If each species had not been specially created as part of the harmonious design of nature, then there might remain no immediate need for a Designer. Similarly, if humankind consituted a mere species randomly evolved in nature, then there might remain no guarantees of human privilege, no material basis for religious faith, in the struggle for existence.

Yet underneath the formal ideology that dominated the first scholarly reviews of *The Origin of Species*, some deeper questioning did occur. While E. J. Chapman (1821–1904), professor of mineralogy and geology at University College, Toronto, shared Dawson's preference for "the Creator's plan" over Darwin's theory, he conceded that "few persons have ever made the close contemplation of nature their study for any time, without having experienced, at one period or another, the visitation of sundry hauntings" about distinctions between species and varieties. And although Dawson had recently developed in his *Archaia: Studies of the Cosmogony and Natural History of the Hebrew Scriptures* (1860) a rigidly antitransformist view that precluded by definition any possibility of variations transcending the realm of their species, Chapman felt unable to "go with the author to the full extent of his argument."[27]

In 1866 John Macoun (1831–1920), who became one of Canada's most influential field botanists with the Geological Survey, revealed to the evolutionist J. D. Hooker at Kew Gardens his own perplexity over Darwin's challenge to his Presbyterian assumptions: "Mr. Darwin is most likely right in his opinion," he wrote, "but I doubt it." Even for Daniel Wilson, who understood his entire intellectual paradigm to be in flux, the scientific case for Darwin's theory remained open for the time being. While under his presidency a copy of *The Origin* was immediately procured for the library of the Canadian Institute, Wilson himself struggled toward a gradual acceptance of evolution in a limited form that excluded human origins.[28]

J. W. Dawson, meanwhile, hoped to close the case permanently with an authentication of the notorious "fossil" that he named *Eozoon canadense*, the dawn animal of Canada. Discovered in the Precambrian Shield of the Ottawa Valley by a staff member of the Geological Survey of Canada in 1858, the curious markings were thought by Dawson to represent evidence of the existence of a relatively sophisticated organism that predated simpler forms. *Eozoon* offered Dawson an apparent counterexample to the theory of natural selection and Canadian romantics a dramatic image of their northern homeland as the very cradle of life on earth. William Logan cautiously referred to *Eozoon* as a "supposed fossil" in his magnum opus, the *Geology of Canada* (1863), while his staff palaeontologist Elkanah Billings privately maintained his own utter rejection of the specimens as fossils. In deference to Dawson's palaeontological authority, however, Logan nevertheless presented *Eozoon* to the British Association for the Advancement of Science in 1864, including it in further publications to 1867. *Eozoon* then quickly and quietly dropped from Logan's embarrassed cognizance when critical evidence against its biological origins piled up soon thereafter. Dawson, in contrast, clung to his conviction for decades longer, undaunted even after it was clear that he stood increasingly alone.[29]

In the meantime, Canadian botanists enjoyed new biogeographical opportunities opened indirectly by Darwin's theory. Unlike the geological record, the geographical distribution of species offered an important source of positive evidence for Darwin's theory, and botany quickly blossomed into a modern cluster of ecological and palaeobotanical approaches to the study of plants. In 1862 the British botanist J. D. Hooker's "Outlines of the Distribution of Arctic Plants" applied Darwin's theory of "creation by variation" to explain the distribution patterns of northern plant forms, in terms of glacial migrations of species from a single center in Scandinavia to Asia and America. Appreciated by Darwin as "the real engine to compel people to reflect on the modification of species," Hooker's theory confirmed the geographical importance of British North America to an overall understanding of plant distribution patterns. It thereby raised international interest in Canada as a northern territory adjacent to two of Hooker's five circumpolar regions of Arctic floras and pointed Canadian botanical attention northward, away from border regions with the United States. The theory encouraged Canadian botanists to collect crucial evidence far beyond the bounds of settlement, and permitted them to do so without having to declare themselves openly on the question of natural selection. Perhaps the most surprising example of such participation was Abbé L.-O. Brunet (1826–76), professor of natural history at Laval University, whose botanical collections from Anticosti Island and other remote places contributed to Asa Gray's Darwinian efforts at Harvard University.[30]

The most widely influential example, however, was George Lawson (1827–95), a Scottish botanist trained by J. H. Balfour at the University of Edinburgh. Lawson came to Queen's University, Kingston, in 1858, and founded the Botanical Society of Canada in 1860. Dismayed by serious infighting at Queen's, he left for Dalhousie University, Halifax, in 1863. Declaring Darwin's case on its own not strong enough to be compelling, Lawson nevertheless recognized Hooker's theory as a boon to his own field of scientific endeavor, and used it as a scientific link between his work in the Province of Canada and his move to Nova Scotia. Lawson adopted Hooker's theoretical framework, recasting his own botanical researches over large northern geographical expanses in terms of the variation, adaptation, and survival of plants under changing climatic conditions. He mobilized a far-reaching network of British North American collectors to contribute specimens while he collated distribution patterns of northern plant forms in their geographical and ecological relations.

In particular, Lawson verified Hooker's biogeographical theory that northern plant forms spread farther south along the Atlantic coast than inland, and that Canadian swamp plants flourished along east-coast hillsides. Although he never declared himself a Darwinian scientist (and only hinted at it in his presidential address to the Royal Society of Canada in 1888), Lawson in 1865 explicitly agreed (incorrectly, as it turned out) with Hooker and Gray – and against Dawson's objections that the plant was introduced by Scottish immigrants – that heather (*Calluna* vulgaris) was native to Nova Scotia and had once spread widely over North America.[31] And while the *Canadian Naturalist and Geologist* duly issued Dawson's critique of Hooker's Darwinian excesses (penned, once again, before he saw Hooker's paper), the journal also included botanical support for Darwin's case even during these early years.[32]

An equally important broader application of Darwinian themes manifested itself in the Canada First movement. With the achievement of Confederation in 1867, this group of young romantic nationalists sought a common source of Canadian identity and pride in the distinctiveness suggested by *Eozoon canadense*. Expanding the myth of the north beyond ideals of harmony to those of dominance, they forged a powerful ideological taproot of expansionist Canadian nationalism as a natural evolutionary product of British imperialism. The movement, which included William McDougall, now an influential expansionist member of the Dominion's first federal (Conservative) government; a few former students of George Lawson; and other admirers of modern science, including the Nova Scotian author and businessman Robert G. Haliburton (1831–1901), fortified its arguments by explicit conceptual links to biological theories of the variation and geographical distribution of species. In a remarkable speech in 1869 Haliburton reminded Canadians that

> We are the sons and heirs of those who have built up a new civilization, and though we have emigrated to the Western world, we have not left our native land behind, for we are still in the North, in the home of the Old Frost Giant, and the cold north wind that rocked the cradle of our race, still blows through our forests, and breathes the spirit of liberty into our hearts, and lends strength and vigour to our limbs.

No longer simply Britons but "sons of the New World," Canadians were thus encouraged to take heart that "as long as the north wind blows, and the snow and the sleet drive over our forests and fields, we may be poor, but we must be a hardy, a healthy, a virtuous, a daring, and if we are worthy of our ancestors, a dominant race." This aggressive nationalism purported to justify Canada's northwestward and transcontinental expansion at the expense of its native peoples, a fulfilment of its manifest destiny, in the view of the Nova Scotian R. G. Haliburton, as a new "Norland" that would succeed Great Britain as the core of the Empire. McDougall, for his part, applied his aggressive expansionist outlook to guide a near-disastrous federal policy of acquiring the Hudson's Bay Company's northwestern lands. His approach provoked an armed resistance by Métis and native peoples in 1869, forcing the concession of full provincial status to the proposed Crown colony of Manitoba.[33]

Second round

Canadians in growing numbers and varieties pursued the implications of evolutionary theory with more concerted intensity, and from more varied directions, after 1871. This greater alacrity resulted not just from the publication of Darwin's *Descent of Man* that year, or from the proliferation of national literary magazines that promoted discussion of its implications for Christian morality. In addition, the rise of "materialism" in the physical sciences, epitomized by John Tyndall's Belfast address to the British Association for the Advancement of Science in 1874, which traced the origins of life back to inorganic compounds and even single atoms, made Darwin appear moderate.

The growing body of ideas associated with scientific materialism signalled a restless questioning that challenged the traditional cultural consensus far more fundamentally than any single revolutionary theorist alone could have accomplished. At the University of Toronto, it was said, students undertook pilgrimages to the lectures of philosophers who, they hoped, could safeguard their religious faith from doubts visited by Darwin and other scientists upon their generation. Professors of physical science at McGill and Dalhousie Universities too felt called to counter a growing "agitation" and "uneasiness" in "the mind of the nation in general," as well as among students who learned of Darwin, Spencer, Huxley, and Tyndall in their classes.[34] In this way, Darwin's evolutionary theory became subsumed by larger evolutionary issues, not all of which Darwin himself had raised.

Yet while there was more room for discussion outside of the realm of science, there was also more room for dissent within – dissent, that is, from the moral imperative, which no longer permeated the outlook of the generation of scientists who came to prominence after 1870. Belonging to the same general age cohort that informed the Canada

First movement with its outspoken youthful idealism, these scientific scions had imbibed much that was modern in their professional training. At Toronto James Loudon (1841–1916), the institution's first Canadian appointee as professor of mathematics and physics in 1875, publicly challenged the traditional assumptions embodied by Daniel Wilson's generation. Loudon disavowed scientists' purported responsibility to reconcile their work with religion or metaphysics; two years later he apprised the Royal Canadian Institute of his commitment instead to the German research ideal, pursuing knowledge for its own sake.[35]

At about the same time Loudon's colleague in natural history, Robert Ramsay Wright (1852–1933), succeeded the Reverend William Hincks and H. Alleyne Nicholson. Hincks, a devoted quinarian taxonomist, had held to idealist morphological theories that classified the natural world by analogy, in interlocking circles of groups. Himself considered a fossil by generations of students who endured while he taught natural history by rote, Hincks cherished beliefs in an elaborate grand Design without which, he told the Canadian Institute in 1870, choas must result. In contrast, Wright in 1876 extolled the unmatched fertility of *The Origin of Species* in opening new fields for scientific research. Untroubled by either Darwin's methodology or his application of the theory to humankind (a mere "deduction from the general inductive law of descent"), the Edinburgh-trained zoologist considered "the theoretical nature of species" to be "definitely settled." He set about during the 1870s teaching modern biology and establishing a research laboratory; during the 1880s he set about developing a research program based upon evolutionary approaches.[36]

Loudon and Wright might have recognized a kindred spirit in George Mercer Dawson (1849–1901) of the Geological Survey of Canada. The son of J. W. Dawson, George returned to Canada in 1872 a star graduate of the Royal School of Mines. As Thomas Huxley's prize student in natural history, he trained in a modern evolutionary approach that underlay his enormous later contributions as a leading figure among the second generation of professional Canadian geologists.

Evolutionary theory proved particularly useful for Dawson's anthropological studies. After joining the Geological Survey in 1875, he spent several seasons in British Columbia. An important result was the publication in 1878 of Dawson's remarkable report on the Haida people of the Queen Charlotte Islands, whom he appreciated for their highly evolved culture, threatened with extinction in its deserted villages and decreasing population. Dawson's evolutionary understanding of aboriginal peoples as lineal descendants of prehistoric peoples spurred him to value and preserve information about their culture as a laboratory of human history – a task whose time, he feared, was limited in the face of the serious challenges for survival facing them. He recommended policies both to encourage the Haida's social evolution and integration into European culture and industry through education and, just as importantly, to respect their "fully developed" concepts of property ownership during forthcoming railway-land negotiations with the Canadian government. Dawson's advice was eclipsed, however, by more aggressive politicians, including William McDougall, who more willingly used science to ride roughshod over native societies while promoting the advance of "civilization," which even Dawson saw as inevitable.[37]

The irony of George Dawson's wide-ranging accomplishments as a field geologist must have struck the elder Dawson, who never seems to have confronted his son over Darwinian differences between them. Yet William Dawson felt himself fighting an uphill battle against growing numbers of popular articles and textbooks that presented organic evolution as a scientific given. "In this way," he recognized, a process of intellectual selection was taking place in education, as "young people are being trained to be evolutionists without being aware of it, and," he regretted, they "will come to regard nature wholly through this medium." Indeed, scientific critics of J. W. Dawson's increasingly discordant searches for counterevidence to natural selection had begun to wish aloud, lest his books "fall into the hands of commencing biological students, who would find it difficult to shake off the false associations that, in it, surround the facts which are discussed." Quoting an "eminent naturalist," one public lecturer pointedly told a Montreal audience in 1883 that, "with one exception, every man in the world who can properly be called a naturalist has accepted the theory of Evolution."[38]

Daniel Wilson too departed from the scientific mainstream when *The Descent of Man* challenged his assumption of humanity's special place in Creation. As his aesthetic temper gained ascendancy, Wilson's extraordinary response took the form of *Caliban: The Missing Link* (1873), which the anthropologist Bruce Trigger recently described as "distinctly post-modernist" in approach. Under the guise of literary criticism, the work explored the philosophical and psychological implications of Darwin's theory by representing a character from Shakespeare's *The Tempest* as an imaginary missing link whose existence Darwinians had consistently failed to prove.[39]

In Quebec the ultramontane Abbé Léon Provancher (1820–92), an entomologist at Laval University, also faced a dilemma. While wishing to encourage popular interest in natural history and science among French Canadians, he felt a responsibility to combat on its own ground the "absurdity and impiety" of modern scientific materialism, including Darwinism. As a result, Provancher founded *Le Naturaliste canadien* in 1872 to serve this dual and sometimes contradictory purpose. His main preoccupation with Darwin, however, remained more religious than scientific. As late as 1887, he introduced a course of critical lectures on the subject, "by popular demand."[40]

In the maritime provinces, too, natural-history societies and universities witnessed some support for Darwinian themes during the 1870s. In essays to the Nova Scotia Institute of Natural Science in Halifax, Angus Ross and Andrew Dewar worked to reconcile modern evolutionary theories with Christian teachings. Ross not only accepted the evidence in nature that supported Darwin's theory of the evolution of species, but also argued that humanity's inclusion in the process served only to explain scientifically its exalted place on the "Tree of Life." Dewar, a Halifax architect, revived the idea of spontaneous or "predestinated" generation, to soften the implications of both scientific materialism and natural selection: "a God which endowed matter from the beginning with properties which enabled it when in a certain condition to form new life," he felt assured, "is certainly greater than one who had to interpose in every new creation." A student at Mount Allison University, a Methodist college in Sackville, New Brunswick, went further in suggesting that Darwin's "grand modifying principle" functioned "as a principle of change to a greater or less degree, not alone in the organic world, but in all the various social institutions that are found as products of civilization in its various stages." It operated, for example, on languages, in which the principles of natural selection revealed the patterns of their development and modification over time.[41]

Missing from these opinion pieces was new scientific research, as Dewar was well aware. But he for one had arrived at the opinion that "so far as experiments are valuable, no new ones can be performed that would materially alter the position of affairs, or give a further insight into the beginnings of life." No microscope, he regretted, could ever "enable us to see the very evolution of life," which he assumed occurred at the atomic level. This limitation consoled Dewar in his regrettable conviction that "no one would put faith in experiments performed in such a benighted country as Nova Scotia." George Lawson at Dalhousie University and his colleague John Sommers of the Halifax Medical College were nevertheless still busily pursuing original research on geographical distribution patterns of northern plant forms, and applying Darwinian concepts of competition for ecological dominance among the various types.[42]

Individual entomologists too began during the 1870s openly to support Darwin's theory of the origin of species. William Couper (?–1886), a specialist in botany and entomology, addressed the Entomological Society of Montreal in affirmative terms in 1875. While Couper had been investigating the role of wind and water in the geographical distribution of plants and insects since the late 1850s, his earliest response to

The Origin of Species is unknown. A fellow entomologist, the beekeeper and free-thinking controversialist Allen Pringle (1834–96) of Selby, Ontario, supported Darwinian evolution at least since 1874.[43] In contrast to the strong program of combined laboratory and field work in botany that George Lawson had brought with him from Edinburgh, it would be another decade before organized Canadian entomologists followed Couper and Pringle along similar Darwinian lines.

Meanwhile, an important catalyst for the reorganization of entomologists to pursue Darwinian approaches could be found in Dr. William Brodie (1831–1909), a Toronto dentist. In 1877 Brodie founded the Toronto Entomological Society (renamed the Natural History Society of Toronto in 1878 and amalgamated with the Royal Canadian Institute in 1885) as an alternative to the older Entomological Society of Ontario, which he believed had grown moribund in a neglect of its scientific responsibilities. An internationally recognized authority on entomology and botany, especially in the field of insect galls and their remedies, Brodie developed an ecological approach to the study of nature. While remaining sceptical of Darwin's theory of natural selection, Brodie nevertheless accepted both the principle of evolution in nature and "the theory that specific life forms are a product of Environment, that relation to Environment is their historic life." In 1898 he explained to a popular audience that

> Nature is said to be a dear old nurse, and we love to rest on her lap and listen to her wonderful tales and enchanting music; but she is many-sided, and on one side is implacable and cruel as the grave. Her constant demand is "conform," and what does not is ruthlessly tramped down.[44]

An important transitional figure between the older natural history and modern biology in Canada, Brodie exerted a powerful influence there not only upon the development and organization of entomology, but also upon the tradition of nature writing that came to be epitomized by Ernest Thompson Seton (1860–1946). A close family friend, Seton spent many youthful hours during the early 1880s on local nature walks with Brodie, who generated charming wildlife stories out of these rambles along the Don River valley and into the local woods near Toronto. Brodie exhibited an impressive receptiveness to nature's offerings: when he went out to "interview" the animals, he once observed, he invariably found it was the animals that were interviewing him. His environmental perspective was further enriched by investigations of insect life, pioneering contributions to the study of intelligence and consciousness as they related to Darwinian psychological theory in Brodie's day.[45]

This second phase also brought David Pearce Penhallow (1854–1910), whom Asa Gray recommended to J. W. Dawson, to occupy the long-vacant chair of botany at McGill University in 1883. Penhallow, an American, had been trained by William Smith Clark at the Massachusetts Agricultural College in the "new school" of botanical science, with its Darwinian emphasis on anatomy, physiology, and ecology, with experimental and phylogenetic approaches to these subdisciplines. Penhallow carefully followed the work of Julius von Sachs (Würzburg) in plant physiology, Carl Nägeli (Munich) in plant cytology, and Anton de Bary (Strasbourg) in the developmental history of plants, and under Dawson's influence he included the study of palaeobotany. Here he followed Heinrich Goeppert (Breslau) in detailing microscopic analyses of fossil specimens and in tracing their genetic relations to modern types. Correcting even some of Dawson's taxonomic errors, Penhallow became a leading authority on fossil plants from the Cretaceous and Tertiary eras, including lignite coals. One of the first to interpret the evolutionary sequence of conifers in physiological terms, Penhallow also applied microscopic analyses of woods to explain their varying strengths and other practical qualities.[46]

The growing receptiveness toward evolutionary science by the 1880s occurred within a particular, and still conservative, intellectual context. Historians of English-Canadian thought emphasize the development during the 1870s of a dominant school of philosophical idealism, derived from Scottish adaptations of Kantian and Hegelian ideas, which facilitated intellectual accommodations between modern evolutionary science and Christian religion. The most prominent representative of this school was perhaps John Watson (1847–1939) of Queen's University, but George Paxton Young (1818–1889) of the University of Toronto, John Clark Murray (1836–1917) of McGill University, and Jacob Gould Schurmann (1854–1942) of Acadia University, Wolfville, Nova Scotia, were also active contributors. While perhaps wisely unwilling to engage in debate over the scientific subject matter per se, these philosophical idealists expressed concern over the doctrine of evolution in its implications for human morality and related issues. Even proof of a developmental relationship among instinct, consciousness, thought, conscience, and morality, they concluded, would not invalidate a traditional moral philosophy based on Christian assumptions. At best it would be irrelevant; for, as Watson insisted, a veil still separated "the holy place of Morality from the outer court of Nature."[47]

Readers of the *Canadian Monthly and National Review* found entertainment in rebuttals of idealists' efforts to preclude human mental and moral development from the realm of natural science. Although hopelessly outnumbered and outranked by the academic bulwark of idealists, Joseph A. Allen (1814–1900), a former Anglican clergyman who had taught history at Queen's University, and William Dawson LeSueur (1840–1917), an Ottawa journalist and civil servant, challenged this limitation as arbitrary and unwarranted. As one anonymous reader put it, "nowadays if the Church disagrees with Science, so much the worse for the Church." This conclusion reflected precisely the fear of Goldwin Smith (1823–1910), the liberal critic, who accepted natural selection in principle yet abhorred its implications for the social order.[48]

The growth in Canada after 1880 of industrialization and urbanization, with their concomitant social problems, intensified the perceived social role of the doctrinal synthesis wrought by philosophical idealism in the country's universities. One important result was the development on a broader popular level of a "practical idealism" that contained a modernized but equally powerful version of the traditional "moral imperative." According to one case study, the teachings of John Watson and his colleagues instilled in future teachers and public servants a set of moral evolutionary principles intended to correct the materialist path taken by modern science. The idealist spirit encouraged a strong commitment to community and what has also been called an "ideology of service" for the public good. In stark contrast to the rampant individualism generally associated with Social Darwinism in the United States and Great Britain at the time, Watson offered freedom in the form of duty, "the joyous doing of what one ought," for the moral evolution of society and the nation. Extramural lectures for local citizens reinforced in university students their moral duty to direct public opinion on important social issues, so as to promote this evolution more widely.[49]

Public educational systems furthermore provided a convenient vehicle for this ideology of applied idealism. In Canada the high school had not yet become a center of mass secondary education, but remained rather a "gatekeeper for the occupational and social order," a role well suited to the idealist cause. What better way to promote these elevated social evolutionary themes than to transmit through public schools multiple versions of the evolutionary creed of practical idealism? This powerful intellectual trend helps to explain the facility with which Ramsay Wright was able to disseminate his evolutionist *Introduction to Zoology for the Use of High Schools* (1889) throughout public educational systems in most Canadian provinces from the 1890s on.[50]

Diffusion

One ought not to underestimate the remarkable profusion in Canada of references to Darwin in a fanning-out process that began slowly during the 1880s, increased apace during the 1890s, and laid the groundwork for a sea change that came, albeit stealthily, earlier than 1900. By the time of J. W. Dawson's death in 1899, his younger geological colleagues charitably differentiated between Dawson the "pure scientist," and Dawson the "Presbyterian of the old school," whose works on evolution retained "but a transitory value." It was difficult for them to comprehend Dawson's continued insistence that Darwin "makes science dry, barren, and repulsive, and diminishes its educational value." (Yet, in the final analysis, what are we to make of the anti-Darwinian who sent his eldest son to study with Thomas Huxley, and appointed a protege of Asa Gray to succeed him?)[51]

That same year the noted Quebec architect and engineer Charles Baillairgé (1827–1906), addressing the Royal Society of Canada, urged that evolutionary biology had come to appear not as a slippery slope to materialism, but rather as a pillar of faith in the divine Creator of the germ of life. Under the impact of European intellectual developments, including Henri Bergson's philosophy of creative evolution, even the hard-line editor of *Le Naturaliste canadien* conceded in 1905 that the Catholic Church had never expressly pronounced itself on the issue of evolution in natural history.[52]

Nor was the question of an exact turning point in Canadian universities more clearcut. An impressionistic survey of course calendars suggests that in most cases before the late 1880s, professors of natural history who did not openly oppose Darwin's theory left little indication of what exactly they were teaching. Curricular transitions that explicitly included evolutionary biology, beginning generally during the 1890s, seem contingent upon the end of long incumbencies of professors of natural history. William Hincks, for example, was ensconced at the University of Toronto, 1853–1871; his successor, Ramsay Wright (after a brief stint by H. Alleyne Nicholson), recalled later that teaching evolution during the 1870s "was still regarded as dangerous."[53] Wright accordingly remained careful not to do so too overtly until about the time of the establishment of the medical school at Toronto in 1887. J. W. Dawson dominated at McGill, 1855–1899; D. P. Penhallow, appointed in 1883, included evolution in the curriculum only implicitly until 1901, after Dawson had retired. George Lawson at Dalhousie University taught botany and natural history, 1863–1895; evolutionary biology entered the calendar more overtly in 1898.

Lawson left Queen's University in 1863, too early to be replaced by a Darwinian biologist. Among his successors, Robert Bell worked for the Geological Survey of Canada under the direction of Sir William Logan; and the Rev. James Fowler, appointed to teach natural history in 1881 and botany after 1892, worked within Lawson's biogeographical framework, introducing Julius Sachs's evolutionary textbook at Queen's only in 1900. More importantly, however, Fowler was joined in zoology in 1893 by A. P. Knight, a graduate of both Queen's in arts and Victoria

College, University of Toronto, in medicine. Knight lost little time in announcing his intention of teaching modern biology, introducing *The Origin of Species* as a textbook at Queen's in 1895.[54]

Among the remaining maritime institutions, Mount Allison University, with its strong connections to northeastern American universities, introduced evolution into its curriculum in 1891. At the University of New Brunswick, Fredericton, the Harvard-trained naturalist Loring Woart Bailey (1839–1925) taught natural history, 1861–1907; he adopted Wright's evolutionary zoology textbook also in 1891. Acadia University, a Baptist college, added evolution overtly to its curriculum in 1908.[55]

More generally in English Canada, by the 1890s the acceptance of evolutionary models facilitated various popular permutations of Darwinian, neo-Lamarckian, and Spencerian iterations of both Canada's history and the growth of its political institutions. John George Bourinot (1836–1902) heralded a constitutional school of historians and political scientists who traced Canada's evolution from colonial status toward "a fuller freedom in an imperial association" with Great Britain, incorporating "one of the most elemental themes in Canadian history, analogous to a scientific 'law.' " For Bourinot, as Carl Berger has shown, "the idea of the progressive development of liberty seemed analogous to the purposeful progress of evolution in nature. Indeed, the process by which Canada achieved self-government was referred to as 'Nature's methods of evolution.' " The "dynamic engine" of this inevitable progress toward liberty in Canadian history, according to this view, was struggle,[56] including a fundamental struggle for existence in a northern climate "peculiarly fitted to rear a people whose northern vigour will give them weight in the world, and will add strength and character to the [British] nation of which they form a part."[57] In 1909, after almost two decades of steady industrialization, urbanization, and immigration under the Liberal government of Wilfrid Laurier, the imperialist George Parkin (1846–1922) predicted with confidence that Canada's cold climate would "secure for the Dominion and perpetuate there the vigour of the best northern races," protecting the country from

> that submerged tenth which in the slums of [America's] great cities is deteriorating our own race. A stern nature takes hold of this type of man when he comes to Canada and imposes upon him a relentless discipline. It either develops in him the virtues of energy, prudence and foresight which he lacks, or it kills him. . . . Apply that process for a century to a country and you will have a survival of the fit which ensures to it, . . . the traditional strength of the northern races.[58]

In stark contrast to these idealizations of Canadians' relationship to their environment, J. W. Dawson had feared instead that Darwinian evolution justified human behavior as "the enemy of wild nature," the exterminator of species. In *Modern Ideas of Evolution* (1890), Dawson foresaw as inevitable not the northwestward march of empire, but rather "the westward march of exhaustion" of both the land and its resources.[59]

In Canadian literature, too, the popular genre of nature writing now drew inspiration from the Darwinian outlook. Developed by Ernest Thompson Seton and the nationalist and wilderness writer Charles G. D. Roberts (1860–1943), these animal stories followed the chronological pattern of small beginnings during the 1880s to full flowering during the 1890s and after. Roberts, the better writer, grew up in the New Brunswick wilderness and shared with Seton and some contemporary poets, including Bliss Carman (1861–1929), strong themes of the amorality of nature and the struggle for survival that expressed their experience of nature in the Canadian context. Seton, the more scientific in approach, read both *The Origin of Species* and *The Descent of Man* during the 1870s, and maintained regular contact with William Brodie after leaving Toronto. His publication of *Wild Animals I Have Known* (1898), and the many animal stories that followed it, depicted nature as governed by a stern ecological ethic of survival rather than by Christian morality. The stories are scientific more than literary in appeal, and more disturbing than entertaining in effect. Seton was also concerned to incorporate the Darwinian ideas of Brodie and other pioneers on animal instinct and intelligence. Sympathy for the animals in his stories, as in Brodie's before him, was intended to convey his complete identification of humankind with them, as opposed to an anthropomorphic interpretation of nature.[60]

Developments in Darwinian psychology also confirmed for specialists in education as well as in industry that humankind was "not apart from but a part of nature." The sooner one proceeded to investigate human nature "as part of the grand whole," argued Wesley Mills (1847–1915), professor of physiology at McGill University, the better it would be "for man and all other animals." In particular, these researches bore enormous significance for educators, in Mills's view "first, last, and always, developer[s]." In this broader evolutionary context, Mills held, teachers who recognized the complete interdependence of mind and body could facilitate students' learning by considering the importance of environment, both physical as well as social.[61] As the British Columbian minister of education told the Mainland Teachers' Institute in Vancouver in 1896,

> you will find the study of evolution a great aid and assistance to you in the noble work of moulding the immature minds which are committed to your charge. It will help you to cultivate, not only the three R's,

> but the more important qualities, the two P's, Pity and Patience – pity for inherited faults – patience to mould them into virtues, and when you meet with aggravating eccentricities of character, it will help you to look upon them, not with anger and impatience, but with a curious interest, as manifestations of inherited tendencies, which it is your duty and your privilege to correct and to reform for the advancement of the human race.[62]

In an age of growing monopoly, capitalism, and intensified competition, these social Darwinian principles of learning were extended in *Industrial Canada*, the trade journal of the Canadian Manufacturers' Association, to the industrial arts, to explain the process of "industrial evolution." The engineering profession, too, scrupulously applied them to analyze the process of invention as an undertaking that promised "a MODERN success" if understood correctly.[63]

Conclusions

By the time the Royal Society of Canada marked the fiftieth anniversary of *The Origin of Species* in 1909, principles of evolution by natural selection were being applied in Canada to human pathology; to the improvement of field crops; and to the biochemical origins of life itself.[64] A. B. Macallum (1858–1934), who as professor of the new science of biochemistry at the University of Toronto was himself a virtual product of the Darwinian revolution, judged in hindsight that Darwin's theory had appeared in "such a degree of completeness as to parallel the birth of Minerva, armed with shield and spear, from the brain of Jupiter." Darwin's thoroughness, Macallum observed, "intensified the keenness of the discussion and precipitated into the struggle all the representatives of the older schools of thought." Such critical attention in turn tested the theory's mettle, "served to clear ideas," and diffused it sufficiently widely to "silence its opposition." Macallum confessed himself part of a generation that took "evolution more or less for granted."[65]

Two years later Macallum's mentor, Ramsay Wright, ventured into the more controversial social Darwinian offshoot of eugenics. Acknowledging recent criticisms of natural selection as a mechanism for evolutionary change, Wright pointed out to the Royal Society of Canada that organic evolution had nevertheless "fertilized every branch of human knowledge since its general acceptance by the scientific world." He highlighted heredity and variation as twin operative factors in the origin of "new forms," and called for eugenics to control undesirable elements in human societies. For unlike George Parkin, Wright reasoned with admitted caution that modern provisions for "philanthropic schemes of various kinds, improvements in modern hygiene and the progress of medical science," effectively eliminated "natural selection, which would otherwise remove the unfit in the struggle for existence." As a result, he feared, a growing "decadence" was "observable in the slum populations of great cities and elsewhere." Wright stressed the "urgent import" of heredity studies for Canada, "emigration to which has now attained such magnitude." The country, he predicted, would "soon be an immense experimental ground in miscegenation, so various are the elements of our new population." There could be "no question about the desirability of fostering as far as possible immigration of a high quality from the British Isles and Northern Europe."[66]

At roughly the same time, however, children in some rural Ontario Sunday schools were still being taught to vilify pieces of coral along Great Lakes shorelines as the horns of the devil ("devils' decorations, and we always smash 'um").[67] These continued differences in outlook underline the fact that the reception of Darwinian evolution in Canada proceeded as part of a large matrix of related ideas, along numerous paths and at many different rates, as it became clear over time that evolution and natural selection could be dealt with on quite separate terms (or in some cases not at all). The reception and diffusion of Darwinian ideas in Canada occurred within a particular historical and geographical context, and was governed to a large extent by Canadians' changing images and experiences of the northern land they inhabited.

Indeed, the long-standing centrality of environmental perceptions and relationships in the Canadian case predated Darwin's theory, and suggests an exception to Peter Bowler's assumption that "only when Darwin made the basic idea of evolution acceptable did the possibility of rehabilitating Lamarck's idea [of acquired characteristics in response to environment] emerge."[68] The Canadian reality appears to have been more complex, as Lamarckian and other older myths recurrently found expression from the pioneer era on, yet shifted in tone to more aggressive postures of dominance within the Darwinian context. As industrialization, urbanization, and immigration proceeded to distance many Canadians from the natural environment – as wilderness became cottage country – the moral imperative was adapted to meet changing conditions. The results amounted to more than mere variations in traditional forms; by the 1890s new species both intellectual and cultural were evolving.

Notes

1 By the late 1850s British North America comprised the United Province of Canada, Nova Scotia, New Brunswick, Prince Edward Island, Newfoundland, Rupert's Land, the North West Territories, and British Columbia. In

1867 confederation united the first three to form the Dominion of Canada; by 1873 all but Newfoundland belonged to the Dominion.

2 The quote is from John Fenlon Cornell, "Sir William Dawson and the Theory of Evolution," unpublished M.A. thesis, McGill University, 1977, p. 21. On Dawson, see also Charles F. O'Brien, *Sir William Dawson: A Life in Science and Religion* (Philadelphia: American Philosophical Society, 1971); Robert John Taylor, "The Darwinian Revolution: The Responses of Four Canadian Scholars," unpublished Ph.D. diss., McMaster University, 1976; T. H. Clark, "John William Dawson," in *Dictionary of Scientific Biography*, ed. Charles C. Gillispie, 16 vols. (New York: Scribner's, 1970–80), vol. 3, pp. 607–09; Peter R. Eakins and Jean Sinnamon Eakins, "John William Dawson," in *Dictionary of Canadian Biography* [*DCB*], ed. Francess Halpenney and Ramsay Cook, 13 vols. (Toronto: University of Toronto Press, 1966–1994), vol. 12, pp. 230–37; and Susan Sheets Pyenson, *John William Dawson: Faith, Hope and Science* (Montreal and Kingston: McGill-Queen's University Press, 1996).

3 Suzanne Zeller, *Inventing Canada: Early Victorian Science and the Idea of a Transcontinental Nation* (Toronto: University of Toronto Press, 1987), pp. 63–64, 77, 93–94, 101–02; and Carl Berger, *Science, God, and Nature in Victorian Canada* (Toronto: University of Toronto Press, 1983), p. 69. See also Morris Zaslow, *Reading the Rocks: The Story of the Geological Survey of Canada 1842–1972* (Ottawa: Macmillan, 1975), pp. 38–82 passim.

4 Berger, *Science, God, and Nature*, pp. 69–70; P. Roome, "The Darwin Debate in Canada, 1860–1880," in *Science, Technology and Culture in Historical Perspective*, eds. L. A. Knafla, M. S. Staum, and T. H. E. Travers (Calgary: University of Calgary, 1976), p. 186.

5 Roome, "Darwinian Debate," pp. 183–86.

6 A. B. McKillop, *A Disciplined Intelligence: Critical Inquiry and Canadian Thought in the Victorian Era* (Montreal: McGill-Queen's University Press, 1979), p. 110.

7 Ibid., especially pp. 99–110; J. W. Dawson, review in *Canadian Naturalist and Geologist* [*CN&G*], 5 (1860):100–22; Daniel Wilson, "President's Address," *Canadian Journal of Industry, Science, and Art* [*CJ*], 5 (1860):99–127; both reprinted in J. D. Raab, ed., *Religion and Science in Early Canada* (Kingston: Ronald P. Frye, 1988), sec. 11; and Clifford Holland, "First Canadian Critics of Darwin," *Queen's Quarterly*, 88 (1981):102. Dawson reiterated his presumption in advice to scientists to "spare themselves the trouble of looking for any such transition from apes to men in any period," in "On the Antiquity of Man," *Edinburgh New Philosophical Journal* [*ENPJ*], 19 (1864):55.

8 Berger, *Science, God, and Nature*, pp. 63–64; Dawson quoted in Zeller, *Inventing Canada*, p. 7; and Richard Jarrell, "L'ultramontanisme et la Science au Canada français," in *Science et médécine au Québec*, ed. Marcel Fournier, Yves Gingras, and Othmar Keel (Quebec: Institut québecois de Recherche de la Culture, 1987), pp. 41–68. For a broader discussion of the "fortress mentality" in Canadian culture, see Gaile McGregor, *The Wacousta Syndrome: Explorations in the Canadian Landscape* (Toronto: University of Toronto Press, 1985).

9 Luc Chartrand, Raymond Duchesne, and Yves Gingras, *Histoire des Sciences au Québec* (Montreal: Boréal Press, 1987), pp. 167–79.

10 Berger, *Science, God, and Nature*, pp. 55, 70–72.

11 Chartrand et al., *Histoire des Sciences*, p. 160.
12 Roome, "Darwinian Debate," pp. 183–87.
13 Gerald Killan, *David Boyle: From Artisan to Archaeologist* (Toronto: University of Toronto Press, 1983), pp. 47–49, 59–61.
14 McKillop, *Disciplined Intelligence*, p. 134; Roome, "Darwinian Debate," pp. 192ff. McKillop is supported in his broader discussion by Ramsay Cook, *The Regenerators: Social Criticism in Late Victorian English Canada* (Toronto: University of Toronto Press, 1985), pp. 9–14. See also McKillop, ed., *A Critical Spirit: The Thought of William Dawson LeSueur* (Toronto: McClelland and Stewart, 1977); John A. Irving, "The Development of Philosophy in Central Canada from 1850 to 1990," *Canadian Historical Review*, 31 (1950):264; Leslie Armour and Elizabeth Trott, *The Faces of Reason: An Essay on Philosophy and Culture in English Canada, 1850–1950* (Waterloo, Ontario: Wilfrid Laurier University Press, 1981); and Raab, *Religion and Science*, Section 11.
15 William Westfall, *Two Worlds: The Protestant Culture of Nineteenth Century Ontario* (Kingston: McGill-Queen's University Press, 1989), p. 189; and David B. Marshall, *Secularizing the Faith: Canadian Protestant Clergy and the Crisis of Belief, 1850–1940* (Toronto: University of Toronto Press, 1992), pp. 54–60. See also Barry H. Moody, "Breadth of Vision, Breadth of Mind: The Baptists and Acadia College," in *Canadian Baptists and Christian Higher Education*, ed. G. A. Rawlyk (Kingston: McGill-Queen's University Press, 1988), pp. 3–30; and Jerry N. Pittman, "Darwinism and Evolution: Three Nova Scotia Religious Newspapers Respond, 1860–1900," *Acadiensis*, 22 (1993):40–60.
16 Michael Gauvreau, *The Evangelical Century: College and Creed from the Great Revival to the Great Depression* (Kingston: McGill-Queen's University Press, 1991), Chap. 4; for a succinct preliminary version, see his "Baconianism, Darwinism, Fundamentalism: A Transatlantic Crisis of Faith," *Journal of Religious History*, 13 (1985):443.
17 Gauvreau, *Evangelical Century*, pp. 128–37.
18 Berger, *Science, God, and Nature*, p. 75.
19 See Graeme Wynn, "Notes on Society and Environment in Old Ontario," *Journal of Social History*, 3 (1979):59.
20 Alec Lucas, "Nature Writers and the Animal Story," in *Literary History of Canada*, ed. Carl F. Klinck (Toronto: University of Toronto Press, 1967), p. 367; Northrop Frye, *The Bush Garden: Essays on the Canadian Imagination* (Toronto: Anansi Press, 1971), pp. 225, 243; and George Altmeyer, "Three Ideas of Nature in Canada, 1893–1914," *Journal of Canadian Studies*, 11 (1975):21–36.
21 John Sheridan Hogan, *Canada: An Essay* (Montreal: John Lovell, 1855), p. 24; italics mine. See also McGregor, *Wacousta Syndrome*; Marcia Kline, *Beyond the Land Itself: Views of Nature in Canada and the United States* (Cambridge: Harvard University Press, 1970); and Carl F. Klinck, ed., *Literary History of Canada*, 2nd ed., 4 vols. (Toronto: University of Toronto Press, 1976–90).
22 George Barnston, "Remarks on the Geographical Distribution of the Cruciferae throughout the British Possessions in North America," *CN&G*, 4 (1859):1–12.
23 Zeller, *Inventing Canada*, pp. 171–73; when asked by the British parliamentary select committee on the Hudson's Bay Company, in 1857, whether his

experiences in British North America could support the impression created by Gibbon that Canada's climate would be ameliorated by settlement, J. H. Lefroy, director of the Toronto Magnetic and Meteorological Observatory, replied in the negative.

24 Zeller, *Inventing Canada*, pp. 257–59. On the earlier impact of Buffon's theories in the American colonies, see Joseph Kastner, *A World of Naturalists* (London: John Murray, 1978), pp. 122–27. On Lamarckian evolution, see Peter J. Bowler, *Evolution: The History of an Idea* (Berkeley: University of California Press, 1984), Chap. 3; and L. J. Jordanova, *Lamarck* (Oxford: Oxford University Press, 1984).

25 Alexander Morris, *Canada and Her Resources* (Montreal: John Lovell, 1855); *The Hudson's Bay and Pacific Territories* (Montreal: John Lovell, 1858); and *Nova Britannia* (Montreal: John Lovell, 1858).

26 For a broader discussion of the background, context, and ramifications of these myths, see Zeller, "Classical Codes: Biogeographical Perceptions of Environment in Victorian Canada," *Journal of Historical Geography*, 24 (1998): 20–35.

27 Cornell, "Dawson and Evolution," pp. 34–35; E. J. Chapman, review of *Archaia*, *CJ*, 5 (1860):61.

28 W. A. Waiser, *The Field Naturalist: John Macoun, the Geological Survey, and Natural Science* (Toronto: University of Toronto Press, 1989), pp. 9–11, follows Berger: "Darwin's ideas were alien to the Canadian tradition of natural science, and were thus largely ignored"; "Annual Report of the Council," *CJ*, new ser., 6 (1861):201. On Wilson: Raab, *Science and Religion*, pp. 327–28; Holland, "First Canadian Critics," pp. 104–05; Bruce Trigger, "Daniel Wilson and the Scottish Enlightenment," *Proceedings of the Society of Antiquaries of Scotland*, 122 (1992):68–69; and Zeller, " 'Merchants of Light': The Culture of Science in Sir Daniel Wilson's Ontario, 1853–1892," in Elizabeth Hulse, ed., *Daniel Wilson* (Toronto: University of Toronto Press, 1999).

29 Zaslow, *Reading the Rocks*, pp. 87–88; A. R. C. Selwyn, "On the Origin and Evolution of Archaean Rocks," Royal Society of Canada [RSC], *Proceedings and Transactions* [*P&T*], 2nd ser., 2 (1896):LXXXV. See also O'Brien, *Dawson*, Chap. 6; and his "*Eozoon canadense* 'The Dawn Animal of Canada,' " *Isis*, 61 (1970):206–23. Dawson published *The Dawn of Life: The History of the Oldest Known Fossil Remains* in 1875.

30 J. D. Hooker, "Outlines of the Distribution of Arctic Plants," Linnaean Society, *Transactions*, 23 (1862):251–310; British North America lay adjacent to Hooker's Western American region, from Bering Strait to the Mackenzie River, and his Eastern American region, from the Mackenzie River to Baffin Bay. Letter, Darwin to Hooker, 25 February, 1862, #354, in *More Letters of Charles Darwin*, ed. Francis Darwin, 2 vols. (London: Appleton, 1903), vol. 1. p. 465; J. H. Faull, "The Influence of Darwin on Botanical Science," *Ontario Natural Science Bulletin*, 5 (1909):33–34; and Chartrand et al., *Histoire des Sciences*, p. 167.

31 Zeller, *Inventing Canada*, pp. 243–57. Lawson was explaining the evolutionary approach of Wesley Mills (a physiologist at McGill University) to comparative psychology: "we aim at the construction of a ladder by which we may

climb from the simplest manifestations of consciousness to the highest performances of the most gigantic human intellect," in Lawson, "Presidential Address," RSC, *P&T*, 6 (1888):XXVIII. See also Zeller, "George Lawson: Victorian Botany, the Origin of Species, and the Case of Nova Scotian Heather," in *Profiles of Science and Society in the Maritimes*, ed. Paul Bogaard (Fredericton: Acadiensis, 1989), pp. 51–64.

32 Dawson, review of Hooker, *CN&G*, 7 (1862):334–44; Harland Coultas, "Origin of Our Kitchen Garden Plants," ibid., n.s. 2 (1865):33–42; and letter, Hooker to Darwin, 2 November, 1862, #355, in Darwin, ed., *More Letters*, p. 466.

33 Zeller, *Inventing Canada*, pp. 262–67; R. G. Haliburton, *The Men of the North and Their Place in History* (Montreal: n.p., 1869); see also Carl Berger, *The Sense of Power: Studies in the Ideas of Canadian Imperialism, 1867–1914* (Toronto: University of Toronto Press, 1970), p. 53; and his "The True North Strong and Free," in *Nationalism in Canada*, ed. Peter Russell (Toronto: University of Toronto Press, 1966), pp. 3–26.

34 Irving, "Philosophy in Central Canada," p. 264. See also Cook, *Regenerators*; Alexander Johnson (McGill University), *Science and Religion* (Montreal: Dawson Brothers, 1876), p. 6; William Lyall, address in *Halifax Reporter and Times*, 4 November, 1874, clipping, Dalhousie University Archives; and Charles MacDonald, "Evolution," [1883] and "Old Lesson in Metaphysics," [n.d.], mss., Dalhousie University Archives.

35 See McKillop, "The Research Ideal and the University of Toronto," in *Contours of Canadian Thought*, ed. A. B. McKillop (Toronto: University of Toronto Press, 1987), p. 85.

36 William Hincks, "The Gorilla,"*CJ*, 8 (1863):319; Hincks, "Presidential Address to the Canadian Institute," *Nature*, 2 (1870):108; H. Alleyne Nicholson, "Sexual Selection in Man," *CN&G*, 6 (1872):449–54; McKillop, "Research Ideal," p. 85; Wright, "Haeckel's Anthropogenie," *CJ*, n.s. 15 (1876–78):235–36; and Sandra F. McRae, "The 'Scientific Spirit' in Medicine at the University of Toronto, 1880–1910," unpublished Ph.D. diss., University of Toronto, 1987, Chap. 3.

37 On Dawson, see Zeller and Gale Avrith, "George Mercer Dawson," in *DCB*, vol. 13, pp. 257–61; Avrith, "George Dawson, Franz Boas, and the Origins of Anthropology in Canada," in *Dominions Apart: Reflections on the Culture of Science and Technology in Canada and Australia, 1850–1945*, eds. Roy MacLeod and Richard Jarrell, special issue of *Scientia Canadensis*, 17 (1994):185–203; and Dawson's report reprinted in *To the Charlottes: George Dawson's 1878 Survey of the Queen Charlotte Islands*, ed. Douglas Cole and Bradley Lockner (Vancouver: University of British Columbia Press, 1993). The larger impact of Social Darwinism on perceptions of (and polices toward) native peoples in Canada has not yet come under close scholarly scrutiny, but see J. R. Miller, *Skyscrapers Hide the Heavens: A History of Indian-White Relations in Canada*, rev. ed. (Toronto: University of Toronto Press, 1989, 1991), pp. 96–98; Miller, *Shingwauk's Vision: A History of Native Residential Schools* (Toronto: University of Toronto Press, 1996), pp. 154–55, 185–86, 414–15; and Zeller, *Inventing Canada*, pp. 266–67. George Dawson may have been attempting to integrate his modern evolutionary training with older theories of cultural evolution

from the Scottish Enlightenment; see Trigger, "Daniel Wilson," especially pp. 66–67; and Zeller, " 'Merchants of Light.' "

38 Dawson, "On the Bearing of Devonian Botany on Questions as to the Origin and Extinction of Species," *American Journal of Science*, 102 (1871): 410–16; Dawson, presidential address to the Natural History Society of Montreal, *CN&G*, 7 (1875):281; Dawson, "Annual Address," ibid., 7 (1875): 1–3; Dawson, "The Present Aspect of Inquiries as to the Introduction of Genera and Species in Geological Time," Victoria Institute, *Journals and Transactions*, 7 (1873–74):388; review of Dawson, *The Story of the Earth and Man*, in *Nature*, 8 January, 1874, p. 180; Robert C. Adams, *Evolution: A Summary of Evidence* (New York: Putnam, 1883), pp. 40–41; and Ronald L. Numbers, *The Creationists* (New York: Knopf, 1992), pp. 7, 352.

39 On Wilson, see McKillop, *Disciplined Intelligence*, pp. 129–32; Holland, "First Canadian Critics," p. 104; Trigger, "Wilson," p. 58.

40 Chartrand et al., *Histoire des Sciences*, pp. 174–77; Provancher, "Etude de l'histoire naturelle," *Le Naturaliste canadien*, 8 (1875):48; "Le Catholicisme et le Science," ibid., 8 (1876):87–92; "Le Darwinisme," ibid., 16 (1887):167–75, 17 (1887):28–57, 90–167.

41 Angus Ross, "On Evolution," Nova Scotia Institute of Natural Science [NSINS], *P&T*, 3 (1871–74):410–35; Andrew Dewar, "Spontaneous Generation, or Predestinated Generation," ibid., 4 (1875–78):34–42; and "Darwinism in Language," Mount Allison University, *The Argosy*, 3 (1875):70–72, 89–90.

42 Dewar, "Spontaneous Generation," p. 35; John Sommers, "Introduction to a Synopsis of the Flora of Nova Scotia," NSINS, *P&T*, 4 (1875):181–221; Sommers, "On a Correspondence between the Flora of Nova Scotia, and that of Colorado, and the Adjacent Territories," ibid., pp. 122–31.

43 Chartrand et al., *Histoire des Sciences*, p. 166; Zeller, *Inventing Canada*, p. 216; Cook, *Regenerators*, pp. 48, 51–52; Pringle, "A Little Afraid of His Own Logic," *The Canadian Bee Journal*, 8 February, 1888, 932–33, and "Those Bee Glands and Evolution," ibid., 14 March, 1888, 1031–32.

44 William Brodie, transcript, Meeting of the Brodie Club, William Brodie Collection, SC20A, Royal Ontario Museum, Toronto, 1925; and untitled Brodie ms., "Home Study Club," *Toronto News* [1898].

45 See also J. T. H. Connor, "Of Butterfly Nets and Beetle Bottles: The Entomological Society of Canada, 1863–1960," *HSTC* [History of Science and Technology in Canada] *Bulletin*, 6 (1982):151–71; John Hamilton, "The Survival of the Fittest Among Certain Species of Pterostichus as Deduced from their Habits," *The Canadian Entomologist*, 16 (1884):73–77; A. R. Grote, "The Origin of Ornamentation in the Lepidoptera," ibid., 19 (1887):114–16; and Grote, "On So-Called Representative Species," ibid., 176–77. Contrast J. Alston Moffat, "Origin and Perpetuation of Arctic Forms," in *Annual Report* (Toronto: Entomological Society of Ontario, 1890), pp. 59–61. Also see Brodie, "Signs of Intelligence in Insects," clipping, *Toronto News* [1898], SC20A, Royal Ontario Museum; and Edith L. Marsh, "Romance of the Past"; Zeller, "William Brodie," in *DCB*, vol. 13, pp. 112–14; and John Henry Wadland, *Ernest Thompson Seton: Man in Nature and the Progressive Era, 1880–1915* (New York: Arno Press, 1978), pp. 72–76.

46 Zeller, "David Pearce Penhallow," in *DCB*, vol. 13, pp. 827–28; see also Margaret Gillett, "Carrie Derrick (1862–1941) and the Chair of Botany at McGill," in *Despite the Odds: Essays on Canadian Women and Science*, ed. Marianne Gosztonyi Ainley (Montreal: Véhicule Press, 1990), pp. 74–87; and Eugene Cittadino, *Nature as the Laboratory: Darwinian Plant Ecology in the German Empire, 1880–1900* (Cambridge: Cambridge University Press, 1990).

47 John Watson, "Darwinism and Morality," *Canadian Monthly and National Review* [*CMNR*], 10 (1876):644; the relevant literature includes Roome, "Darwinian Debate;" McKillop, *Disciplined Intelligence*; Raab, ed., *Religion and Science*; and Armour and Trott, *Faces of Reason*. Schurmann went on to become president of Cornell University.

48 J. A. Allen, "The Evolution of Morality," *CMNR*, 11 (1877):491–501; Watson, "The Ethical Aspect of Darwinism: A Rejoinder," ibid., pp. 638–44; LeSueur, "Science and Materialism," ibid., pp. 22–29; "Darwin and His Work," *CMNR*, 8 (1882):541; and Goldwin Smith, "The Prospect of a Moral Interregnum," *CMNR*, 14 (1879):651. See also Roome, "Darwinian Debate," pp. 196–97; McKillop, ed., *Critical Spirit*; Cook, *Regenerators*, pp. 28–32. Not one among these advocates seemed available, however, to inform the editors of a magazine for workers, who waited in vain to witness "any one species in the act of transforming itself into any other;" see "Charles Robert Darwin," *Scientific Canadian: Mechanics' Magazine and Patent Office Record*, 10 (1882):129–30.

49 See McKillop, "The Idealist Legacy," in *Contours of Canadian Thought*, pp. 96–110; and S. E. D. Shortt, *The Search for an Ideal: Six Canadian Intellectuals and their Convictions in an Age of Transition, 1890–1930* (Toronto: University of Toronto Press, 1976), especially pp. 20–21. The case study is B. Anne Wood, *Idealism Transformed: The Making of a Progressive Educator* (Kingston: McGill-Queen's University Press, 1985), pp. 26–29. See also John English, *The Decline of Politics* (Toronto: University of Toronto Press, 1977); and Doug Owram, *The Government Generation* (Toronto: University of Toronto Press, 1986).

50 R. D. Gidney and W. P. J. Millar, *Inventing Secondary Education: The Rise of the High School in Nineteenth-Century Ontario* (Montreal: McGill-Queen's University Press, 1990), p. 316; and Wood, *Idealism Transformed*, pp. 21–25.

51 Frank Dawson Adams, "In Memoriam – Sir John William Dawson," RSC, *P&T*, 9 (1901):12–13; see also E. W. MacBride, "Zoological Problems for the Natural History Society of Montreal," *Canadian Record of Science*, 8 (1899):10; Berger, *Science, God, and Nature*, p. 75; Sheets Pyenson, *Dawson*, leaves this question essentially unanswered.

52 Charles Baillairgé, *La Vie – L'Evolution – Le Matérialisme*, reprinted from RSC, *P&T*, 23 May, 1899, Canadian Institute for Historical Microreproductions [CIHM], #2504 (microfiche), 37 pp.; see also Chartrand et al., *Histoire des Sciences*, pp. 179–82; and "Opinions sur le Transformisme," *Le Naturaliste canadien*, 32 (1905):14.

53 Ramsay Wright, "The Progress of Biology," RSC, *P&T*, 3rd ser. 5 (1912):XLVI. Where not otherwise stated, the information for these paragraphs is taken directly from the relevant university calendars.

54 Hilda Neatby, *Queen's University*, eds. Frederick W. Gibson and Roger Graham, 2 vols. (Montreal: McGill-Queen's University Press, 1978), vol. 1,

pp. 164, 218, 229; James Fowler, "Arctic Plants Growing in New Brunswick, with Notes on their Distribution," RSC, *P&T*, 5 (1887):189–205; A. P. Knight, "Animal Biology," *Queen's Quarterly*, 1 (1893–94):135–39; and "Species," ibid., pp. 214–22. See also B. N. Smallman, H. M. Good, and A. S. West, *Queen's Biology* (Kingston: Queen's University, 1991), Chap. 5.

55 L. W. Bailey's (1839–1925) research interests centered largely on geology, and then on diatoms. He graduated from Harvard University in 1858 and taught in the traditions of both Gray and Agassiz; see Joseph Whitman Bailey, *Loring Woart Bailey: The Story of a Man of Science* (Saint John, N.B.: J & A McMillan, 1925).

56 Berger, "Race and Liberty: The Historical Ideas of Sir John George Bourinot," in *Annual Report* of the Canadian Historical Association (1965), pp. 94–95, 97, 101. This theme is developed more fully in Berger, *The Writing of Canadian History: Aspects of English-Canadian Historical Writing since 1900* (Toronto: University of Toronto Press, 1986), Chap. 2.

57 George R. Parkin, *The Great Dominion: Studies of Canada* (London: Macmillan, 1895), p. 8. See also Charles G. D. Roberts, *A History of Canada* (Boston: Page, 1897); and Charles R. Tuttle, *Our North Land* (Toronto: Blackett Robinson, 1885), p. 1.

58 George R. Parkin, "The Railway Development of Canada," *The Scottish Geographical Magazine*, 25 (1909):249–50; on Parkin, see Terry Cook, "George R. Parkin and the Concept of Britannic Idealism," *Journal of Canadian Studies*, 10 (1975):15–31; and Berger, *Sense of Power*.

59 Dawson, *Modern Ideas of Evolution* [1890], reprint ed. (New York: Prodist, 1977), p. 220.

60 Wadland, *Seton*, pp. 167, 196–99, 204; see also Lucas, "Nature Writers," pp. 364–88. For the poetic response, see S. R. MacGillivray, "Darwin Among the Canadian Poets," in *Religion and Science*, pp. 297–308.

61 Wesley Mills, "The Evolution and Development of Animal Intelligence," *The Ottawa Naturalist*, 10 (1896–97):178–83; Mills, "Heredity in Relation to Education," *The Popular Science Monthly*, 44 (1894):472–80; Mills, *The Nature and Development of Animal Intelligence* (London: Unwin, 1898).

62 Lt.-Col. Hon. James Baker, *Evolution of Mind* (Vancouver, n.p., 1896), p. 13; #841 (microfiche), CIHM.

63 "The Evolution of Society," *Industrial Canada*, 8 (1908):494; Egerton R. Case, "'Natural Selection' or the Law of the Survival of the Fittest, Viewed as applied to the Development of the Industrial Arts," *The Canadian Engineer*, new ser., 6 (1905):286–87:

> The Law of Industrial-Arts' Development implies the preservation of those elements and process-steps that are beneficial to the invention; the discarding of those elements and process-steps that are injurious thereto, and the combining with the beneficial elements or process-steps new elements or process-steps that will make the invention a MODERN success.

64 J. J. Mackenzie, "Human Evolution and Human Disease," RSC, *P&T*, 8 (1909): 535–47; A. B. Macallum, "On the Origin of Life on the Globe," pp. 423–41;

L. H. Newman, "Certain Biological Principles and their Practical Application in the Improvement of the Field Crops of Canada," *The Ottawa Naturalist*, 23/5 (1909):85–110.

65 Macallum, "The Semi-Centennial of the Origin of Species," RSC, *P&T*, 3rd ser., 3 (1909):180–81.

66 R. Ramsay Wright, "The Progress of Biology," RSC, *P&T*, 3rd ser., 5 (1911): App. A, pp. XLVII–XLVIII; on eugenics in Canada, see Angus McLaren, *Our Own Master Race: Eugenics in Canada, 1885–1945* (Toronto: McClelland and Stewart, 1990).

67 Col. C. C. Grant, "Our Critics Answered," *Journal and Proceedings of the Hamilton Association*, 12 (1893–94):152.

68 Peter J. Bowler, *Evolution: The History of an Idea* (Berkeley: University of California Press, 1984), pp. 243–44.

5

Darwinism in the American South

RONALD L. NUMBERS AND LESTER D. STEPHENS

No region in the world has won greater notoriety for its hostility to Darwinism than the American South. Despite the absence of any systematic study of evolution in the region, historians have insisted that southerners were uniquely resistant to evolutionary ideas. Rarely looking beyond the dismissals of Alexander Winchell from Vanderbilt University in the 1870s and James Woodrow from Columbia Theological Seminary in the 1880s – or beyond the Scopes trial in the 1920s – they have concluded in the words of Monroe Lee Billington that "Darwinism as an intellectual movement ... bypassed southerners." W. J. Cash, in his immensely influential *The Mind of the South*, contended that "the overwhelming body of Southern schools either so frowned on [Darwinism] for itself or lived in such terror of popular opinion that possible heretics could not get into their faculties at all or were intimidated into keeping silent by the odds against them." Darwin's few southern converts either "took the way of discretion" by moving to northern universities or so qualified their discussions of evolution as to render the theory "almost sterile."[1]

Historians of religion and of science have generally concurred with the judgment of southern historians. Uncompromising antievolutionism, says the American church historian George M. Marsden, "seems more characteristic of the United States than of other countries and more characteristic of the South than of the rest of the nation." Because the region was more religiously conservative and less well educated than the North, such differences were only to be expected. The historian of science David N. Livingstone echoes Marsden, describing antievolutionism as "mainly a southern phenomenon."[2]

We have no desire to discount southern resistance to organic evolution. The evidence for that is ample – and no doubt in greater quantities

This chapter is reprinted from Ronald L. Numbers, *Darwinism Comes to America* (Cambridge, Mass.: Harvard University Press, 1998) and appears with the permission of the President and Fellows of Harvard College.

than for other regions in the United States. Jon H. Roberts is probably correct when he suggests in *Darwinism and the Divine in America* that "although ardent Biblicists could be found in every geographical region in the United States, a slightly disproportionate number of them resided in the southern and border states." Besides, southern intellectuals undoubtedly accepted evolution more slowly than their northern counterparts. Unlike so many northerners who eagerly embraced Darwinism, observed A. T. Robertson in 1885, "the more plodding scholars of the South advanced slowly, seeing well their way, and being firmly convinced of the reality of the claims of evolution before accepting it."[3] Our point is that the South was far less uniform in its opposition to Darwinism than most scholarly accounts suggest. In fact, the very success of Darwinism in the South contributed significantly to the outburst of antievolutionism in the 1920s. Although far from definitive, our survey of southern responses to evolution shows that focusing mainly on Winchell, Woodrow, and Scopes gives a highly distorted picture of southern attitudes toward evolution in the years from the early 1860s to the late 1920s.

The South's reputation for being inhospitable to science dates back to the pre-Darwinian period, when, according to one eminent historian, "the cotton kingdom . . . killed practically every germ of creative thought." Certainly the antebellum South's overall record for scientific achievement was no match for the Northeast's, but in urban centers such as Charleston and New Orleans and in college towns scientific activity often flourished. In the decades before the American Civil War the slave states of the South and border regions supported as many leading scientists per 100,000 white residents as the free states to the north and west.[4] And in responding to scientific developments of concern to many Christians – such as the nebular origin of the solar system, the antiquity of the earth, and the plural creation of human races – southerners could be found on both sides of the debate.[5]

When the young American Association for the Advancement of Science met in Charleston in 1850, local fears about maintaining the harmony between Genesis and geology prompted the northern visitor Alexander Dallas Bache to comment that the same wave of opposition to geology that had swept over the North twenty years earlier was now washing over the South. Perhaps there was some justification for Bache's observation, but southern intellectuals were not nearly as hostile to geology as his reaction suggests. Even the hidebound Presbyterian theologian Robert Lewis Dabney, one of the region's most vocal critics of geological efforts to stretch the history of the earth beyond Eden, stopped short of damning Christian geologists who sought to harmonize the Bible with the testimony of the rocks. He did, however, express resentment over scientists' "continual encroachments . . . upon

Scripture teachings." At first they had requested only a pre-Adamite earth; now they were demanding acceptance of the nebular hypothesis, a local flood, ancient humans, and even organic evolution.[6]

Numerous southerners in the years before 1859 openly pushed for reinterpreting Genesis in the light of modern geology – and apparently suffered few ill effects for their boldness. For example, Michael Tuomey in his 1848 *Report on the Geology of South Carolina* sought to accommodate the findings of geology by inserting an immense span of time between the creation "in the beginning" and the much later Edenic creation. Similarly, the paleontologist and physician Robert W. Gibbes assured members of the South Carolina Institute that Mosaic silence on the date of the original creation permitted Christians to accept the notion that "the earth has been inhabited by animals and adorned with plants during immeasurable cycles of time antecedent to the creation of man." Richard T. Brumby of South Carolina College also took this position. At first Brumby felt intellectually isolated from friends who clung to the doctrine of a recently "finished" earth, created in six literal days, but before his death in the mid-1870s he rejoiced to see that that teaching had finally been "exploded." Indeed, by that time it seemed that "most intelligent Christians" agreed that Adam and Eve had arrived on earth long after other members of the animal kingdom.[7]

In his study of the "gentlemen theologians" of the Old South, E. Brooks Holifield found that these literate ministers generally displayed great enthusiasm, not hostility, toward natural science; yet he detected "signs of strain" beginning to appear in the 1850s. Indicative of the growing ambivalence of some southerners toward science on the eve of the Civil War was the creation of the Perkins Professorship of Natural Science in Connexion with Revealed Religion, established at the Presbyterian Columbia Theological Seminary in South Carolina. In 1857 the Tombeckbee Presbytery in Mississippi had unanimously adopted a resolution calling for a professorship on science and religion "to forearm and equip the young theologian to meet promptly the attacks of infidelity made through the medium of the natural sciences." The Presbyterians of Mississippi had grown alarmed by recent scientific theories, such as those espoused by the still-unknown author of *Vestiges of the Natural History of Creation*, and they worried that an inexperienced minister, posted to a backwoods parish, might have his faith tested by some impertinent infidel, perhaps a physician who had simultaneously learned how to heal the body and "kill the soul." But in addition to protecting young preachers from such attacks, the Tombeckbee Presbyterians wanted to provide seminarians "with such enlarged views of science, and its relationship to revealed religion, as will prevent them from acting with indiscreet zeal in defending the Bible against the supposed assaults of true science."[8]

The Columbia seminary selected as the first occupant of the chair an English-born scientist-cleric, James Woodrow, a former student of Louis Agassiz's at Harvard who had gone on to the University of Heidelberg for a doctorate, presumably in chemistry. While teaching at the Presbyterian Oglethorpe College after returning from Europe, he had been ordained into the Presbyterian ministry. In 1861, the year he assumed the Perkins chair, he took over the editorship of the *Southern Presbyterian Review*, the leading quarterly of southern Presbyterianism, and four years later he added the weekly *Southern Presbyterian* to his editorial responsibilities. The latter magazine, he sometimes said, gave him a constituency of 4,000 readers. From 1869 to 1872 and again from 1880 to 1897 he served concurrently on the faculty of the University of South Carolina.[9]

Woodrow at first adopted a cautious stance in his new position. He was, after all, a social and theological conservative who professed to believe in "the absolute inerrancy" of the Bible. In his inaugural address he affirmed his acceptance of the antiquity of the earth and allowed that the deluge might have been a local affair, but he left no doubt about his rejection of the plurality of the human races. For twenty-four years he taught that evolution "probably was not true," but that "even if true it did not contradict or in any way affect the truth of the Scriptures." However, in preparing for a requested address to the alumni association in 1884, he reviewed the evidence in favor of evolution and became convinced that it was "probably true."[10] "We cannot go back to the beginning," he told the assembled alumni,

> but we can go a long way. The outline thus obtained shows us that all the earlier organic beings in existence, through an immense period, as proved by an immense thickness of layers resting on each other, were of lower forms, with not one as high or of as complex an organization as the fish. Then the fish appeared, and remained for a long time the highest being on the earth. Then followed at long intervals the amphibian, or frog-like animal, the reptile, the lowest mammalian, then gradually the higher and higher, until at length appeared man, the head and crown of creation.[10]

Although Woodrow had come to believe that divinely guided evolution had produced man's body, he insisted that his soul had been "immediately created." And because of "insurmountable obstacles" connected with the biblical story of Eve, he continued to hold, as a biblical inerrantist, that both body and soul of the first woman had been specially created, a concession derided by creationist critics as "unscientific." To harmonize his evolutionary views with the Mosaic account of creation, he adopted the day-age reading of Genesis 1 formulated by the Princeton geographer Arnold Guyot.[11]

The publication of Woodrow's address in both the *Southern Presbyterian Review* and in pamphlet form touched off a controversy that raged within southern Presbyterianism for four years. "At once a vehement attack upon him was begun," reported the Woodrow-edited *Southern Presbyterian*, "not apparently for his own views as given in the address, but on account of . . . the whole brood of Evolutionists from the beginning, especially the atheistic part of it, most of his assailants seeming not to have read the address at all." Leading the charge against Woodrow was his own colleague at the seminary, the powerful theologian John Lafayette Girardeau, who feared that enrollments would decline if Columbia became known as the "Evolution Seminary." In view of the recent closing of the institution for two years because of a shrinking student body, Girardeau's fears were not unfounded. Besides, at least one financial backer of the seminary had complained that "the Church did not give you money to have Darwinism taught."[12]

It quickly became clear that most southern Presbyterians had little use for "tadpole theology." Yet Woodrow was not without support. When the seminary trustees met in September, 1884, to respond to the developing controversy, they voted eight to three to back Woodrow on the grounds that "the Scriptures, while full and clear in asserting the fact of creation, are silent as to its mode." This decision inflamed Woodrow's opponents, who quickly succeeded in reconstituting the board and obtaining a call for Woodrow's resignation. When he declined, the board fired him. Two of Woodrow's friends on the faculty quit in support of their beleaguered colleague, leaving Girardeau and his associates free, as the *Southern Presbyterian* put it, to form "a new 'Anti-Evolution Seminary.' "[13]

For several more years, however, the evolution question continued to preoccupy "the upper circles" of southern Presbyterians. At times it seemed as though the church was devoting "more zeal and attention to discussing the origin of Adam's body than to the interest of the souls of Adam's descendants." In 1886 alone, three different levels of church courts wrestled with it: the Presbytery of Augusta, which tried Woodrow for heresy, the four synods responsible for the seminary, and the General Assembly of the Presbyterian Church in the United States, which had received official complaints from no fewer than eight different presbyteries scattered throughout the South. By a vote of fourteen to nine, Woodrow won acquittal in Augusta, but the General Assembly adopted a hard-line position against human evolution:

> The Church remains at this time sincerely convinced that the Scriptures, as truly and authoritatively expounded in our confession of Faith and Catechisms, teach that Adam and Eve were created, body and soul, by

> immediate acts of Almighty power, thereby preserving a perfect race unity.
>
> That Adam's body was directly fashioned by Almighty God, without any natural animal parentage of any kind, out of matter previously created of nothing.
>
> And that any doctrine at variance therewith is a dangerous error . . . . [14]

But even this declaration did not bring the "long and violent warfare on Dr. Woodrow" to an end. His enemies appealed his acquittal to the Synod of Georgia, which overturned the lower ecclesiastical court in a decision upheld in 1888 by the General Assembly. Although Woodrow took some pride in his role as a latter-day Galileo, and at times seemed to find exhilaration in combat, he anguished over having not only his orthodoxy but also his spirituality brought into question. At times he was forced to defend himself against charges ranging from "neglecting the sanctuary" (a result of poor personal and family health) to living an "almost wholly secularized" life (because of his involvement with numerous businesses). For years he fought to save his job and his reputation, but in the end even his friends conceded that "our little company [was] completely routed." Worse yet, at a time when John William Draper's best-selling *History of the Conflict between Religion and Science* (1874) was still on everyone's mind, the church's treatment of Woodrow had put "another javelin . . . in the hands of future John W. Drapers to hurl at the Church."[15]

At first glance the Woodrow affair seems strikingly to confirm accusations of southern antipathy to evolution, but a closer examination of this episode reveals the need to qualify this assessment. A majority of southern Presbyterians may have condemned Woodrow and his beliefs, but the embattled professor did win two key decisions, and over a hundred sympathizers at one time or another voted for him in the church trials. One Presbyterian pastor estimated that the pro-Woodrow faction constituted "at least one-tenth of our Church." Many of Woodrow's defenders would no doubt have described themselves as anti-Darwinists, but they did support his right to advocate a theistic version of evolution. At the height of the controversy an elderly alumnus of the Columbia seminary estimated that a "strong minority" of southern Presbyterians regarded evolution "as a mere scientific deduction, which may be true or not, but which can in no way imperil the interest of true religion, inasmuch as the Bible teaches the simple fact that God made Adam, but does not explain the mode or the particular substance out of which his body was formed." This same minister surmised that there were "scores of intelligent Christian men and women in almost all parts of the Church, who after studying the subject in the light of the discussions that have been going on for more than a year past have come to the conclusion . . . that Adam's body was 'probably evolved' from organic matter."[16]

It is also important not to generalize about southern attitudes towards evolution generally from the Woodrow controversy over *human* evolution. "Neither party," declared one participant, "denies that descent with modification is probably the law of the successive appearances of the *animal* tribes on this globe from the beginning until we come down to man . . . . We differ only upon one point, viz., the creation of the body of Adam." Another writer noted the same tendency to distinguish between animal and human evolution: "The point of discussion is . . . not Evolution in general. For life below man this is conceded generally, and one newspaper pronounces it 'harmless.' The controversy begins when the doctrine is applied to man."[17]

Because the above opinions appeared in Woodrow's *Southern Presbyterian*, one might suspect that the writers were painting an overly positive picture of toleration for prehuman evolution in order to make all evolution seem less threatening. But the same distinction appears in the writings and speeches of the anti-Woodrow George D. Armstrong, southern Presbyterianism's leading voice on matters of science and religion. As a sometime science professor turned minister, he claimed to reject all forms of evolution "on grounds purely scientific." Nevertheless, he readily conceded that if evolution excluded the transition from inorganic to organic, at the beginning of the process, and from animals to humans, at the end, it was neither atheistic nor irreconcilable "with the Bible account of the origin of plants and animals in the world." Speaking before the General Assembly in 1886, he warned Christian evolutionists in the audience not to let evolution carry them "to the belief that it refers to man made in the image of God. It will necessitate giving up the doctrine of the fall." To ripples of laughter, he explained that "According to evolution, man was at his lowest stage, just evolved from a brute – how could he fail? he was already low as he could get."[18]

Southern Presbyterians were not alone in distinguishing between animal and human evolution. As they sometimes noted, even such leading northern lights as Princeton Theological Seminary's James McCosh, Francis Patton, and Archibald Hodge made the same distinction. "About the lower animals," explained Hodge, apparently speaking for his colleagues as well, "we are willing to leave it to the scientists as outside of immediate theological or religious interest."[19] In light of this widespread feeling, we should not assume in the absence of direct evidence that Woodrow's southern critics opposed all forms of organic evolution. And before we take Woodrow's dismissal from the faculty of a theological seminary as representative of southern intolerance of Darwinism, we should keep in mind that Woodrow, as the South's most notorious evolutionist, continued to serve undisturbed on the faculty of the University of South Carolina until his retirement in 1897, spending his last six years there as president.

In trying to account for why antievolution became "a standard test of the faith among southern evangelicals earlier than it did among northern fundamentalists," George M. Marsden draws on the Woodrow case to suggest that "the most likely principal explanation was that their northern counterparts had been infected by a liberal spirit, evidenced in the first instance in their unbiblical attacks upon slavery." Although this argument may have some merit, it finds little support in the actual debates over Darwinism. Southern Presbyterians were well aware that their opposition to evolution distinguished them from many northern brothers and sisters. Some feared that the northern church would call them "heretics" if they did not condemn evolution. Others, such as one of Woodrow's female correspondents, grieved "that the Northern Church should have occasion to comment upon a want of union among ourselves, with some little unchristian exultation."[20]

Indeed, even the Presbyterian press in the North seemed to relish reporting the monkey business disturbing fellow believers in the South. The *Presbyterian Journal* of Philadelphia, for example, called the attack on Woodrow "the ecclesiastical blunder of this generation" and accused his persecutors of yielding "to a spasm of terror." The *Interior*, a Presbyterian paper in Chicago, wondered editorially if there was "ever in the world such a thundering fiasco as the Woodrow business in the Southern Presbyterian Church!" Invoking meteorological metaphors, the editor concluded that "Southern cyclones do not have the faculty of catching on." Northerners congratulated themselves that the internecine struggle in the South could never occur in their region. But despite the intense and at time acrimonious interregional rivalry, noted one of Woodrow's staunchest supporters, the northern church's *"toleration of Evolution has never been named"* in the long list of errors charged against it.[21]

Woodrow's fate at the Columbia seminary tells us as little about the overall reception of Darwinism in southern institutions of higher learning as does the equally celebrated expulsion of Alexander Winchell from the Methodist Vanderbilt University in Nashville, Tennessee. Winchell, a respected geologist and Methodist layman from the North, was already well known as a theistic evolutionist when the Vanderbilt trustees invited him in 1876 to join the faculty as a part-time lecturer, which required his spending only two months in the South each spring. All went smoothly until 1878, when Winchell published a little book titled *Adamites and Pre-Adamites*, in which he argued that humans had populated the earth long before the appearance of Adam in the Garden of Eden. Even more provocatively, he insisted that Adam had "descended from a black race, not the black races from Adam." When Winchell's views began circulating in the racially sensitive South, the trustees responded to criticism by abruptly abolishing his position. Publicly they justified their action on "purely economic grounds"; privately the Methodist bishop who headed the trustees informed Winchell that he had lost his job because of his opinion on "Adamites and Pre-Adamites," a position that did not require belief in evolution.[22]

The Methodist's Tennessee Conference applauded the trustees for courageously confronting the "arrogant and impertinent claims of ... science, falsely so called," while at least one local newspaper sprang to Winchell's defense. The anticlerical *Popular Science Monthly*, published in the North, railed against the "stupid Southern Methodists" who used their "power to muzzle, repress, silence, and discredit the independent teachers of scientific truth." The editor, like many later commentators, overlooked the fact that evolution had had little to do with Winchell's brief stay at Vanderbilt, where by the 1890s "Christianized versions of Darwinian evolution" were again welcome on campus.[23]

Racism may have motivated some of Winchell's critics, but racial concerns rarely manifested themselves in the antievolutionary literature we examined. One southern historian, writing about the evolution controversies of the 1920s, has noted that although objections to evolution "were usually religious in nature, the frequent references to the theory's heretical implications regarding man's ancestry suggested that much of the anti-evolution sentiment may have been related to traditional concepts of race held by a majority of the citizens of the state." However, the only evidence he cites in support of this claim is W. J. Cash's personal recollection that "one of the most stressed notions which went around was that evolution made a Negro as good as a white man – that is, threatened White Supremacy." This may have been true, but it is difficult to document from the publications of the time.[24]

With the exception of Woodrow, no southern professor before World War I seems to have lost a job over Darwinism. And during that time evolution frequently appeared in the classrooms of both state and sectarian colleges. Henry Clay White's experience at the University of Georgia illustrates the extent to which evolutionists could survive, at times even thrive, in the intellectual atmosphere of the New South. A professor of chemistry and geology, White joined the faculty in Athens in 1872. In 1875 he first cautiously endorsed evolution, and within a few more years he was freely teaching it to his students. Although Darwinism remained "heavily under fire from all sides," as one of his former students put it, White continued to expose his classes to it, and in 1887, without incident, he publicly declared himself to be an evolutionist. That same year a senior orator at the university delivered a positive address on evolution – stripped of any mention of its religious implications because of the school's ban on "any references to politics or religion upon the college stage."[25]

During the late 1870s or early 1880s White (or perhaps one of his colleagues in geology) commissioned a fresco depicting evolution for the ceiling of the geology lecture room. The artist created what a contemporary described as "a beautifully painted design, representing the evolution of life through all the geologic or zoologic ages." In 1909, when Darwin would have been a hundred years old, White planned a special birthday celebration in honor of the revered scientist. In deference to the concerns of the chancellor of the university, who feared criticism from evangelical antievolutionists, White hosted the event in his own home. Three of White's colleagues at the university – a historian, a biologist, and a classicist – joined him in honoring Darwin with prepared papers, as did a noted Episcopal bishop. Despite such eye-catching enthusiasm for evolution, White remained in the good graces of the university until his death in the 1920s.[26]

At Tulane University in New Orleans, theistic evolutionists also flourished, especially in the medical school, where there were three or four known Darwinists in the mid-1880s. J. W. Caldwell, Tulane's professor of geology and mineralogy, had resigned his position at Southwestern Presbyterian University (now Rhodes College) in Tennessee when the controlling synods in the wake of the Woodrow affair declared, as one cynic put it, "that Evolution is inconsistent with Synodical natural science." At Tulane, Caldwell suggestively advertised in the school catalog that he aimed "to consider organic life, as it is expressed in the fossils of the various strata, and to discover, if possible, the connexion between the successive fauna and flora."[27]

John B. Elliott, professor of the theory and practice of medicine at Tulane, used his platform as president of the New Orleans Academy of Sciences in the late 1880s to promote evolution as a "great natural law" and to defend it against the charge that it led to unbelief. The son of an Episcopal bishop, he could at times barely contain his fervor for this new revelation from science. "The effect of the theory of natural selection upon the human mind," he declared on one occasion, "has been vivifying in the extreme, so bold, so clear-cut, and so simple; accounting for so much that can be accounted for in no other way." Despite his praise of natural selection, Elliott, like virtually all other scientific evolutionists in the South – and the vast majority in the North – preferred non-Darwinian modes of evolutionary development.[28]

At the University of Mississippi, George Little, who had earned a Ph.D. at the University of Berlin, openly taught evolution, even though at the time of his arrival in 1881 "the controversy between evolution and orthodoxy was at its height." Never one to conceal his preferences, he liked to boast that his library contained "the works of Darwin, Huxley, Tindall [sic] and Spencer, alongside of those of Dr. McCosh of Princeton and Hugh Miller of Scotland." When a visiting general prayed that God would save the students at the University of Mississippi from "the teachings of 'science falsely so called' " and a short time later pointedly sent Little an article on evolution, the professor returned the piece with the terse comment "Evolution is a workable hypothesis, like Newton's law of gravitation and Dalton's Atomic Theory." In his opinion, evolution was also theologically harmless. In support of that claim he pointed out that many of his former students and one of his own sons had gone on to become ministers of the Gospel.[29]

At other public universities around the South – at Alabama, North Carolina, South Carolina, Virginia, and presumably elsewhere – the story was much the same: theistic evolution could be taught with relative impunity.[30] This does not mean that southern universities flaunted the teaching of evolution or that university administrators never advised caution. We know, for example, that in 1909 the dean of the University of South Carolina warned a prospective lecturer that, "though we have made progress toward ... evolution since Dr. Woodrow's time ... we have hardly reached the point where we could make the subject too prominent." The dean expressed confidence that thoughtful persons would not object, but he worried that narrow-minded religionists, particularly those associated with denominational colleges, would criticize the university if the visiting scholar touted evolution.[31]

Surprisingly, many religiously affiliated colleges in the South were themselves teaching evolution by the last decades of the nineteenth century. As Woodrow and his friends loved to point out, this was true even of a number of Presbyterian schools, such as Davidson College in North Carolina, Hamden-Sydney College in Virginia, Southwestern Presbyterian University in Tennessee, and Central University (which later merged with Centre College) in Kentucky, all of which at least taught evolution in "a purely expository manner" or used evolutionary texts. In fact, by the mid-1880s it seemed "doubtful whether any college deserving the name in the United States, North or South, uses a text-book on geology or biological science whose author is *not* an evolutionist, and in which Evolution is not taught."[32]

The fortunes of evolution at Southwestern Presbyterian are particularly interesting. Caldwell, the professor of the natural sciences there in the early 1880s, had come to accept evolution as "God's ordinary mode of working" even before Woodrow and had begun teaching theistic evolution in a nondogmatic manner. Because his "teaching did not assume a prominent or offensive form," explained the chancellor, he had seen no reason to bring Caldwell's activities to the attention of the board of directors. When the Woodrow bomb burst on the southern Presbyterian church, Caldwell voluntarily submitted his resignation and ran for cover to Tulane. Ironically, many of the same church leaders who opposed Woodrow's teaching in a seminary saw no reason for Caldwell to resign because, as one of Woodrow's supporters put it, "*he was A Professor in a* COLLEGE, [and] *there was no objection to him as a scientific man holding or teaching these views in a* COLLEGE!" This created an "almost incredible" situation, in which "our young men . . . may be taught by our scientific professors, even in our church schools, that Evolution is true; and then our professors in the theological schools *must* tell these same young men . . . 'Evolution is false.' "[33]

Evolutionists could also be found on Southern Baptist campuses. William Louis Poteat, a German-trained biologist on the faculty of Wake Forest College in North Carolina, helped to pioneer the teaching of scientific evolution in the South in the 1880s. Shortly after the turn of the century, Baylor University in Texas called one of Poteat's students, John Louis Kesler, to organize that school's biology department. He in turn recruited Lulu Pace, a theistic evolutionist, to join him in the department. Few other Baptist colleges at the time could afford to hire a professional biologist. On the eve of World War I, Kesler could think of only two Southern Baptist biologists, Poteat and Pace, who "would be thought of out of their own neighborhood when biology is mentioned." Both were widely known to be evolutionists, and neither had yet heard more than murmurings about their teaching Baptist students the theory of evolution.[34]

The Methodists' Wofford College in South Carolina and the Quakers' Guilford College in North Carolina were likewise exposing their students to evolution before the end of the century. During the 1890s two Guilford biologists, Joseph Moore and T. Gilbert Pearson, repeatedly and openly advocated organic evolution. Pearson, who went on to become a distinguished ornithologist, explained that life on earth had begun with a single cell – a product of "the divinely appointed agencies of heat, gravity, chemical affinity, water, air and organic life" – which had evolved over "millenniums on millenniums" into the diversity of life seen today. At one point a committee of Guilford trustees investigated his orthodoxy, but Pearson presented such a compelling case for his Christian beliefs that he shortly thereafter received a fifty percent raise. He continued to teach evolution during his tenure at Guilford "and never heard any further complaint about my unorthodox views."[35]

Some southern scientists, however, did criticize Darwin, especially during the first fifteen years or so after the publication of his *Origin of Species*, before a scientific consensus in favor of evolution developed. For example, the Louisville chemist and mineralogist J. Lawrence Smith, one of the region's most distinguished scientists, took a parting shot at Darwin in his address as the retiring president of the American Association for the Advancement of Science. But though he condemned Darwin for departing from "true science" in his "purely speculative studies," he professed to harbor no religious concerns. "If it be grounded on truth, it will survive all attempts to overthrow it," he declared; "if based on error, it will disappear, as many so-called scientific facts have done before." Similar sentiments also appeared from time to time in southern medical journals.[36]

Perhaps the most ardent scientific opponent of Darwinism in the South was John McCrady, a native of Charleston who spent the last years of his life as professor of biology and the relation of science and religion at the Episcopal University of the South in Tennessee. Before the Civil War McCrady had studied with Agassiz at Harvard and had developed a special interest in embryology and the Radiata. He had even, in 1860, published a preliminary paper on what he termed "the law of development by specialization," in which he hinted at the possibility of organic evolution and cautiously applauded Darwin's recent work. However, after the war, in which he suffered terrible losses, he began bitterly opposing "the current erroneous views of (so-called) 'Evolution,'" which he associated with Darwin. In 1873, at the invitation of Agassiz, McCrady joined the faculty of Harvard's Museum of Comparative Zoology, where he remained until the university president forced him to resign in 1877, ostensibly for inadequate teaching and research but perhaps, as McCrady suspected, also because of his social and religious conservatism and his refusal to teach "Darwinism, Huxleyism and Spencerism." Returning to the South, he devoted the last years of his life to working on his law of development, which he grandiosely hoped would harmonize "the apparently antagonistic views of Agassiz & Darwin." However, because of his reticence to publish his view on evolution, he exerted relatively little influence on southern thought.[37]

During the 1850s and 1860s, McCrady had taught at the College of Charleston and had become an active participant in the circle of naturalists in the city, who, especially before the war, earned a remarkable reputation for their scientific work. Within the circle, opinion regarding Darwinism was divided, with several prominent members refusing to be drawn into debate. The elderly John Bachman, a well-known Lutheran pastor and naturalist, condemned Lamarckian evolution but apparently remained silent about Darwin's theory. Francis S. Holmes, an Episcopalian who taught at the College of Charleston until 1869, also avoided the subject, though he sold Darwin's works in his bookstore. Lewis R. Gibbes, a fellow Episcopalian and longtime professor at the college, seems to have commented on evolution only once. Responding in 1891 to a request for his opinion on Joseph LeConte's efforts to reconcile evolution and Christianity, he jotted a note at the bottom of the letter saying that it was "best to avoid discordant discussion."[38]

At least two members of the Charleston circle looked favorably on Darwinism. Henry W. Ravenel, a noted mycologist, found some aspects of Darwin's theory unconvincing but nevertheless judged it "an attractive doctrine." Gabriel Manigault, a prodigious osteologist and longtime curator of the Charleston Museum, observed late in life that evolution was irrefutable. He thought it "strange" that so many contemporaries feared its "atheistic consequences." Besides McCrady, only J. H. Mellichamp, a botanist from Beaufort, South Carolina, with close ties to the circle, openly criticized Darwin. "The great Mr. Darwin, whom most worship," he informed a correspondent, is "the most inconclusive of all writers." Mellichamp wondered where Darwin would "be 50 years hence? – I wonder! – perhaps quite forgotten."[39]

As we have seen, scientific opposition to Darwinism did exist in the South before World War I, but for every McCrady and Mellichamp, there were perhaps several disciples of Darwin such as Silas McDowell and Moses Ashley Curtis, botanists in North Carolina and associates of Asa Gray, Darwin's chief American ally.[40] Outside of scientific circles there was considerably more anti-Darwinian sentiment, though some literate southerners appreciated Darwin for delivering "the final *coup de grace* to the untenable doctrine of the extreme rigidity of species and absolute invariability of types."[41] Even in the churches of the South one could sometimes hear voices urging toleration if not acceptance. "Let the scientific men grapple with the hypothesis, they will deal with it according to its true merits," wrote a somewhat fearful Presbyterian; "and if it be true, it will ultimately take its place as an accepted scientific theory, in spite of the fulminations of the Vatican or the artillery of Protestant divines and metaphysicians." The Baptist New Testament scholar A. T. Robertson put it somewhat more colorfully in stating his openness to evolution with " 'God' at the top." "I can stand it if the monkeys can," he would tell his students at the Southern Baptist Theological Seminary in Louisville.[42]

The relative tranquility evolutionists in the South enjoyed in the late-nineteenth and early-twentieth centuries declined rapidly in the years after World War I, when angry fundamentalists, convinced that the teaching of human evolution was causing many of the nation's social ills, tried to dislodge evolutionists from their professorships and to ban the offending doctrine in public schools. Evolutionary biologists in Southern Baptist colleges suffered particularly harsh treatment during the witchhunts of the 1920s. Poteat at Wake Forest and C. W. Davis of Union University in Jackson, Tennessee, both survived efforts to oust them, but colleagues at Baylor, Mercer University in Georgia, and

Furman University in South Carolina all fell victim to the fundamentalist frenzy. Apparently no southern Presbyterians lost their professorships over evolution in the 1920s, perhaps because their denomination had learned a lesson from the embarrassing Woodrow affair, and the other major Protestant bodies generally remained aloof from the debates over evolution.[43]

Although state legislatures across the land debated the wisdom of antievolution laws, only three states – Tennessee, Mississippi, and Arkansas – actually outlawed the teaching of human evolution, and two other states – Oklahoma and Florida, respectively – prohibited the adoption of evolutionary textbooks and condemned the teaching of Darwinism as "improper and subversive." The fact that all these states were in the South enhanced the region's reputation for hostility to evolution and encouraged even some fundamentalists to believe that the South was "the last stronghold of orthodoxy on the North American continent." The hoopla surrounding the Scopes trial in 1925 seemed dramatically to confirm the South's distinctiveness.[44]

Writers who describe the South as a bastion of antievolutionism typically neglect to mention that most southern legislatures refused to outlaw the teaching of evolution in the 1920s. This was true in Alabama, Georgia, North Carolina, South Carolina, Kentucky, Louisiana, and Texas, as well as in Florida and Oklahoma, where legislators declined to make the teaching of evolution a crime. In Virginia antievolutionists failed even to find a sponsor for their bill. Newspapers throughout the South, from the *Louisville Courier-Journal* and the *Atlanta Constitution* to the *Richmond News-Leader* and the *Raleigh News and Observer*, played a decisive role in helping to turn back the fundamentalist onslaught. After the Alabama legislature defeated an antievolution bill in 1923, the editor of the *Montgomery Advocate* boasted that "every daily newspaper in the state, city daily and small daily also, opposed the bill, while the right to teach evolution has never been an issue in any college in Alabama – that is, not openly."[45]

A generally overlooked factor that contributed to the outburst of militant antievolutionism in the South was the growing popularity of Darwinism among the educated classes in the region. "Practically all biological teachers in the high schools and colleges (including Baptist colleges) believe in some form of biological development," wrote a Baptist educator from Kentucky in 1921. He estimated that "nine out of every ten of our Baptist preachers who are full college and seminary graduates believe in some form of biological development" and that "practically all of our grammar school, high school, and college graduates either believe as true or at least accept as plausible some sort of theory of biological development." These estimates may have been on the high side, but they would not have surprised J. Frank Norris, the fiery fundamentalist preacher from Fort Worth. "Contrary to what most people here in the South think," he informed William Jennings Bryan in 1923, "evolution has already made tremendous gains in our schools." For confirming evidence, he needed to look no farther than nearby Waco, where Baptist biologists at Baylor had been teaching organic evolution for years and where one sociologist had recently published a description of primitive man as "a squat, ugly, somewhat stooped, powerful being, half human and half animal, who sought refuge from the wild beasts first in the trees and later in caves." Exposed and vilified by Norris, the social scientist submitted his resignation. But the fact that a Southern Baptist professor had felt free openly to advocate human evolution tells us much about the degree to which Darwinism had penetrated the "mind" of the South.[46]

Notes

1 Monroe Lee Billington, *The American South: A Brief History* (New York: Scribner, 1971), pp. 301–02; and W. J. Cash, *The Mind of the South* (New York: Knopf, 1941), pp. 140–41. For other negative views by historians of the South, see William B. Hesseltine, *A History of the South, 1607–1936* (New York: Prentice-Hall, 1936), p. 340; Clement Eaton, *Freedom of Thought in the Old South* (Durham, N.C.: Duke University Press, 1940), pp. 312–14; Thomas D. Clark, *The Emerging South*, 2nd ed. (New York: Oxford University Press, 1968), pp. 248–52; John Samuel Ezell, *The South since 1865*, 2nd ed. (New York: Macmillan, 1975), pp. 348–52; and Carl N. Degler, *Place over Time: The Continuity of Southern Distinctiveness* (Baton Rouge: Louisiana State University Press, 1977), p. 23. C. Vann Woodward's classic *Origins of the New South, 1872–1913* (Baton Rouge: Louisiana State University Press, 1951) remains surprisingly silent about evolution in the New South.

2 George M. Marsden, *Understanding Fundamentalism and Evangelicalism* (Grand Rapids: Eerdmans, 1991), pp. 168–73; and David N. Livingstone, *Darwin's Forgotten Defenders: The Encounter between Evangelical Theology and Evolutionary Thought* (Grand Rapids: Eerdmans, 1987), p. 124.

3 Jon H. Roberts, *Darwinism and the Divine in America: Protestant Intellectuals and Organic Evolution, 1859–1900* (Madison: University of Wisconsin Press, 1988), p. 222; and A. T. R[obertson], "Darwinism in the South," *Wake Forest Student*, 4 (1885):205–06. James Moore brought this article to our attention.

4 S. E. Morison, *The Oxford History of the United States, 1783–1917*, 2 vols. (London: Oxford University Press, 1927), vol. 2, p. 24; and Ronald L. Numbers and Janet S. Numbers, "Science in the Old South: A Reappraisal," in *Science and Medicine in the Old South*, ed. Ronald L. Numbers and Todd L. Savitt (Baton Rouge: Louisiana State University Press, 1988), pp. 9–35.

5 On southern responses to the nebular hypothesis, see Ronald L. Numbers, *Creation by Natural Law: Laplace's Nebular Hypothesis in American Thought* (Seattle: University of Washington Press, 1977), pp. 37–38, 63–64, 86.

6 A. D. Bache, "Remarks upon the Meeting of the American Association at Charleston, S. C., March, 1850," in *Proceedings, Fourth Meeting... 1850* (Washington, D. C.: AAAS, 1851); Robert L. Dabney, "Geology and the Bible," *Southern Presbyterian Review*, 14 (1861):246–74; and Dabney, "A Caution against Anti-Christian Science," in *Discussions by Robert L. Dabney, D. D., LL. D.*, ed. C. R. Vaughan, 4 vols. (Richmond, Va.: Presbyterian Committee of Publication, 1892), vol. 3, pp. 116–36. On the AAAS meeting, see also Sally Gregory Kohlstedt, *The Formation of the American Scientific Community: The American Association for the Advancement of Science, 1848–60* (Urbana: University of Illinois Press, 1976), p. 116.

7 Michael M. Tuomey, *Report on the Geology of South Carolina* (Columbia, S.C.: A. S. Johnston, 1848), pp. 58–59; Robert W. Gibbes, *The Present Earth the Remains of a Former World: A Lecture Delivered before the South Carolina Institute, September 6, 1849* (Columbia, S.C.: A. S. Johnston, 1849), p. 31; R. T. B[rumby], "The Pre-Adamite Earth: Relations of Geology to Theology," *Southern Quarterly Review*, 19 (1852):420–55; [Brumby], "Relations of Science to the Bible," *Southern Presbyterian Review*, 25 (1874):1–31; and [Brumby], "Gradualness Characteristic of All God's Operations," ibid., pp. 524–55, quotation on p. 540.

8 [James A. Lyon], "The New Theological Professorship – Natural Science in Connexion with Revealed Religion," *Southern Presbyterian Review*, 12 (1859): 181–95; E. Brooks Holifield, "Science and Theology in the Old South," in *Science and Medicine*, pp. 127–43, quotation on p. 142. See also Holifield, *The Gentlemen Theologians: American Theology in Southern Culture, 1795–1860* (Durham, N.C.: Duke University Press, 1978).

9 J. William Flinn, "James Woodrow, A. M., D. D., M. D., LL. D.," *Columbia [SC] State*, 18 January, 1907, pp. 10–11. This biographical account, as well as many pamphlets related to the Woodrow affair, can be found in the John William Flinn Collection, Department of History, Presbyterian Church, Montreat, N.C. The reference to 4,000 constituents appears in "The Trial of Professor Woodrow," *Southern Presbyterian*, 9 September, 21 (1886):2.

10 James Woodrow, Editorial Note, *Southern Presbyterian*, 28 May, 20 (1885):2.

11 Woodrow, *Evolution: An Address Delivered May 7th, 1884, before the Alumni Association of the Columbia Theological Seminary* (Columbia, S.C.: Presbyterian Publishing, 1884), pp. 17–18, 23, 29. A more colorful version of Woodrow's depiction of evolution from fish to man appeared as a direct quotation from Woodrow in a speech by William Adams, reprinted in "The General Assembly," *Southern Presbyterian*, 31 May, 23 (1888):2. For criticism of Woodrow's views on the creation of Eve, see George D. Armstrong, *The Two Books of Nature and Revelation Collated* (New York: Funk & Wagnalls, 1886), p. 94. On Woodrow as a biblical inerrantist, see T. Watson Street, "The Evolution Controversy in the Southern Presbyterian Church with Attention to the Theological and Ecclesiastical Issues Raised," *Journal of the Presbyterian Historical Society*, 37 (1959):234. On Woodrow's endorsement of Guyot, see "Professor Woodrow's Speech before the Synod of South Carolina," *Southern Presbyterian Review*, 36 (1885):55. For Guyot's views, see Numbers, *Creation by Natural Law*, pp. 91–100.

12 "The Seminary Board Question before the Synod," *Southern Presbyterian,* 30 July, 20 (1885):2; John L. Girardeau, *The Substance of Two Speeches on the Teaching of Evolution in Columbia Theological Seminary, Delivered in the Synod of South Carolina, at Greenville, S. C., Oct. 1884* (Columbia, S.C.: William Sloan, 1885), p. 35; and Ernest Trice Thompson, *Presbyterians in the South,* 3 vols. (Richmond, Va.: Knox Press, 1973), vol. 2, p. 464. Thompson offers perhaps the best account of the Woodrow affair, but see also Street, "The Evolution Controversy in the Southern Presbyterian Church," and the largely derivative discussion in Frank Joseph Smith, "The Philosophy of Science in Later Nineteenth Century Southern Presbyterianism," Ph.D. diss., City University of New York, 1992. On Girardeau, see George A. Blackburn, *The Life Work of John L. Girardeau, D. D., LL. D.* (Columbia, S.C.: State Company, 1916). On October 24, 1884, the *Greenville Daily News* credited Girardeau with firing "the first shot in the evolution controversy."

13 Flinn, "Evolution and Theology: The Consensus of Science against Dr. Woodrow's Opponents," *Southern Presbyterian Review,* 36 (1885):510; "The Seminary Board Question before the Synod," p. 2; "Columbia Theological Seminary," *Southern Presbyterian,* 18 December, 19 (1884):2; and "Professor Woodrow's Removal," Ibid.

14 "A Sure Enough Subject for the Charleston 'Inquisition,' " *Southern Presbyterian,* 8 November, 23 (1888):2; and George D. Armstrong, *A Defence of the "Deliverance" on Evolution, Adopted by the General Assembly of the Presbyterian Church in the United States, May 26th, 1886* (Norfolk, Va.: John D. Ghiselin, 1886), pp. 3–5.

15 "The Outcome," *Southern Presbyterian,* 5 July, 23 (1888):2; and "The General Assembly," ibid., 31 May, 23 (1888):1–3. On Woodrow's identification with Galileo, see "Professor Woodrow's Speech before the Synod," pp. 56–58. On his alleged secularity, see "The Perkins Professor's Case," *Southern Presbyterian,* 10 September, 20 (1885):2; and *The Examination of the Rev. James Woodrow, D. D., by the Charleston Presbytery* (Charleston, S.C.: Lucas & Richardson, 1890), p. 1. The quotation about being "routed" appears in Thompson, *Presbyterians in the South,* vol. 2, p. 489. The reference to Draper appears in Flinn, "Evolution and Theology: The Logic of Prof. Woodrow's Opponents Examined," *Southern Presbyterian Review,* 36 (1885):268–304, quotation on p. 270.

16 Flinn, "Evolution and Theology: Consensus," p. 508; and J. Leighton Wilson, "The Evolution Difficulty," *Southern Presbyterian,* 17 September, 20 (1885):2.

17 John B. Adger, "The Synod at Cheraw," *Southern Presbyterian,* 4 November, 21 (1886):2; "Evolution in the Church," ibid., 16 October, 19 (1884):2.

18 Armstrong, *The Two Books,* pp. 86, 96; and "The General Assembly," *Southern Presbyterian,* 27 May, 21 (1886):2. On Armstrong's standing, see Thompson, *Presbyterians in the South,* vol. 2, p. 477.

19 "Drs. Patton and Hodge on Evolution and the Scriptures," *Southern Presbyterian,* 6 May, 21 (1886):2. See also "Sir William Dawson on the Relations of Evolution to the Bible," ibid., 13 May, 21 (1886):2.

20 Marsden, *Understanding Fundamentalism,* pp. 168–73; Flinn, "Evolution and Theology: Consensus," p. 579; and letter, Emma M. Barnett to James Woodrow, 5 September, 1884, quoted in Smith, "The Philosophy of Science," p. 316.

21 "The Ecclesiastical Blunder," *Southern Presbyterian*, 11 December, 19 (1884):2; "More Work for the 'Inquisition,'" ibid., 29 November, 23 (1888):2; "Evolution in the South," *New York Times*, 5 April, 1885, reprinted in the New Orleans *Daily Picayune*, 14 April, 1885, from a copy in James Woodrow's scrapbook, Department of History, Presbyterian Church; Flinn, "Evolution and Theology: Consensus," pp. 578–79 (emphasis in the original).

22 Leonard Alberstadt, "Alexander Winchell's Preadamites – A Case for Dismissal from Vanderbilt University," *Earth Sciences History*, 13 (1994):97–112; Edwin Mims, *History of Vanderbilt University* (Nashville: Vanderbilt University Press, 1946), pp. 100–01; Paul Conkin, *Gone with the Ivy: A Biography of Vanderbilt University* (Knoxville: University of Tennessee Press, 1985), pp. 50–51, 60–63; and Livingstone, *Darwin's Forgotten Defenders*, pp. 86, 91. On pre-Adamism, see Livingstone, *The Preadamite Theory and the Marriage of Science and Religion*, Transactions of the American Philosophical Society, vol. 82, part 3 (Philadelphia: APS, 1992).

23 Conkin, *Gone with the Ivy*, pp. 63, 97; "Religion and Science at Vanderbilt," *Popular Science Monthly*, 13 (1878):492–95; "Vanderbilt University Again," ibid., 14 (1878):237–39.

24 William B. Gatewood, Jr. *Preachers, Pedagogues & Politicians: The Evolution Controversy in North Carolina, 1920–27* (Chapel Hill: University of North Carolina Press, 1966), p. 154; and Cash, *Mind of the South*, p. 339.

25 Lester D. Stephens, "Darwin's Disciple in Georgia: Henry Clay White, 1875–1927," *Georgia Historical Quarterly*, 78 (1994):66–91.

26 Ibid., pp. 66–91.

27 Flinn, "Evolution and Theology: Consensus," p. 545; "Professor Caldwell at Tulane University," *Southern Presbyterian*, 23 July, 20 (1885):2; and "What Is It?," ibid., 19 January, 23 (1888):2. Flinn pastored a Presbyterian church in New Orleans from 1878 to 1888.

28 John B. Elliott, "President's Address before the New Orleans Academy of Sciences," in *Papers Read before the New Orleans Academy of Sciences*, vol. 1, No. 2 (1886–1887), pp. 5–18; and Elliott, "The Deeper Revelations of Science: Annual Address before the Academy of Sciences, Feb. 7, 1888," ibid., vol. 1, No. 2 (1887–1888), pp. 8–19.

29 George Little, *Memoirs of George Little* (Tuscaloosa, Ala: Weatherford, [1924]), p. 101. See also Flinn, "Evolution and Theology: Consensus," p. 544.

30 On evolution at the University of Virginia and the University of North Carolina, see "Professor Woodrow's Speech before the Synod," pp. 34–35. For the University of Alabama, see James B. Sellers, *History of the University of Alabama* (University: University of Alabama Press, 1953), pp. 540–41. Woodrow taught evolution in his geology classes at the University of South Carolina. The standard histories of other state universities in the South say virtually nothing about the teaching of evolution.

31 Daniel Walker Hollis, *University of South Carolina*, Vol. 2: *College to University* (Columbia: University of South Carolina Press, 1956), pp. 165–80, 245–46.

32 "Professor Woodrow's Speech before the Synod," p. 34; "What Is It?," p. 2; "Inaccurate Reports," *Southern Presbyterian*, 20 November, 19 (1884):2; and Flinn, "Evolution and Theology: Consensus," pp. 545–46.

33 "Professor Caldwell at Tulane University," p. 2; "The Southwestern Presbyterian University and Evolution," *Southern Presbyterian*, 4 December, 19 (1884):2; and Flinn, "Evolution and Theology: Consensus," pp. 544–45.

34 This paragraph is taken in large part from Ronald L. Numbers, *The Creationists* (New York: Knopf, 1992), p. 40.

35 Regarding Wofford, see "Professor Woodrow's Speech before the Synod," p. 34. Regarding Guilford, see Joseph Moore, "The Greatest Factor in Human Evolution," *Guilford Collegian*, 6 (1894):240–44; T. Gilbert Pearson, "Evolution in Its Relation to Man," ibid., 8 (1896):107–11; Pearson, *Adventures in Bird Protection: An Autobiography* (New York: Appleton-Century, 1937), pp. 58–59; and Oliver H. Orr, Jr., *Saving American Birds: T. Gilbert Pearson and the Founding of the Audubon Movement* (Gainesville: University Press of Florida, 1992), pp. 19, 42, 46, 49, 81.

36 J. Lawrence Smith, "Address," in *Proceedings of the American Association for the Advancement of Science* (Portland, Maine: AAAS, 1873), pp. 14–16. On medical opinion, see, e.g., J. C., Review of *On the Origin of Species by Means of Natural Selection*, by Charles Darwin, *Richmond and Louisville Medical Journal*, 9 (1870):84–100; and F. M. Robertson, "President's Address," in *Transactions of the South Carolina Medical Association* (1880–81), Appendix, pp. 1–15. We are indebted to the late Patricia Spain Ward for these last two references.

37 John McCrady, "The Law of Development by Specialization: A Sketch of Its Probable Universality," *Journal of the Elliott Society of Natural History*, 1 (1860):101–14; Lester D. Stephens and Dale R. Calder, "John McCrady of South Carolina: Pioneer Student of North American Hydrozoa," *Archives of Natural History*, 19 (1992):39–54; Numbers, *The Creationists*, pp. 8–9.

38 John Bachman, "An Investigation of the Cases of Hybridity in Animals, Considered in Reference to the Unity of the Human Species," *Charleston Medical Journal*, 5 (1850):168–97, especially p. 186; Stephens, *Ancient Animals and Other Wondrous Things: The Story of Francis Simmons Holmes, Paleontologist and Curator of the Charleston Museum* (Charleston, S.C.: Charleston Museum, 1988); Lewis R. Gibbes, marginalia on letter from John L. Girardeau to Gibbes, 5 March, 1891, in the Lewis R. Gibbes Papers, Manuscript Division, Library of Congress. See also Stephens, "Lewis R. Gibbes and the Professionalization of Science in Antebellum South Carolina," unpublished paper presented at the annual meeting of the Southern Historical Association, 13 November, 1980; and Stephens, "A Sketch of Natural History Collecting in Charleston, South Carolina: The Golden Age, 1820–1865," unpublished paper presented at the second North American meeting of the Society for the History of Natural History, 24 October, 1986.

39 Tamara Miner Haygood, "Henry Ravenel (1814–1887): Views on Evolution in Social Context," *Journal of the History of Biology*, 21 (1988):457–72; Gabriel E. Manigault, manuscript autobiography written ca. 1887–1897, in the Manigault Family Papers, Manuscripts Department, Library, University of North Carolina at Chapel Hill; and letter, J. H. Mellichamp to George Englemann, 26 August, 1872, in the South Carolina Collection, Charleston Museum Library.

40 Silas McDowell, undated (ca. 1865 or later) manuscript, entitled "Evolution," in the Silas McDowell Papers, Manuscripts Department, Library, University of North Carolina at Chapel Hill; Edmund Berkeley and Dorothy Smith Berkeley, *A Yankee Botanist in the Carolinas: The Reverend Moses Ashley Curtis, D. D. (1808–72)* (Berlin: J. Cramer, 1986).

41 "The Evolution Hypothesis," *Southern Review*, 3 (1868):408–40, quotation on p. 419. For somewhat more critical assessments in the same journal, see "The Origin of Species," ibid., 9 (1871):700–28; "Darwinism," ibid., 12 (1873):406–23; and "Philosophy versus Darwinism," ibid., 13 (1873):253–73.

42 [W. S. Bean], "The Outlook of Modern Science," *Southern Presbyterian Review*, 25 (1874):331–38, quotation on p. 335. The Robertson quotation appears in James Moore, *The Darwin Legend* (Grand Rapids: Baker Books, 1994), p. 119.

43 Numbers, *The Creationists*, pp. 46–49.

44 Numbers, *The Creationists*, p. 41; and "Fighting Evolution at the Fundamentals Convention," *Christian Fundamentals in School and Church*, 7 (1925):5.

45 Maynard Shipley, *The War on Modern Science: A Short History of the Fundamentalist Attacks on Evolution and Modernism* (New York: Knopf, 1927), pp. 75–186, quotation on p. 141. See also Norman F. Furniss, *The Fundamentalist Controversy, 1918–31* (New Haven: Yale University Press, 1954), pp. 76–100.

46 F. D. Perkins, "Evolution Theory Taught in American Schools," *Western Recorder*, 96 (1921):4; letter, J. Frank Norris to William Jennings Bryan, 28 December, 1923, Box 38, Bryan Papers, Manuscript Division, Library of Congress; and Numbers, *The Creationists*, p. 47.

6

Darwinism, American Protestant thinkers, and the puzzle of motivation

JON H. ROBERTS

The sustained labor of historians during the past six decades has provided us with a relatively clear understanding of the response of American Protestant intellectuals to Darwinism. We know, for example, that at the time that Charles Darwin published his *Origin of Species* in 1859, most Protestant thinkers in the United States assumed that interpretations of the history of life were inextricably bound up with beliefs that lay at the very heart of Christian theology. We now also know, however, that prior to the middle of the 1870s, virtually all spokespersons in the American Protestant community, recalling the fate of earlier transmutation hypotheses and aware that many reputable scientists were fiercely hostile to Darwin's work, concluded that the Darwinian hypothesis was a false system of metaphysics masquerading as science. Accordingly, insofar as they addressed themselves to that hypothesis at all, they tended to focus on its scientific deficiencies and its metaphysical affinities with the heretical works of Thomas Huxley, Herbert Spencer, and John Tyndall. Only after 1875, when it had become abundantly clear that the scientific community had endorsed the theory of organic evolution, did American Protestant thinkers feel compelled to

During the course of writing this piece I incurred a number of intellectual debts that I am happy to acknowledge. I am grateful to Ronald L. Numbers and John Stenhouse for the invitation to participate in the conference that served as the inception for the volume in which this article appears. While writing this paper I received institutional support from the Institute for Research in the Humanities at the University of Wisconsin-Madison and the University Personnel Development Committee at the University of Wisconsin-Stevens Point. Several scholars read this paper in its formative stage and made significant contributions in improving the final product: Paul S. Boyer, Matthew Kramer, Ronald L. Numbers, and William B. Skelton. I profited from the comments of all the participants at the Dunedin Conference, especially Peter J. Lineham and Jim Moore. In 1995 I presented a version of this essay before the Committee on the History and Philosophy of Science at the University of Colorado-Boulder. I am grateful to Fred Anderson for issuing the invitation and to the participants for their helpful responses. As always my wife, Sharon (ILYS), and my son, Jeff, gave me loving support and wise counsel.

engage in a sustained assessment of its theological implications. During the next quarter of a century most of them concluded that judicious modifications of traditional formulations of Christian doctrine would enable them to accept the transmutation hypothesis. Not all Protestant opinion leaders, however, endorsed that view. A minority concluded that the theory of evolution was so irrevocably antagonistic to Christian theology that reconciliation was simply impossible.[1]

Yet, despite being able to describe the responses of members of the American Protestant community to Darwinism, we are still only beginning to unearth the many layers of motivation that prompted these responses. At first glance, this might well seem rather surprising. After all, students of scientific and religious thought characteristically analyze the views of articulate people at pains to explain themselves. Unfortunately, however, it is rarely possible to accept the explanations of these people at face value. For one thing, as even the most determined introspection suggests, human beings are seldom entirely conscious of the factors influencing their decisions. Further complicating the picture is the work of such "masters of suspicion" as Karl Marx, Sigmund Freud, and Michel Foucault, who have convinced us to view the accounts that people give of their behavior with skepticism. Self-deception now plays an integral role in the way we understand the human condition. Finally, as if these problems were not enough, sociologists of knowledge and their votaries have advanced persuasive arguments that the status of ideas within cultural communities cannot be entirely understood in terms of the internal dynamics of intellectual discourse; they are "socially constructed" as well.

In view of the difficulties involved in gaining a purchase on motivation, it is perhaps not altogether surprising that most historians who have assessed the responses of religious thinkers to the transmutation hypothesis have eschewed trying to explain those responses in favor of the more modest goal of simply describing them. Although some attempts have been made to grapple with the issue of motivation, agreement remains elusive. The dialogue continues, however, for the question of why thinkers adopted the positions they did is as potentially illuminating as it is problematic.

In an effort to contribute to this dialogue, I shall assess the variables that historians have most frequently advanced to account for the responses of American Protestant leaders to the theory of organic evolution between 1875 and 1930. I then offer a hypothesis of my own. Before proceeding, however, several methodological issues should be addressed. First, the focus of this paper is limited to clergy and theologians within the American Protestant community. This is not entirely unexceptionable, for scientists and a variety of other lay people (William

Jennings Bryan comes to mind) were "religious thinkers" in a culturally important sense. Moreover, even this broader group does not begin to encompass all the Protestants in the United States who responded to the theory of evolution. Still, the confines of a paper make selectivity necessary, and because professional religious spokespersons were, other than possibly scientists, the individuals for whom the issues raised by Darwinism were most compelling and relevant, it seems entirely appropriate to concentrate on their responses.

More troublesome is the problem of making sense of the variety of ostensibly proevolutionary positions that arose in the wake of the Darwinian controversies. The manner in which historians deal with this difficult problem may well affect their assessment of which variables were particularly decisive in motivating religious leaders to take the positions they did. One distinction that has seemed crucial to some historians is that which separates religious thinkers who embraced the general theory of organic evolution from those who went even further to accept Charles Darwin's more specific account of how the evolutionary process operated.[2] I am convinced, however, that this distinction is unnecessary and even misleading. Most American Protestant intellectuals tended to equate "Darwinism" with the more general theory of evolution rather than with the more specific mechanism of natural selection, and there do not appear to be salient differences between this majority and the minority committed to drawing a sharp distinction between them.[3] Just as importantly, though most American Protestant thinkers who embraced the transmutation hypothesis did not accord natural selection as great a role as Darwin would have liked, neither did most evolutionists within the scientific community, who characteristically subordinated natural selection to other mechanisms.[4] This suggests that the hostility or indifference of Protestant thinkers to natural selection signifies little more than a determination to be "scientifically rigorous" in accordance with contemporary understandings of that phrase.

A more relevant and difficult challenge confronting historians concerned with religious responses to Darwinism is that of characterizing provisional and limited conversions to the theory of organic evolution. It seems appropriate to denominate Protestant leaders who embraced the idea of organic evolution by resident forces, even when they interpreted those forces as manifestations of God's immanent divine "energy," as evolutionists. It also seems fitting to lablel Protestants who rejected the concept of transmutation in favor of a continued commitment to the fixity of species as antievolutionists. But how should we categorize religious leaders who claimed to be evolutionists or refused to exclude the likelihood that some transmutation had occurred, yet continued to insist on the need for periodic supernatural interventions during the course of the history of life? Is it appropriate to place these leaders within the evolutionists' camp? I have attempted to resolve this classificatory difficulty by invoking a third term to describe the position of individuals who embraced this middle way: progressive creationism.[5]

Estimating the relative size of these three groups is highly problematic. Many Protestant intellectuals who participated in the Darwinian controversy addressed themselves to individual issues without indicating precisely where they stood with regard to more fundamental questions concerning the nature, scope, and even the validity of evolution. Moreover, some altered their position over time. Given due allowance for these difficulties, it would appear that by 1900 approximately thirty-five percent of American Protestant leaders had wholeheartedly embraced the theory of organic evolution, roughly twenty-five percent had rejected it, and about forty percent had endorsed some version of progressive evolutionism.

For a long time historians' assessments of the differing responses of American Protestant opinion leaders to the theory of organic evolution between 1875 and 1930 appear to have reflected the unfavorable publicity attending the fundamentalist movement of the 1920s. Prior to 1970, historians frequently treated the controversy over evolution as another skirmish in the protracted confrontation between enlightenment and ignorance. One of the earliest chroniclers of the Darwinian controversy, Bert James Loewenberg, thus chided nineteenth-century opponents of the transmutation hypothesis for offering "rationalizations unworthy of a sophomoric intelligence, . . . superstitions more reasonable in an African tribesman than a nineteenth-century American." Similarly, in a work published in 1954, Norman F. Furniss ascribed much of the Protestant opposition to the hypothesis during the 1920s to "complete misunderstanding" and "distorted opinions."[6]

Such assessments, which seem to exemplify partisanship masquerading as interpretation, now (thankfully) enjoy somewhat less support than they once did. Today one seldom encounters suggestions that complex controversies can be adequately understood in terms of the relative intellectual rigor of the participants. This may indicate that historians are more wary of hubris (at least in print). Or it may have something to do with the advance of historical scholarship, which has disclosed that opponents of the theory of organic evolution were as committed to "reason" as its proponents.[7] Finally, it probably reflects the recognition that normative judgments issued *ex cathedra* violate an important canon of the sociology of knowledge: that neither truth nor error is a sufficient explanation for the adherence of groups and individuals to a particular idea.[8]

Moving beyond such normative judgments, the historian pondering the problem of motivation discovers that much of the early work discussing the American Protestant community's response to the theory of organic evolution focused on two variables: denominational affiliation and geography. Denominational affiliation assumed an analytical importance as early as 1936, when Windsor Hall Roberts made use of it in his pioneering study of American Protestant responses to Darwinism. Roberts's work was followed by numerous others, including many that focused on the reactions of individual denominations to the transmutation hypothesis.[9]

In order to sustain the claim that denominational affiliation was salient (much less decisive) in shaping Protestant leaders' responses to Darwinism, it is necessary to establish significant patterns of correlation between the stances that those leaders took and their denominational affiliations. Surprisingly, historians have not yet used the sophisticated instruments of statistical analysis to examine this relationship systematically. Until this is done, an impressionistic view must suffice. It would appear that within "mainline" Protestantism, embracing such groups as the Congregationalists, the Presbyterians, the Methodists, the Baptists, the Episcopalians, and the Disciples of Christ, denominational affiliation is not very useful in predicting the responses of religious leaders to the theory of organic evolution. In fact general patterns seem strikingly similar in all major denominations: the initial, almost universally hostile response in the 1860s and early 1870s gave way by the mid-1870s to the emergence of all three of the major positions: proevolutionism, antievolutionism, and progressive creationism. This is hardly surprising, for none of the mainline denominations took a formal stand on the theological acceptability of the transmutation hypothesis. This freed individuals to decide the matter in accordance with their own peculiar sets of priorities.[10] And because none of the positions that emerged succeeded in becoming the self-evidently "Christian view," agreement proved elusive.

A more thorough investigation may disclose significant differences in the percentages of spokespersons within mainline denominations who embraced the various intellectual options. Some historians have suggested, for example, that denominations that placed greater emphasis on religious experience as a source of insight and authority expressed somewhat less concern with evolution than did those denominations that drew their theological authority more self-consciously from the Scriptures.[11] It is also possible that denominations that placed less emphasis on formal education and the value of reading secular sources might have been less familiar with the transmutation hypothesis and thus devoted less attention to it.[12] At this point, we simply do not know enough to make well-grounded generalizations.

If the analysis of the relationship between denominational affiliation and the responses to the theory of organic evolution is extended beyond the mainline, a more meaningful pattern begins to emerge. Although a systematic investigation is also needed here, it seems clear that spokespersons within denominations to the theological "right" of the mainline, such as Missouri Synod Lutherans, Seventh-day Adventists, and churches within the American holiness movement, were far more likely than opinion leaders within mainline denominations to unite in opposition to the theory. Because they prided themselves on being people of "one book," they were more apt to find the transmutation hypothesis a source of provocation. On the other side of the theological spectrum, it would appear that once the scientific community embraced the theory of organic evolution, Unitarian clergy and theologians, as well as leaders of other groups generally viewed as theologically to the left of mainline Protestantism, almost unanimously followed suit.[13]

The behavior of Protestant leaders outside the mainline suggests that in at least some cases, denominational affiliation partially predicts one's response to Darwinism. It is not altogether clear, however, that strong correlations between denominational affiliation and views of Darwinism, even when they do exist, actually tell us very much. Denominations represent clusters of more general theological orientations, creedal commitments, and sociological allegiances. Any effort to move beyond description to explanation will therefore necessitate a further unpacking of those characteristics. Denominational affiliation, in short, does not constitute an appropriate terminus of explanation.

This is also true of geography, another variable often cited as particularly salient in shaping the response of American Protestants to Darwinism. A number of studies, possibly reflecting an assumption that Dayton, Tennessee, was microcosmic, have suggested that critics of evolution tended to hail from rural areas, especially in the South.[14] These studies imply that rejection of the transmutation hypothesis provides a regrettable, albeit colorful, example of "cultural lag."[15]

David N. Livingstone has argued persuasively (Chapter 1) that certain correlations exist between geographical location and the responses of American Protestants to the theory of organic evolution.[16] His analysis can also be extended. For example, it seems clear that as time went on, the number of supporters of the theory within northern seminaries increased, while few seminaries in the South harbored outspoken proponents. Moreover, the agitation for legislation outlawing evolution in the public schools centered (though it was not exclusively located) in the South.[17] To be sure, most southern legislatures rejected such legislation.

Moreover, as Ronald L. Numbers and Lester D. Stephens have shown (Chapter 5), proponents of the transmutation hypothesis in the South certainly existed. Still, it would appear that southern Protestants who opposed evolution significantly outnumbered those who supported it.[18]

When attention is limited to the geographical location of American Protestant leaders, however, the waters become somewhat murkier. Robert A. Wenger inferred from a careful study of forty fundamentalist and liberal clergy and theologians that leaders in *both* groups were most likely to come from the East and serve in northern and eastern urban areas; a disproportionately fewer number came from the South and West.[19] A larger study, focusing on 126 liberal and 100 conservative Protestant leaders found that conservatives were considerably more likely than liberals to come from the South, but the difference was not sufficiently dramatic to be predictive.[20] Although additional investigation concentrating more explicitly on leaders who played an active role on all sides of the evolutionary controversy is sorely needed, it seems provisionally clear, at least, that the great majority of Protestant leaders who rejected the transmutation hypothesis, like those who supported it, were not southern in either background or residence.

There is also good reason to view claims about the "rural" character of opposition to the theory of organic evolution with some suspicion. To be sure, many of the Protestant clergy and theologians who rejected the Darwinian hypothesis grew up on farms and in small towns, but this was hardly unusual in the late-nineteenth and early-twentieth centuries; Protestant leaders who embraced the transmutation hypothesis had similar backgrounds.[21] Just as significantly, supporters of the hypothesis were not the only Protestant leaders to pursue careers in urban areas. Many who rejected Darwinism, such as Luther Tracy Townsend, De Witt Talmage, William Bell Riley, John Roach Straton, and J. Frank Norris, also ministered to urban congregations or served urban seminaries.

In the event that correlations between geographical location and the responses of clergy and theologians to the theory of organic evolution ultimately prove significant, it will be because they point to the existence of more fundamental factors. These factors might include differences in the degree to which the various regions felt compelled to engage "modernity," differences in educational levels among residents of those regions, and class differences. In short, our cursory examination of geography has yielded the same lesson as our examination of denominational affiliation: in order to make judgments about motivation, historians must move beyond descriptive variables to the underlying context in which those variables are embedded.

They still need to decide, however, *which* context is the appropriate object of analysis. Historians have proposed several promising candidates:

social interests, cultural-psychological strain, and theological considerations. It is, of course, possible that the construction of a master narrative of the Darwinian controversies will require the deployment of more than one of these contextual elements; there are, after all, worse sins than overdetermination.[22] At the very least, it seems unlikely that any explanation of motivation will suffice that is not consistent with each element of the contextual landscape. This does not mean, however, that one must accord those elements parity, and the explanations thus far advanced by historians to account for the responses of religious thinkers to Darwinism have not done so. The hypothesis I advance will be no exception.

The claim that ideas "do work" has become one of the reigning assumptions of contemporary intellectual and cultural history. One of the consequences of this functionalist assumption has been the appearance of concerted efforts to draw connections between ideas and the social context in which they emerge and flourish. Indeed, social constructionism has become a cottage industry in most fields of humanistic scholarship during the past decade. At one level the idea that social factors play an integral role in creating boundary conditions for the development and articulation of ideas seems quite unexceptionable. The nature of one's educational background, for example, doubtless helps to determine one's range of ideas, conceptual apparatus, and vocabulary.

If, however, it is clear that in this relatively weak sense intellectual positions are socially constructed, this does not mean that ideas are largely, or even primarily, expressions of social interests. An early application of this stronger position can be found in the work of Richard Hofstadter and others, who attributed the motivation of Protestant intellectuals, in common with other middle-class Americans, at least partly to considerations of social status. Although this view received a barrage of criticism, the claim that social interest played a fundamental role in shaping the contours of the post-Darwinian landscape continues to find favor.[23] James Moore presents perhaps the clearest statement of this claim in his 1985 article, "Herbert Spencer's Henchmen: The Evolution of Protestant Liberals in Late Nineteenth-Century America." Moore maintains that during the half-century after 1882, evolution served an "ideological function" for "many of the most influential spokesmen for Protestant liberalism." These spokespersons, Moore asserts, embraced a Spencerian view of cosmic evolution largely because they believed that it could be used to justify normative judgments concerning the appropriate means and pace of social change. Evolutionism served as an "adjustive, palliative, and gradualist" theodicy justifying "law and order, and hope," rather than "revolutionary social change." Although Spencer was the primary architect of this theodicy, Moore quite rightly emphasizes that nonscientists in the late-nineteenth and early-twentieth centuries frequently conflated Spencerianism with Darwinism.[24]

Moore has accounted for the Protestant antievolutionist position in somewhat different terms. Beset as they were by a chaotic social environment, he argues, opponents of the evolutionary hypothesis within the American religious community found the "majestic vision of a fixed natural order summoned into existence immediately by God and subject to unswerving moral judgment" enormously appealing. This prompts Moore to conclude that "just as liberal Protestants found justification in evolution for functional social relationships that would foster industrial progress without revolution, so those who became Fundamentalists found in Genesis a sanction for stable social forms that would conserve eternal values."[25]

The most influential proponent of the view that responses to Darwinism were ideological expressions of social interests is Robert M. Young. Young, whose work focuses on England rather than the United States, believes that social context is of vital importance in constituting knowledge. He maintains that during the nineteenth century, participants in discussions of science and theology shared a "common context"; disputes were quarrels among individuals of similarly conservative social views as to whether a secular or a theological theodicy better justified the status quo.[26] Hence, in the Darwinian debates, evolutionists within the religious community sanctioned the continued "mystification" of the social order and the justification of existing class relations by emphasizing that inequality, the existing hierarchical and exploitative division of labor, and progress through struggle were "natural," and hence privileged, means of ensuring progress. By contrast, opponents of Darwinism (for reasons that Young does not carefully spell out) refused to embrace this "secular theodicy" and continued instead to advance a more traditional, theologically oriented defense of the status quo.[27]

However successful the interpretations by Young and Moore might be in collapsing the distinction between scientific naturalism and religious thought, they fail to show the relevance of social interest in accounting for why individual Protestants responded to the theory of organic evolution in the way that they did. If rejection of revolutionary social change characterized all sides in the controversy, it cannot be used to distinguish Protestant evolutionists from their opponents. Moreover, though it is quite true that few American Protestant leaders supported revolutionary change, they hardly needed the evolutionary hypothesis as a prop for their conservatism. Beyond this, however, it is not entirely clear that relationships actually existed between theological position and social interest. Whereas some supporters of the evolutionary hypothesis within the American religious community did maintain that it reinforced or even justified a conservative view of the social structure, others invoked the theory of evolution to justify the need for significant efforts to change society. The Christian socialist Walter Rauschenbusch, for example, declared in 1912: "Translate the evolutionary theories into religious faith, and you have the doctrine of the Kingdom of God."[28]

Opposition to evolution has also been compatible with a number of different assessments of society and politics.[29] Even if it could be established that opponents of the theory of organic evolution were less inclined than supporters to embrace change in the social structure, this would hardly prove that those Protestants opposed the theory of organic evolution *because* they were hostile to social change. It might well be that their opposition to both evolution and social change had a common source: a conviction that the universe is controlled by a sovereign, gracious God who does not always act in accordance with human definitions and standards. In any case, it should be apparent that the link between social interest and theological positions cannot merely be assumed; it must be shown to exist. This is a difficult assignment. Indeed, if Young is correct in claiming that "ideology is pervasive but often implicit," a troublesome question arises: What would a knockdown argument for the existence of a link between responses to Darwinism and social interest actually look like?[30]

More generally, few historians have carefully studied the socioeconomic background of American Protestants during the late-nineteenth and early-twentieth centuries. The studies that do exist suggest that while this variable may play at least some role in predicting the way that lay people interpret theology, it is not very useful in drawing distinctions within the Protestant leadership itself.[31] No less importantly, even if correlations were found, their mere existence would certainly not be sufficient to prove that responses to Darwinism were a function of social class. In view of the above considerations, it is difficult to avoid the conclusion that so far, at least, claims that responses to Darwinism represent ideological expressions of social interest are largely driven by dogma rather than evidence.

Self-interest, however, can be construed in a variety of ways, and in dealing with the attitude of American Protestant leaders toward the theory of organic evolution, studies that center on the interaction of culture and individual psychology have been more fruitful, in my judgment, than those that focus on social interest. Two studies published in the 1970s exemplify the emphasis on efforts that individuals made to resolve psychic tensions brought about by new ideas: William R. Hutchison's "Cultural Strain and Protestant Liberalism" and James R. Moore's *The Post-Darwinian Controversies*.[32]

Hutchison argues that during the period between 1875 and the end of World War I, American Protestant thinkers experienced a good deal of tension and even a "loss of meaning" in the face of challenges to traditional formulations of Christian theology. Their efforts to cope with this "cultural strain" impelled them to examine some of their most fundamental assumptions. Ultimately Protestant thinkers arrived at much different conclusions concerning the nature and sources of religious truth. In accounting for the differences between liberals and conservatives, Hutchison emphasizes the salience of "distinctive patterns" of family life, conversion experience, education, and personality.[33]

James R. Moore's analysis of the interaction of individuals with the larger culture focuses somewhat more narrowly on Protestants who played a prominent role in the Darwinian controversies. Using Leon Festinger's concept of cognitive dissonance, he has assessed the impact of the theory of organic evolution on twenty-eight Protestant leaders in Great Britain and America. Moore concludes that the efforts of Protestants to come to grips with the theory of evolution culminated in the emergence of three major groups: Christian anti-Darwinians, for whom dissonance was "negligible"; Christian Darwinians, largely comprising orthodox Christians who "understood Darwin's theory and left it substantially intact"; and "Christian Darwinisticists" (Moore now prefers the term "Christian evolutionists"), composed of liberal Protestants who "misunderstood, misinterpreted, or modified Darwin's theory" in their efforts to reconcile transmutation with their theological commitments.[34]

Although cultural strain and cognitive dissonance may be useful in describing the psychological context in which Protestant leaders in the United States rendered their judgments about Darwinism, these concepts shed little light on the question of why some Protestants chose to embrace the transmutation hypothesis while others opted to reject it. Recognizing this, Moore has called attention to the role of prior theological considerations. He contends that whereas orthodox Protestants were likely to accept the Darwinian version of the theory, liberal Protestants typically adhered to more optimistic renditions of evolutionary thought. Indeed, he argues, "it was only those who maintained a distinctly orthodox theology who could embrace Darwinism; liberals were unable to accept it."[35]

This is not the place to engage in an extended critique of Moore's position. It is important to emphasize, however, that relatively few American Protestant leaders who remained committed to orthodox formulations of Christian theology actually embraced Darwinism. More commonly they rejected the theory of organic evolution altogether.[36] In practice, Calvinism and other categories used to describe American theology prior to 1875 are only modestly useful as a predictor of individual Protestants' view of Darwinism. By the last quarter of the nineteenth century, opposition to Calvinism, far from being a bold step, was sufficiently common to render it relatively useless as a predictor of precisely how Protestant leaders would respond to the evolutionary hypothesis.[37] Although one can be reasonably sure that believers who remained committed to Calvinism would find the immanentist orientation of the liberals' "New Theology" unacceptable, one could not predict with similar confidence whether they would choose to endorse progressive creationism or reject the theory of organic evolution altogether.[38] And while virtually all liberal Protestants accepted the transmutation hypothesis, the terms of their acceptance ranged from a belief in evolution by resident forces to the halfway house of progressive creationism.

These caveats, however, are not intended to dismiss the value of theological considerations in shaping the responses of American Protestant leaders to the theory of organic evolution. To the contrary, I would argue that little progress in making sense of those responses will be made until we appreciate the theological stakes that those religious thinkers believed were at issue and the differing theological presuppositions that they embraced.

In order to understand the dynamics of the decision-making process that occurred in the wake of scientists' endorsement of the transmutation hypothesis, it is necessary to realize that during the first three quarters of the nineteenth century, views concerning the nature of the relationship between science and the Bible remained considerably fuzzier than a superficial examination might suggest. To be sure, there was widespread agreement that God had revealed himself in both nature and the Scriptures and that, when properly interpreted, those two sources of revelation would prove consistent. Protestant intellectuals also agreed that although both sources of divine revelation were sufficiently clear to be understood, fallible human beings did not always immediately apprehend their real meaning.[39] This enabled them to account for the conflicts that had periodically arisen between the conclusions of scientists and the prevailing understanding of the scriptural message: they indicated that either nature or the Bible had been erroneously interpreted. Convinced that God had provided human beings with the Bible in order to reveal the scheme of redemption rather than to provide them with a detailed knowledge of natural history, most Protestant leaders reasoned that as long as the findings of science were not irreconcilable with doctrines central to Christian faith and practice, those findings could be accepted and biblical interpretations could be revised accordingly. Prior to about 1875 this reasoning led to the periodic alteration of their interpretation of scriptural passages to bring them into accord with the consensus of science.

These principles seemed to provide clear guidelines for reconciling science and the Scriptures. In reality, however, they did not. They provided no clear criteria for determining which biblical texts conveyed "essential" doctrines. Nor did they furnish guidance for interpreting such texts. Perhaps because American Protestant thinkers were confident that tension between science and the Bible would remain outside the realm of essentials, they devoted little attention prior to 1875 to the task of determining the precise nature and scope of biblical inspiration or the relative authority to be conferred on science and scriptural testimony. The source and ultimate basis of religious knowledge, in short, remained lamentably vague.[40]

The scientific community's endorsement of the theory of organic evolution forced American Protestant intellectuals to confront these troublesome issues. By challenging the credibility of the special divine creation of the human species, the Fall of Adam, the meaning of Christ's atonement, and other themes central to the biblical narrative, the theory challenged major elements of the orthodox understanding of Christianity. Accordingly, scientists' acceptance of that theory impelled Protestant leaders to make critical choices concerning the relative authority they were prepared to accord to modern science and the Scriptures. Stated rather crudely, Protestants who embraced the theory of organic evolution conferred primary authority on the findings of the scientific community and assumed that the task confronting them was to show that it was possible to accept the theory of evolution while embracing a world view that remained essentially Christian. Those who rejected the theory, in contrast, characteristically did so in the name of fidelity to scriptural revelation. For their part, progressive creationists attempted to carve out a compromise along lines similar to those that had been made by Protestant intellectuals during the first three quarters of the nineteenth century.

Underlying and even determining these decisions were fundamentally different presuppositions about the nature of both God and human beings. A skeletal description of those presuppositions is in order.[41] American Protestant clergy and theologians who embraced the theory of organic evolution proceeded from the conviction that truth could most reliably be acquired by subjecting human experience to rigorous examination.[42] Although they acknowledged that knowledge (which they often associated with divine revelation) was accessible to all, by 1875 they had come to assume that the acquisition of that knowledge was most frequently achieved through the patient investigation of specialists.[43] There is little evidence that their esteem for specialized expertise was motivated by an estimate of their own prospects for career enhancement; to the contrary, it would appear that the outcome of specialization – professionalism – actually had the effect of undermining the cultural authority of religious elites.[44] Instead their deference to expertise appears to have been rooted in their assessment that the history of thought was moving in the direction of increasing specialization and their eagerness to equate historical change with divinely sanctioned progress.[45]

The respect of Protestant evolutionists for professional expertise conditioned, and was in turn heightened by, the emphasis they placed on intelligibility in their conceptualizations of human nature and divine activity. As individuals living in a century notable for its advances in knowledge and as heirs of a religious tradition committed to the doctrine that humanity had been created in the image of God, these Protestants confidently assumed that the Creator had empowered human beings to understand both the natural world and human history. At the same time, the progress already attained by scientific investigation and historical scholarship convinced them that God characteristically chose to act in a manner comprehensible to his human children. This view enabled them to welcome with greater equanimity descriptions of natural history that employed intelligible natural processes rather than periodic, arbitrary acts of divine interposition.[46]

Protestant opinion leaders who endorsed the theory of organic evolution were also influenced by another consideration. They recognized that other educated men and women shared their conviction that scientists were the most reliable judges of theories concerning the natural world. As George Foot Moore, professor of the Old Testament at Andover Theological Seminary, put it in 1888, "The new knowledge of the universe gained by astronomy, geology, and biology, with the philosophy which seeks to unify this knowledge, has created the modern conception of the universe, which is held more or less consistently and clearly by all educated men."[47] Their awareness of the prestige of science convinced them that opposition to Darwinism would not only confirm the charges of those who were intent on tarring Christianity with the obscurantist brush but also impel many believers to abandon their faith. For Protestant leaders who shared this perspective, the development of a recognizably Christian interpretation of the theory of organic evolution seemed to be not only a psychological imperative but also an essential apologetic strategy.[48]

Acceptance of the theory of organic evolution led to a number of important alterations in theological formulations. Recognition of the fact that the transmutation hypothesis could not be reconciled with a "plain sense" reading of the Scriptures was one of the considerations that convinced most Protestant evolutionists to abandon the claim that the Bible was either an infallible or an exclusive source of divine revelation. To be sure, they continued to defend some biblical doctrines – most notably the doctrine that the human species bore the divine image – quite vigorously.[49] Nevertheless, they found themselves increasingly placing the testimony of the Scriptures before the bar of modern scholarship. They continued to insist that the Bible contained the word of God, but they no longer assumed it to *be* that word.[50]

This view of the Scriptures, coupled with the conviction that knowledge was a product of progressive illumination, went a long way toward undermining the allegiance of many American Protestant leaders to biblical authority. Indeed, by the 1920s some of the more liberal American Protestant intellectuals were doing little more than paying lip service to the Scriptures. To those who viewed the entire natural world and all of human history as arenas of divine revelation, it seemed clear that doctrinal change should be welcomed rather than feared, for it was the product of increasing wisdom about the nature of the divine – human encounter.[51]

Protestant clergy and theologians who rejected the theory of organic evolution proceeded from a different set of convictions concerning human nature and divine activity. Though they respected the intellect as the most obvious evidence of humanity's kinship with God, their view of the human condition was informed more profoundly by their belief in the universality of sinfulness and fallibility than by awareness of humanity's past accomplishments or its promises of future potential. From this perspective, the glorification of specialized experts and the exalted visions of their ability to ascertain the nature of reality seemed to exemplify the very pride that served to separate human beings from their Creator.[52]

If, however, these Protestants tended to share a rather bleak view of humanity's nature and possibilities, this does not mean that they believed that God had simply cast human beings adrift.[53] Rather, they emphasized that in keeping with his infinite grace and mercy, the Deity had provided in the Bible a complete, clear, and inerrant guide to His plan of salvation, an "ultimate standard of authority as to religious truth."[54] An interpretation of that divinely inspired text in accordance with "common sense principles," these Protestants maintained, would disclose the essential elements of God's scheme of redemption.[55] This perspective, which was "populist" in the sense of affirming the equal accessibility of religious knowledge to all people, prompted Charles Hodge, an eminent Princeton Seminary theologian, to announce proudly that his institution had never been "charged with originating a new idea."[56] Given Hodge's belief that a merciful God had presented human beings with a unique and perfect rule of Christian faith and practice, it made little sense to alter creeds, for as long as they expressed the scheme of redemption revealed in the Scriptures, their validity, like God Almighty, transcended time itself.[57]

Although anticipations of this "muscular" view of biblical authority can be found long before 1875, its systematic development did not occur until the last quarter of the nineteenth century.[58] In that period a significant minority of American Protestant leaders, confronted by the theory of organic evolution and the higher criticism of the Bible, began to construct a systematic discourse of scriptural infallibility that precluded compromise with rival secular sources. More than any other factor, this discourse and the thinking that informed it shaped their assessment of modern science. Noting that the history of science revealed all too clearly that scientists had frequently erred in their interpretations of the natural world, proponents of biblical infallibility concluded that conflict between the clear message of the Bible and the verdict of scientists was tantamount to conflict between an infallible vehicle of divine truth and the fallible conclusions of human beings.[59]

This pattern of reasoning inevitably led a sizable minority of American Protestant leaders to join Jesse B. Thomas, the pastor of Brooklyn's First Baptist Church of Pierrepont Street, in concluding that "to recast a theological system to the pattern of a shifting and precarious biological hypothesis is madness."[60] These thinkers maintained that the evolutionary hypothesis was irreconcilable with "the whole system of truth, for the revelation of which the Scriptures were given to men."[61] The Methodist theologian Miner Raymond warned his readers that "if the origin of the race be found anywhere else than in the special creation of a single pair, from whom all others have descended, then is the whole Bible a misleading and unintelligible book."[62] For Raymond and many other American Protestant leaders, it seemed patently clear that if a "plain sense" reading of passages relating to the creation of human beings and other species did not disclose the truth about natural history, there was no reason to assume the reliability of the Scriptures on any other subject.[63] Convinced that an untrustworthy Bible was incompatible with the benevolence and mercifulness of God, they concluded that Christians must reject the theory of organic evolution in favor of the faith that "once for all was delivered to the saints."

These thinkers continued to view the natural world as one source of divine revelation, and they continued to deny that any real conflict between valid interpretations of nature and the Bible was possible. They inferred from the scientific community's conversion to the theory of organic evolution, however, that they could no longer trust scientists to interpret nature correctly.[64] Indeed, modern science appeared to be a particularly egregious example of a corrupt humanity's overweening pride – a classic case of individuals who, pretending to be wise, had become fools.[65] The fact that many Protestant leaders joined scientists in their folly suggested that they were giving "that infallibility to secular science which alone belongs to the theology of the Bible."[66] By the early-twentieth century, most proponents of biblical inerrancy had come to view the theory of organic evolution as, in the words of one historian, "something akin to a rival religion" and its supporters as rival communicants.[67]

In spite of their increasing hostility to the tenor of modern science, American Protestant leaders who opposed the theory of organic evolution sometimes supported their claim that it was inimical to "true science" by invoking the names of scientists who rejected it. Between 1875 and 1930, however, this tactic played a subordinate role in opposition strategy.[68] Clergy and theologians who rejected the transmutation hypothesis typically assumed that the clearest evidence of its falsity was that its implications were irreconcilable with the clear testimony of the Scriptures.

It would, of course, be too much to claim that all clergy and theologians who opposed the theory of organic evolution did so primarily in the name of fidelity to the biblical narrative. Nevertheless, this issue did prove more compelling than other considerations that have been advanced by students of the Darwinian controversies. Some historians have asserted, for example, that a philosophical (as opposed to distinctively Christian) commitment to the fixity of species was paramount in accounting for opposition to Darwinism.[69] However, the willingness of all but the most fervent biblical literalists in the period prior to 1930 to accept the idea of fossil sequences and the nebular origin of the solar system would seem to refute the notion that the idea of change over time was itself a source of great difficulty, even to Christians wedded to strong views of biblical authority.[70] Protestants who opposed the theory of organic evolution did commit themselves to the notion that the boundaries of the originally created "kinds" had remained fixed, but the motive underlying that commitment was biblical rather than philosophical.[71]

Historians have also cited concern about the impact of Darwinism on the classic argument from design as a source of major concern.[72] Such concern undoubtedly existed, but Protestants who accepted the theory of organic evolution managed successfully to negotiate changes in their approach to natural theology, and there is no reason to assume that Protestants who rejected the theory could not have done likewise if they had been satisfied on other grounds.[73] Clergy and theologians did not oppose theistic evolutionism because they believed that it was inconceivable that God could have populated the world by means of transmutation or because they assumed that it would have demeaned his glory if he had done so; they opposed it because of a conviction that the Bible clearly indicated that special creation had been the method God had chosen to employ.[74]

Protestants also opposed the theory of organic evolution because they suspected that the animus of the theory was hostility to supernaturalism.[75] Convinced that the defects of a fallen world could only be remedied by means of supernatural intervention, antievolutionists tended to be "catastrophists" in their vision of the nature and mode of divine activity. This can be seen, for example, not only in their approach to creation but also in their view of the nature of conversion as a determinate, sudden experience.[76] To recognize this, however, is simply to acknowledge in slightly altered fashion the vision of human depravity and divine benevolence that grounded their belief in the existence of an infallible source of divine revelation.[77]

The appeal of biblical literalism did not stem from its ability to allay concerns generated by social and intellectual change, as some historians have suggested.[78] To be sure, such change was pervasive and often anxiety provoking, but there is no reason to assume that biblicists were more intent than their opponents on finding sources of certainty. Protestant leaders undoubtedly found the belief that they possessed a clear and authoritative guide to salvation psychologically gratifying. Indeed, at least some of those individuals may well have derived from their conviction that a gracious God had provided human beings with an infallible revelation of his scheme of redemption the psychic and intellectual resources they needed to defy the larger culture's verdict concerning the authority and scope of scientific investigation. Yet, proponents of evolution within the American Protestant community also found the ideal of certainty appealing. Unlike Protestants who opposed the theory of organic evolution, however, they did not believe that this ideal could be achieved by a rigorous adherence to biblical testimony. Rather, they placed their trust in sustained use of the scientific method, the tools of modern historical investigation, the "religion of Jesus," and the testimony of religious experience as a means of ensuring continued spiritual and secular progress.[79]

The conviction that the transmutation hypothesis was incompatible with the Bible underlay a good deal of the rhetoric of fundamentalism during the first half of the twentieth century.[80] In recent decades, however, opponents of the theory of evolution have become increasingly unwilling to rely primarily on biblical considerations. They have opted instead to spar with scientists over the meaning and significance of paleontological and biological data.[81] But despite the recent turn to technical discussions, biblical considerations remain at the heart of the creationist position.

Though it is often difficult to judge the nature and extent of the commitment of many progressive creationists to evolution, the motives of this third group of leaders seem relatively clear. Progressive creationism was a middle position, a grudging compromise akin to the approach that most American Protestant leaders had taken during episodes of tension between science and religion during the first three quarters of the nineteenth century. The individuals who embraced progressive creationism remained committed to the basic framework of the biblical narrative, yet eager to harmonize their position with that of the scientific community. These thinkers were convinced that scientists' favorable verdict concerning the transmutation hypothesis made it not only intellectually advisable but also strategically imperative that Christians arrive at some form of *modus vivendi* with the idea that evolutionary change accounted for much of the history of life. At the same time, they remained more wedded to traditional supernaturalist, providentialist conceptions of divine activity and stronger views of biblical authority than did the group I have labeled Protestant evolutionists. As a result, they refused to abandon the claim that God had periodically intervened in a special manner during the process of creation to create matter, life, the human body, or the human mind.[82] Characteristically they sustained this compromise by maintaining – quite disingenuously from the vantage point of Protestant evolutionists and Protestant antievolutionists alike – that the words of the biblical account of creation could be reinterpreted to make room for the claim that most life-forms were the product of evolution.[83] Progressive creationism represents, in short, a much more attenuated and less decisive embrace of the evolutionary hypothesis.

This theologically oriented interpretation of motivation is clearly vulnerable to the claim that it has simply pushed the puzzle back a step. It is certainly legitimate to ask why individual American Protestant leaders chose to adopt the differing assessments of the nature of humanity, God, and revelation that I have claimed were especially salient in shaping their responses to the transmutation hypothesis. I would nevertheless argue, however, that it is appropriate to view the problem of determining motivation as akin to stratigraphy in that it is necessary to analyze the phenomena at more than one level.[84] If this is the case, it makes sense to try to more thoroughly understand intermediate strata – such as prior theological commitments – before burrowing deeper. In the end, we may discover that unrecorded contingencies and individual idiosyncracies will frustrate efforts to achieve a more complete reconstruction. Although this does not mean that such efforts should be abandoned, it does mean that in dealing with the problem of motivation, modesty is fitting, even essential, if we are to avoid "guesses passing as scholarship."[85]

Notes

1 I have treated this subject at some length in Jon H. Roberts, *Darwinism and the Divine in America: Protestant Intellectuals and Organic Evolution, 1895–1900* (Madison: University of Wisconsin Press, 1988), especially pp. 14–20, 32–87. See also Gary Scott Smith, *The Seeds of Secularization: Calvinism, Culture, and Pluralism in America, 1870–1915* (Grand Rapids: Eerdmans, 1985), p. 97. For a useful summary of previous approaches to the periodization of Protestant responses to Darwinism, see James R. Moore, *The Post-Darwinian Controversies: A Study of the Protestant Struggle to Come to Terms with Darwin in Great Britain and America, 1870–1900* (Cambridge: Cambridge University Press, 1979), pp. 9–10.

2 Moore, *Post-Darwinian Controversies*, pp. 300–07, passim.

3 Roberts, *Darwinism and the Divine*, p. xiv.

4 Peter J. Bowler, *The Non-Darwinian Revolution: Reinterpreting a Historical Myth* (Baltimore: Johns Hopkins University Press, 1988); and Ronald L. Numbers, *Darwinism Comes to America* (Cambridge: Harvard University Press, 1998), pp. 24–48.

5 Members of the group I have termed "progressive creationists," which included such notable figures as George Frederick Wright, Augustus H. Strong, James McCosh, Benjamin B. Warfield, and Archibald Hodge, are discussed in David N. Livingstone, *Darwin's Forgotten Defenders: The Encounter Between Evangelical Theology and Evolutionary Thought* (Grand Rapids: Eerdmans, 1987), pp. 106–23, 125–31, 134–37; Moore, *Post-Darwinian Controversies*, pp. 218, 241–42; Roberts, *Darwinism and the Divine*, passim; and Smith, *Seeds of Secularization*, pp. 98–105, 107–11. I should confess that the position I am taking in this paper is different from the one I took in a previous work, in which I refused to accord the most conservative members of this group evolutionist status and tended instead to interpret individuals as evolutionists or antievolutionists on a case-by-case basis. See Roberts, *Darwinism and the Divine*, pp. 209, 318–19, n. 1. The problem with my former approach is that it failed to establish a meaningful, sharply delineated set of criteria by which individuals could be classified. As a result, it seemed somewhat arbitrary.

6 Bert James Loewenberg, "The Controversy over Evolution in New England 1859–1873," *New England Quarterly*, 8 (1935):257; and Norman F. Furniss, *The Fundamentalist Controversy, 1918–1931* ([1954]; Hamden, Conn.: Archon Books, 1963), pp. 19–20. See, for additional examples of this approach, Andrew Dickson White, *A History of the Warfare of Science with Theology*, 2 vols. (New York: Appleton, 1896); Richard Hofstadter, *Anti-Intellectualism in American Life* (New York: Random House, 1963); Stewart G. Cole, *The History of Fundamentalism* (New York: Richard R. Smith, 1931), p. 328, passim; Frank Hugh Foster, *The Modern Movement in American Theology: Sketches in the History of American Protestant Thought from the Civil War to the World*

War ([1930]; Freeport, N.Y.: Books for Libraries, 1969), pp. 40, 42–43, 47; and Sidney E. Mead, *The Lively Experiment: The Shaping of Christianity in America* (New York: Harper & Row, 1963), pp. 148–49. Perceptive discussions of the role of the "military metaphor" in shaping the historiography dealing with the relationship between science and religion include Moore, *Post-Darwinian Controversies*, 19–100; David C. Lindberg and Ronald L. Numbers, "Beyond War and Peace: A Reappraisal of the Encounter between Christianity and Science," *Church History* 55 (1986):338–54; and Numbers, "Science and Religion," *Osiris*, 2nd ser., 1 (1985):59–80.

7 Robert E. Wenger, "Social Thought in American Fundamentalism, 1918–1933," Ph.D. diss., University of Nebraska, 1973, pp. 33, 87–92; Paul A. Carter, "The Fundamentalist Defense of the Faith," in *Change and Continuity in Twentieth-Century America: The 1920's*, eds. John Braeman, Robert H. Bremner, and David Brody (Columbus: Ohio State University Press, 1968), pp. 206–07; and George M. Marsden, *Fundamentalism and American Culture* (New York: Oxford University Press, 1980), pp. 47–48.

8 David Bloor, *Knowledge and Social Imagery*, 2nd ed. (Chicago: University of Chicago Press, 1991); and Barry Barnes, *Scientific Knowledge and Sociological Theory* (London: Routledge & Kegan Paul, 1974).

9 Windsor Hall Roberts, "The Reaction of the American Protestant Churches to the Darwinian Philosophy, 1860–1900," Ph.D. diss., University of Chicago, 1936, pp. 195–97. For an even earlier discussion of the reception of Darwinism by Protestants that relies on denominational categories, see Foster, *Modern Movement in American Theology*, pp. 38–58. General treatments of denominational responses to evolutionary ideas include Milton Berman, *John Fiske: The Evolution of a Popularizer* (Cambridge: Harvard University Press, 1961), pp. 173–98; Edward Justin Pfeiffer, "The Reception of Darwinism in the United States, 1859–1880," Ph.D. diss., Brown University, 1957, pp. 67–74; Herbert Schneider, "The Influence of Darwin and Spencer on American Philosophical Theology," *Journal of the History of Ideas*, 6 (1964):3–18, especially p. 5; Stow Persons, "Evolution and Theology in America," in *Evolutionary Thought in America*, ed. Stow Persons ([1950]; Hamden, Conn.: Archon Books, 1968), p. 425; and Numbers, *Creation by Natural Law: Laplace's Nebular Hypothesis in American Thought* (Seattle: University of Washington Press, 1977), pp. 119–23. For examples of treatments of individual denominations' response to the theory of evolution, see Dennis Royal Davis, "Presbyterian Attitudes toward Science and the Coming of Darwinism in America, 1859 to 1929," Ph.D. diss., University of Illinois, Champaign-Urbana, 1980; Deryl Freeman Johnson, "The Attitudes of the Princeton Theologians toward Darwinism and Evolution from 1859–1929," Ph.D. diss., University of Iowa, 1968; T. Watson Street, "The Evolution Controversy in the Southern Presbyterian Church with Attention to the Theological and Ecclesiastical Issues Raised," *Journal of the Presbyterian Historical Society*, 37 (1959):232–50; Edward Lassiter Clark, "The Southern Baptist Reaction to the Darwinian Theory of Evolution," Th.D. diss., Southern Baptist Theological Seminary, 1952; Reginald W. Deitz, "Eastern Lutheranism in American Society and American Christianity, 1870–1914: Darwinism – Biblical Criticism – The Social Gospel," Ph.D.

diss., University of Pennsylvania, 1958; and C. David Grant, "Evolution and Darwinism in the *Methodist Quarterly Review*, 1840–1870," *Methodist History*, 29 (1991):175–83.

10 I am not the first historian to make this point. See, for example, Numbers, "Science and Religion," pp. 75–76. In 1886 the General Assembly of the Presbyterian Church in the United States did declare that "any doctrine at variance" with the special divine creation of human beings was a "dangerous error"; George D. Armstrong, *A Defence of the "Deliverance" on Evolution, Adopted by the General Assembly of the Presbyterian Church in the United States, May 26th, 1886* (Norfolk, Va.: John D. Ghiselin, 1886), pp. 3–5. The General Assembly did not condemn the transmutation hypothesis as a whole, however, and its position concerning the creation of human beings did not prevent a number of Presbyterians from attempting to reconcile their faith with the transmutation hypothesis.

11 Numbers, "Creation, Evolution, and Holy Ghost Religion: Holiness and Pentecostal Responses to Darwinism," *Religion and American Culture*, 2 (1992): 127.

12 Some historians have suggested that as a group, nineteenth-century Wesleyan and Baptist clergy and theologians had less formal education than their counterparts in Congregational, Presbyterian, and Episcopal churches; see Roberts, "Reaction of the American Protestant Churches," p. 196; and Robert Mapes Anderson, *Vision of the Disinherited: The Making of American Pentecostalism* (New York: Oxford University Press, 1979), p. 133. For some statistical discussion of the Baptists' relatively uneducated ministry, see Walter Edmund Warren Ellis, "Social and Religious Factors in the Fundamentalist-Modernist Schisms Among Baptists in North America, 1895–1934," Ph.D. diss., University of Pittsburgh, 1974, pp. 53–54. However, although systematic analysis might show that members of denominations characterized by a lack of education were less intent than members of other denominations on confronting Darwin's work, my more impressionistic evaluation of the sources has made me rather dubious about this.

13 For the theological "right's" response, see, for example, Numbers, "Creation, Evolution, and Holy Ghost Religion," pp. 127–58; Numbers, "Science Falsely So-Called: Evolution and Adventists in the Nineteenth Century," *Journal of the American Scientific Affiliation*, 27 (1975):18; Numbers, " 'Sciences of Satanic Origin': Adventist Attitudes Toward Evolutionary Biology and Geology," *Spectrum*, 9 (1979):17–30; and Milton L. Rudnick, *Fundamentalism & the Missouri Synod: A Historical Study of Their Interaction and Mutual Influence* (Saint Louis: Concordia, 1966), pp. 83, 88–89. I deal with the response of Unitarians to the transmutation hypothesis in Roberts, *Darwinism and the Divine*, passim.

14 Perhaps the classic articulation of this view is H. Richard Niebuhr, "Fundamentalism," in *Encyclopaedia of the Social Sciences*, ed. Edwin R. A. Seligman ([1931]; New York: Macmillan, 1937), pp. 526–27. See also Kenneth Bailey, *Southern White Protestantism in the Twentieth Century* (New York: Harper & Row, 1964), pp. 74–75; and William E. Leuchtenberg, *The Perils of Prosperity 1914–32* (Chicago: University of Chicago Press, 1958), p. 222. Marsden has

provided a good discussion of the role of the Scopes trial in shaping views of fundamentalism in his *Fundamentalism and American Culture*, pp. 184–91.

15 Carter, "Fundamentalist Defense," p. 202; Hofstadter, *Anti-Intellectualism*, p. 126; and Furniss, *Fundamentalist Controversy*, p. 129.

16 See also Livingstone, "Darwinism and Calvinism: The Belfast-Princeton Connection," *Isis*, 83 (1992):408–28.

17 Donald Tinder, "Fundamentalist Baptists in the Northern and Western United States, 1920–1950," Ph.D. diss., Yale University, 1969, p. 110; Bailey, *Southern White Protestantism*, especially pp. 74–75, 78, 79, 90. Bailey's study focuses on the Fundamentalist movement as a whole rather than the more specific issue of religious leadership. For a discussion of Baptist fundamentalist clergy that supports Bailey's findings, see Ellis, "Social and Religious Factors," p. 78.

18 Marsden, *Fundamentalism and American Culture*, pp. 103–04; and Livingstone, *Darwin's Forgotten Defenders*, p. 124.

19 Wenger, "Social Thought," pp. 57–62, 30. See also Ernest R. Sandeen, *The Roots of Fundamentalism: British and American Millenarianism, 1800–1930* (Chicago: University of Chicago Press, 1970), pp. xii, 132–269.

20 William R. Hutchison, "Cultural Strain and Protestant Liberalism," *American Historical Review*, 76 (1971):386–411, especially pp. 394, 408–09.

21 Once again, it is necessary to draw inferences about leaders in the evolutionary controversy from studies of liberal and conservative leadership. See, for example, Hutchison, "Cultural Strain," pp. 408–09; and Carter, "Fundamentalist Defense," pp. 203–05.

22 On this subject, see Robert M. Young, "Darwin and the Genre of Biography," in *One Culture: Essays in Science and Literature*, ed. George Levine (Madison: University of Wisconsin Press, 1987), pp. 206–07.

23 Hofstadter, *Anti-Intellectualism*, p. 121; Ellis, "Social and Religious Factors," pp. 30, 82, passim.

24 Moore, "Herbert Spencer's Henchmen: The Evolution of Protestant Liberals in Late Nineteenth-Century America," in *Darwinism and Divinity*, ed. John Durant (Oxford: Basil Blackwell, 1985), pp. 76–100, quotations on pp. 81, 89, 92, 91, and 93.

25 Moore, "The Creationist Cosmos of Protestant Fundamentalism," in *Fundamentalisms and Society: Reclaiming the Sciences, the Family, and Education*, eds. Martin E. Marty and R. Scott Appleby (Chicago: University of Chicago Press, 1993), p. 45.

26 Robert M. Young, quoted in Ingemar Bholin, "Robert M. Young and Darwin Historiography," *Social Studies of Science*, 21 (1991):619; Young, "Natural Theology, Victorian Periodicals and the Fragmentation of a Common Context," in *Darwin to Einstein: Historical Studies on Science and Belief*, eds. Colin Chant and John Fauvel (New York: Longman, 1980), pp. 69–107; and Young, "The Historiographic and Ideological Contexts of the Nineteenth-Century Debate on Man's Place in Nature," in *Changing Perspectives in the History of Science: Essays in Honor of Joseph Needham*, eds. Mikulas Teich and Robert Young (London: Heinemann, 1973), p. 376.

27 Young, "Man's Place," pp. 371–86.

28 Walter Rauschenbusch, *Christianizing the Social Order* (New York: Macmillan, 1912), p. 90.

29 Paul Carter makes the case for identifying Fundamentalism as "the religious version of radical rightism," but then, quite properly in my judgment, rejects it. Carter, "Fundamentalist Defense," pp. 189–90. As he notes, "nearly all of the several dozens of essays in *The Fundamentals* were characterized less by 'right-' or 'left-wing' views than by the absence of political discussion altogether." He concludes that "the major social thrust of Fundamentalism, in its earliest years, at least, was neither liberal nor conservative in the political sense." Ibid., pp. 191, 193.

30 Young, "Science, Ideology and Donna Haraway," *Science as Culture*, 3 (1992): 166.

31 Hutchison, "Cultural Strain," pp. 393–95. I have not done a systematic quantitative study, but I have been unable to discern significant socio-economic differences between proponents and opponents of the theory of organic evolution within the American Protestant intellectual community in the late nineteenth century; see Roberts, *Darwinism and the Divine*, p. xii.

32 Another treatment of Protestant thought in the late-nineteenth and early-twentieth centuries that can legitimately be viewed as an exposition employing cultural strain and crisis is Marsden, *Fundamentalism and American Culture*, p. 3, passim.

33 Hutchison, "Cultural Strain," pp. 398, 403, 389, 399.

34 Moore, *Post-Darwinian Controversies*, especially pp. 111–13, 115–17, 218, 301; and Moore, "Preface to the Paperback Impression," ibid.

35 Ibid., especially pp. 301–07, quotation on p. 303. Since publishing *Post-Darwinian Controversies*, Moore has apparently come to believe that greater emphasis should be placed on the role of social interest and ideology in shaping attitudes toward Darwinism.

36 This is one of the reasons, I suspect, why Moore's "Christian Darwinians were never more than a tiny minority." Moore, *Post-Darwinian Controversies*, pp. 350–51. In his discussion of Christian Darwinism, Moore put forward only two proponents of that position in the United States: Asa Gray and George Frederick Wright. Significantly, both of those thinkers made substantial modifications of Darwin's own views. See Roberts, *Darwinism and the Divine*, pp. 38–41, 80; and Numbers, *The Creationists* (New York: Alfred A. Knopf, 1992), pp. 20–36.

37 Roberts, *Darwinism and the Divine*, pp. 132–33.

38 See, for example, George Frederick Wright's rejection of immanentism in G. F. W[right], "Bad Philosophy Going to Seed," *Bibliotheca Sacra*, 52 (1895): 559–61.

39 Marsden, *Fundamentalism and American Culture*, pp. 14–16.

40 A good summary of attitudes toward the Bible is Marsden, "Everyone One's Own Interpreter? The Bible, Science, and Authority in Mid-Nineteenth-Century America," *The Bible in America: Essays in Cultural History*, eds. Nathan O. Hatch and Mark A. Noll (New York: Oxford University Press, 1982), pp. 79–100. For a discussion of the absence of systematic theological discourse

about the nature and scope of biblical inspiration and authority in the United States prior to 1875, see Sandeen, *Roots of Fundamentalism*, pp. 106, 110; and Norman H. Maring, "Baptists and Changing Views of the Bible, 1865–1918 (Part I)," *Foundations*, 1 (1958):60. I have also dealt with this subject at greater length in Roberts, *Darwinism and the Divine*, pp. 21–30, 235.

41 I have discussed most of the following issues in greater detail and provided documentation for the period between 1875 and 1900 in Roberts, *Darwinism and the Divine*.

42 Willard B. Gatewood, Jr., ed., *Controversy in the Twenties: Fundamentalism, Modernism, and Evolution* (Nashville: Vanderbilt University Press, 1969), p. 15; Marsden, *Fundamentalism and American Culture*, p. 215; and James Turner, *Without God, Without Creed: The Origins of Unbelief in America* (Baltimore: Johns Hopkins University Press, 1985), p. 133. For a representative contemporary view, see A. V. G. Allen, "The Theological Renaissance of the Nineteenth Century," *Princeton Review*, n. s., 10 (1882):280–81.

43 Roberts, *Darwinism and the Divine*, p. 118. For the tendency of evolutionists to equate knowledge with divine revelation, see ibid., p. 157–59, 234; and Moore, *Post-Darwinian Controversies*, p. 219.

44 Turner, *Without God*, pp. 121–25.

45 Hutchison, *Modernist Impulse*, pp. 2, 99–101, 186–93; Grant Wacker, *Augustus H. Strong and the Dilemma of Historical Consciousness* (Macon, Ga.: Mercer University Press, 1985), p. 11; Marsden, *Fundamentalism and American Culture*, pp. 48, 63; Moore, *Post-Darwinian Controversies*, pp. 240, 350; Turner, *Without God*, pp. 86–88, 130–31; Charles D. Cashdollar, *The Transformation of Theology, 1830–1890: Positivism and Protestant Thought in Britain and America* (Princeton: Princeton University Press, 1989), pp. 160, 358–59; and Roberts, *Darwinism and the Divine*, pp. 161–62, 182, 234–35.

46 Roberts, *Darwinism and the Divine*, pp. 124, 136–42, 145.

47 George F. Moore, "The Modern Historical Movement and Christian Faith," *Andover Review*, 10 (1888):333.

48 Roberts, *Darwinism and the Divine*, pp. 118–20, 237.

49 Ibid., pp. 177–79.

50 An excellent general treatment of the attitude of liberal Protestants toward Scripture is Grant Wacker, "The Demise of Biblical Civilization," in *The Bible in America*, pp. 121–38. See also Kenneth Cauthen, *The Impact of American Religious Liberalism* (New York: Harper & Row, 1962), pp. 13–20. For a somewhat more narrowly focused view, see Roberts, *Darwinism and the Divine*, pp. 149, 151–65.

51 Roberts, *Darwinism and the Divine*, pp. 157–61, 171–73, 234–35; Wenger, "Social Thought," pp. 21–22; Cauthen, *Impact of American Religious Liberalism*, pp. 22–24, 29–30, 218–19; and Hutchison, *Modernist Impulse*, pp. 95–105.

52 Tinder, "Fundamentalist Baptists," pp. 45, 91–92; Wenger, "Social Thought," pp. 92–93, 95–96; Hart, "True and the False," pp. ix, 21–22; Rennie Schoepflin, "Anti-Evolutionism and Fundamentalism in Twentieth-Century America," M. A. paper, University of Wisconsin, 1980, p. 19; and James R. Moore, "Interpreting the New Creationism," *Michigan Quarterly Review*, 22 (1983):333.

53 Roberts, *Darwinism and the Divine*, pp. 222–23.
54 John L. Girardeau's undated sermon, "The Signs of the Times – In the Church," *Sermons*, ed. George A. Blackburn (Columbia, S.C.: State Company, 1907), p. 116. See also Roberts, *Darwinism and the Divine*, pp. 213–18, 222–23.
55 Stuart Robinson, "The Pulpit and Skeptical Culture," *Princeton Review*, 4th ser., 3 (1879):147. See also William B. Riley, quoted in Ellis, "Social and Religious Factors," pp. 104–05.
56 Charles Hodge, quoted in Alexander H. Hodge, *The Life of Charles Hodge* (London: T. Nelson, 1881), p. 594. See also Roberts, *Darwinism and the Divine*, pp. 215–20; and Wenger, "Social Thought," pp. 94–96.
57 Hutchison, *Modernist Impulse*, pp. 202–04; Roberts, *Darwinism and the Divine*, pp. 221–22. A superb discussion of the ahistoricism of Protestant orthodoxy is Wacker, *Augustus H. Strong*, pp. 11–12, 23–31.
58 Charles B. Warring, *The Mosaic Account of Creation, The Miracle of To-Day; or, New Witnesses to the Oneness of Genesis and Science* (New York: J. W. Schermerhorn, 1875), p. 21.
59 Roberts, *Darwinism and the Divine*, pp. 213–19, 234; Numbers, *Creationists*, pp. 14–17.
60 Jesse B. Thomas, "A Symposium on Evolution: Is the Darwinian Theory of Evolution Reconcilable with the Bible? If So, with What Limitations?" *Homiletic Monthly*, 8 (1884):534.
61 John T. Duffield, "Evolutionism, Respecting Man and the Bible," *Princeton Review*, 4th ser., 2 (1878):173–74.
62 Miner Raymond, *Systematic Theology*, 3 vols. (Cincinnati: Hitchcock and Walden, 1877), vol. 2, p. 16.
63 Roberts, *Darwinism and the Divine*, pp. 210–11, 212, 220; T. T. Martin, quoted in Gatewood, *Controversy in the Twenties*, p. 19; A. C. Dixon and John Bradbury, quoted in Tinder, "Fundamentalist Baptists," p. 117.
64 Tinder, "Fundamentalist Baptists," p. 93; Roberts, *Darwinism and the Divine*, pp. 230–31, 95–97, 223–24.
65 For condemnations of the notion that human judgment was an adequate guide to the truth, see Robert Cameron, quoted in Hart, "True and the False," p. 40; ibid., pp. 54–55; Romans 1:22; and Roberts, *Darwinism and the Divine*, pp. 97–98, 236–37, 220–24.
66 Henry Darling, "Preaching and Modern Skepticism," *Presbyterian Review*, 2 (1881):763–64. See also Roberts, *Darwinism and the Divine*, pp. 98, 224–30.
67 Tinder, "Fundamentalist Baptists," pp. 111–12.
68 Roberts, *Darwinism and the Divine*, pp. 91–98. The 1920s witnessed an increasing tendency on the part of opponents of evolution to resort to appeals to data from natural history to justify their position. Numbers, *Creationists*, pp. 50–101.
69 Moore, *Post-Darwinian Controversies*, pp. 205–14, 219.
70 Numbers, *Creationists*, p. 17.
71 To be fair, it is important to note that Moore has acknowledged that proponents of the fixity of species used the Bible as well as science to justify their views. Moore, *Post-Darwinian Controversies*, pp. 210, 215.
72 Livingstone, "The Idea of Design: The Vicissitudes of a Key Concept in the

Princeton Response to Darwin," *Scottish Journal of Theology*, 37 (1984):329–57; Livingstone, *Darwin's Forgotten Defenders*, pp. 103–04; Cynthia Eagle Russett, *Darwin in America: The Intellectual Response, 1865–1912* (San Francisco: W. H. Freeman, 1976), pp. 26, 32; R. J. Wilson, *Darwinism and the American Intellectual: A Book of Readings* (Homewood, Ill.: Dorsey, 1967), pp. 39–40; and Berman, *John Fiske*, pp. 109–10.

73 Roberts, *Darwinism and the Divine*, pp. 120–29, 209–10.

74 John Horsch, *Modern Religious Liberalism: The Destructiveness and Irrationality of the New Theology* (Scottdale, Pa.: Fundamental Truth Depot, 1921), p. 233; John Roach Straton, quoted in Tinder, "Fundamentalist Baptists," pp. 112–13; and A. C. Dixon, quoted ibid., p. 115. It is significant, I think, that many clergymen responding to a questionnaire in 1929 who believed that the theory of organic evolution was incompatible with the teachings of the Bible nevertheless acknowledged that the theory was "consistent with belief in God as Creator"; see George Herbert Betts, *The Beliefs of 700 Ministers and Their Meaning for Religious Education* (New York: Abingdon, 1929), pp. 26, 44.

75 Roberts, *Darwinism and the Divine*, pp. 101–03, 211; Horsch, *Modern Religious Liberalism*, pp. 233, 235; and J. C. Massee, quoted in Bruce L. Shelley, *Conservative Baptists: A Story of Twentieth-Century Dissent* (Denver: Conservative Baptist Theological Seminary, 1960), p. 19.

76 Hart, "True and the False," pp. 169, 202; Tinder, "Fundamentalist Baptists," pp. 33, 38; and Wenger, "Social Thought," p. 38. Conversely, Protestants who opposed the transmutation hypothesis commonly rejected the idea that environmental improvement was sufficient to change the human condition. Ibid., p. 50.

77 Roberts, *Darwinism and the Divine*, pp. 103–04. For explicit association of supernaturalism and the status of the Bible after 1900, see Curtis Lee Laws, quoted in Tinder, "Fundamentalist Baptists," pp. 116.

78 For variations or intimations of such an interpretation, see Hart, "True and the False," p. xviii; Ellis, "Social and Religious Factors," p. 30; and Gatewood, ed., *Controversy in the Twenties*, p. 20.

79 Roberts, *Darwinism and the Divine*, p. 222; Wenger, "Social Thought," p. 22; and Moore, *Post-Darwinian Controversies*, pp. 239–40.

80 It is often suggested that evolution played a relatively insignificant role in the emergence of fundamentalism as a discrete movement within American Protestantism. See, for example, Moore, *Post-Darwinian Controversies*, pp. 70–73. It is probably true that proponents of fundamentalism were not entirely in agreement in their attitudes toward the evolutionary hypothesis. See Numbers, *Creationists*, p. 39. On the other hand, it is clear that even a number of spokespersons in *The Fundamentals* who did not focus primarily on that hypothesis nevertheless ascribed higher criticism and other modern heresies to it. See, for example, Franklin Johnson, "Fallacies of the Higher Criticism," in *The Fundamentals: A Testimony to the Truth* (Chicago: Testimony Publishing, n. d.), vol. 2, p. 54; J. J. Reeve, "My Personal Experience with the Higher Criticism," ibid., vol. 3, p. 99.

81 Numbers, *Creationists*. See also Moore, "Interpreting the New Creationism," pp. 321–333, especially pp. 332–33.

82 For valuable discussions of G. Frederick Wright, see Numbers, "George Frederick Wright: From Christian Darwinist to Fundamentalist," *Isis*, 79 (1988):624–45; and Numbers, *Creationists*, pp. 20–36. For Reuben Torrey's views, see ibid., p. 39.
83 Roberts, *Darwinism and the Divine*, pp. 136, 143–44, 147–48, 176.
84 I am indebted for the stratigraphy metaphor to a conversation I had with Lynn Nyhart.
85 Numbers, "Science and Religion," p. 73.

7

Exposing Darwin's "hidden agenda": Roman Catholic responses to evolution, 1875–1925

R. SCOTT APPLEBY

In 1931, not long after the theory of evolution by means of natural selection had gained widespread acceptance in the scientific community, the *Catholic Encyclopaedic Dictionary* rendered the official Roman Catholic teaching on the subject as follows:

> EVOLUTION, EVOLUTIONISM. Transformism, or Evolutionism, means the theory of the transformation of species only, but evolution is a more general and universal theory which is applied to the physical world, to the realm of ethics, to man and to society (Spencer). Absolute evolutionism is not justified by physical science which has established without doubt the stability of species, without ever discovering veritable specific transformations; moreover, it is condemned by metaphysics which refuses to admit that effects can be more perfect than their efficient causes; and "extreme" evolution denies the special act of creation of life, attributing the whole process to a natural development from inorganic matter. The doctrine of the natural development of all the species of the animal and vegetable world from a few primitive types created by God is *moderate* evolution. Catholics are free to believe in moderate evolution, excluding the evolution of man. Animals, as distinguished from man, are devoid of reason. Hence the animal soul, i.e., the principle which gives an animal life, is essentially material. Hence, man's soul, though depending on material things for its activities, being essentially spiritual, the evolution of man *as a whole* from the lower animal is impossible.[1]

This somewhat tendentious appraisal of the status of evolution as both a science and a philosophy epitomized the official gloss on half a century of conflict within the church. Its confident and authoritative tone gave little indication that the Roman Catholic response to Darwinism in the United States, as in Europe, had been varied and ardently contested, reflecting developments both in American Catholicism and in the theory of organic evolution itself.

From roughly 1875 to 1925, Darwinism generated a lively and intense debate in the United States. During the heart of this period, from about 1885 to 1910, the clerical leadership of the American Catholic community found itself divided on certain issues, including the reception of Darwin, into "progressive" and "conservative" camps. These labels can be misleading, for they suggest, erroneously, that the so-called progressives were not "conservative" on matters of faith and church doctrine, and were therefore insufficiently orthodox. Equally misleading is the implication that the conservatives were unthinking obstructionists when it came to making pastoral and intellectual accommodations to mainstream American society, including its burgeoning scientific culture. The two sides differed not over fundamental doctrine or the necessity of adaptation, but over the means and pace of accommodation, with the conservatives less willing to countenance the dissemination of new theological and philosophical perspectives and methods that departed in important respects from Thomism, the official theology of the church. These new perspectives and methods, most of which originated among a handful of European Catholic intellectuals, were part of an attempt to bring Catholic thought more closely into conversation with modern developments and ideas, not least of which was evolutionary theory.

The two camps vied for the attention and loyalty of a growing Catholic middle class, which was beginning to send its children to colleges and universities in significant numbers at the beginning of the twentieth century. The progressives were led by a small group of bishops and priests, sometimes described as "Americanists," who attempted to make room within American Catholic culture for a scientifically informed world view, and who urged greater collaboration between Catholics and non-Catholics in the workplace, the public schools, and the laboratory. John Ireland (the archbishop of St. Paul), Bishop John J. Keane (rector of The Catholic University of America), and John Lancaster Spalding (bishop of Peoria) were among the episcopal leaders of initiatives that might be described as "Americanist." Their most powerful friend and occasional collaborator was James Cardinal Gibbons, archbishop of Baltimore. Gibbons, Ireland, Keane, and Spalding were among those who saw no reason to stifle the new biblical scholarship making inroads in Catholic seminaries and publications (including critical readings of the Book of Genesis); and they positively encouraged the research of progressive Catholic scientists such as John Zahm, a Holy Cross priest at the University of Notre Dame, who was interested in reconciling Catholic doctrine and Darwinism.[2]

The conservative camp, which included the majority of Catholic prelates and priests, was led by Michael A. Corrigan, archbishop of New York, Bernard J. McQuaid, bishop of Rochester, and in the early years of the new century, by a new breed of bishops such as William H. O' Connell, who embodied the trend toward the Romanization of the American hierarchy that developed in the early decades of the twentieth century.[3] Behind the conservatives' resistance to proposals that seemed to advance the rapid cultural and intellectual Americanization of immigrant Catholics stood an appraisal of the religious condition of the Catholic laity that was decidedly less enthusiastic than that of the progressives. The conservatives, in short, feared that the immigrant faith might not survive a direct encounter with liberalism, Protestantism, scientific materialism, and other perceived ideological threats to supernaturalist religion.

While the progressives won some important battles – Gibbons, Ireland, and Keane were successful, for example, in preventing the Vatican from condemning U.S. labor unions – they experienced several setbacks in their efforts to adapt Catholic thought to modernity. In 1897 officials of the Holy Office of the Roman curia warned Zahm to cease publication of *Evolution and Dogma*, his book advocating a position known as "theistic evolution." Two years later, Pope Leo XIII issued *Testem Benevolentiae*, an apostolic letter addressed to Cardinal Gibbons and warning against the threat of "Americanism." It was not until Leo's successor, Pope Pius X, issued a sweeping condemnation of modernism in 1907, however, that progressive priests and the bishops who had gingerly supported their research, finally turned away from original biblical and theological scholarship and scientific inquiry. Some few European and American Catholic priests nonetheless continued to maintain that theistic evolution did not fall under the ban on modernism and that a modified form of Darwinism was therefore acceptable.

The historiographical treatment of American Catholic responses to the Darwinian revolution has been thin, and it has generally failed to explore these ecclesial contexts, which shaped so much of the debate. The most detailed and comprehensive history of the subject, an unpublished 1951 dissertation, is aware of the split between the conservatives and the progressives, but does not link it to the larger question of how Roman Catholicism incorporated modern scientific thought and methods into its self-understanding as a social and religious reality.[4] Nor does the scant literature on the topic set the American Catholic debate over evolution in the equally important context of late-nineteenth-century nativism – the social movement which sought to protect Anglo-Saxon privilege in the United States – and the twentieth-century rise of social engineering. Yet most American Catholics found both movements morally questionable, offensive, and personally threatening. Eventually, these Catholic concerns transcended the progressive-conservative divide, and the priests who helped to shape popular Catholic opinion on evolution came to see Darwinism as the legitimating philosophy for a morally contemptible program of eugenics.

This chapter therefore considers the ways in which the external politics of American social reform, as well as the internal politics of the church itself, shaped American Catholic attitudes toward modern science in general and the reception of Darwinism in particular.

The Irish Catholic baboon

"Darwinism" took many forms after 1859, with its Spencerian version widely influential in North America. Catholics, who initially preferred the "Catholic science" of the French naturalist Jean Baptiste Lamarck, assimilated the elements of Darwin's theory at their own pace and on their own terms. As the debate over evolution unfolded, for example, Catholic opponents of evolution tended to conflate the general theory of evolution that was emerging as a modified form of Darwinism, and Darwin's own account of evolution by natural selection. They tended, further, to depict Darwin as a powerful advocate of modern atheism, with "random, godless" natural selection as his most devasting weapon against religious faith. While it was fashionable among Roman Catholic apologists to claim that true religion had nothing to fear from true science, and vice versa, most commentators, progressives as well as conservatives, expressed concerns about the theological and philosophical implications of *Origin of Species* (1859) and, especially, *The Descent of Man* (1871).

From a theistic point of view, their concerns seem understandable. Proponents of Darwinian natural selection denied that any teleological design was discernible in the natural world; nature, Darwin stressed, harbored no aim at all. The fields of anthropology and comparative religion emerged, under the shadow of Darwin's influence, to tell a tale of the gradual evolution of all societies and religions, lending a decidedly relativist cast to Christianity's – and Roman Catholicism's – truth claims.[5] *The Descent of Man* seemed to confirm the theist's worst fears; it needed and reserved no place for Creation, Providence, or any guiding supernatural force. Victorian-era agnosticism, seeking a new base for morality in place of a discarded theism, found initial sustenance in the "moral content" as well as the intellectual achievements of science as embodied in the persons of Charles Darwin and Herbert Spencer.[6]

In addition to the theological difficulties Darwinism posed, Catholics encountered a problem of social theory as well. Adrian Desmond and James Moore make a compelling case that "social Darwinism" – the extension of the "survival of the fittest" mentality to economic and political realms – was not the misrepresentation or uninvited elaboration of Darwin's thought that many have depicted it to be, for his notebooks indicate that "competition, free trade, imperialism, racial extermination, and sexual inequality were written into the equation from the start – 'Darwinism' was always intended to explain human society."[7]

In order to appreciate the specific nature of the threat social Darwinism posed to Catholicism, one must consider the social context of the time. During the last half of the nineteenth century American Catholicism was primarily a working-class phenomenon, a religion of immigrants struggling to make their way in a foreign land.[8] Cities were clogged with Irish, German, Polish, and Italian immigrants willing to take low-paying jobs and to crowd into urban tenements. The Irish and the Germans were the first to arrive in significant numbers, and the Know-Nothing movement, which flourished immediately prior to the Civil War, was a militant nativist response to the unexpected and unwelcome presence of so many uneducated, "unwashed" foreigners. In the 1880s, as Italian, Polish, Lithuanian, and other immigrants arrived from eastern and southeastern Europe, anti-Catholicism flared up again in the form of the American Protective Association (APA). Roman Catholics observed with growing trepidation the apparent social consequences of Darwinian evolution, including the advocacy of sterilization of the "feebleminded," a notion supported by APA activists, and the move to administer intelligence tests to eastern European immigrants to demonstrate their "inferiority."[9] In 1886, the *New York Evening Telegram* printed a caricature of an Irish servant girl "with the mouth of a baboon and horns upon her."[10]

The writings and public statements of both conservative social Darwinists and progressive social scientists, neither of whom looked with favor on the character or physical constitution of the immigrant Catholic, made it virtually impossible for priests and lay leaders, including a steadily growing number of middle-class Irish-American Catholics, to ignore the new strain of nativism.[11] Conservatives used Darwinism to defend the social and economical hierarchy of nineteenth-century America; liberal reformers and social scientists embraced the Darwinian notion of progress to legitimate a program of eugenics and social engineering. The strands converged in movements such as the Immigration Restriction League (IRL), founded in 1894 as an organization of predominantly white, middle-class men who sought a literacy test as a means to curtail emigration from southeastern Europe.[12] Yale sociologist Henry Pratt Fairchild, a prominent IRL member, vividly portrayed "the immigrant problem" as posing a dire threat to the racial purity, the "virility and authority," of American society.[13] Other IRL activists approvingly quoted Darwin's definition of social progress as "the rearing of the greatest number of individuals in full vigor and health, with all their faculties perfect under the conditions to which they are subjected."[14] The naturalist Joseph LeConte, a Protestant exponent of theistic evolution, could have been speaking for Roman Catholics in 1891 when he worried that "if selection of the fittest is the only method available, if we are to have race improvement at all, *the dreadful law of the destruction of the weak and helpless* must, with Spartan firmness, be carried out voluntarily and deliberately." Against such a course, LeConte stated, "all that is best in us revolts."[15]

Within this general social and cultural context the American Catholic reception of Darwinism occurred on elite and popular levels, although popular awareness of Darwinism was restricted to the relatively small American Catholic middle class, especially to those lay people who attended the Catholic Summer Schools in the 1890s, joined the Catholic Reading Circle, and subscribed to its *Review*, where much of the controversy was reported. Even among this select group of middle-class Catholics, there was little familiarity with the thought of Haeckel, Spencer, or Huxley, for "the American popular mind, when directed to intellectual, scientific, and religious matters, centered on discussions of a more general nature."[16] The supply of priests and other Catholic educators conversant with "the evolution question" was thin, for "seminaries, bound to traditional instruction, were unable to keep up with the breathless pace of evolutionary thought, since they were training priests to become pastors of flocks with little interest in Darwinism."[17]

Nonetheless, a small number of well-informed priests were engaged, along with their Protestant counterparts, in the controversy over the higher biblical criticism, the teaching of evolution, and theological modernism within the seminaries and colleges. The battle over "Americanism" shaped the small Catholic intellectual community's reponses to Darwinism as well. Indeed, the Americanist controversy was the most important "internal context" of the Catholic response to Darwinism, and it accounted for many of the most important divergences of the Catholic response from that of Protestant elites.

The so-called Americanists within the Catholic clergy were attempting to balance what they perceived to be the needs of the immigrant community – specifically, the need to gain access to the mainstream institutions of American society – with the requirements of theological orthodoxy as defined by the Vatican. With the possible exception of John Zahm's brief career as a theistic evolutionist, this balancing act produced an inconsistent, tentative, and sometime confused response to evolution. On the one hand, the progressives sought to demonstrate that Roman Catholics would take a back seat to no one when it came to scientific research and freedom of inquiry. On the other hand, these same men were clergy of a church that perceived a pervasive threat to religious belief in modern scientific thought. When American Catholics engaged Darwinism in its modified form, cited Protestant and secular authorities on the emerging general theory, and began to develop their own theological responses to it, their ecclesial opponents raised charges of "Americanism" and "modernism." Thomas Dwight, a celebrated anatomist at Harvard University and a lay Catholic, was free to deliver a qualified endorsement of the scientific theory, but the clergy was timid after 1910, in the wake of the Vatican's condemnation of modernism. By that date the response to evolutionary theory had become a litmus test for orthodoxy, a marker of allegiances in the contest for influence over Catholic education.

In sum, the American Catholic response to the Darwinian revolution revealed as much about the mind-set and internal struggles of the polyglot, urbanizing, immigrant Catholic community as it did about the scientific theory. Darwinism was engaged critically by only a few elite Catholics, while most Catholic commentators and publicists quickly framed it within ongoing intraecclesial controversies. Two mild ironies resulted from this practice. First, the debate over Darwinism itself, for all its virulence, was actually an occasion for American Catholics to work out a number of identity-defining issues facing the immigrant community. Second, the debate did not leave a strong antievolutionist legacy to future Catholic educators in quite the way Protestant fundamentalism did. The advent of evolutionary theory, in other words, served as a catalyst for the resolution of internal Catholic issues rather than as a sustained evaluation of Darwinism and evolutionary theory in itself. Despite the seeming victory of the conservatives, moreover, the scientific theory of evolution was never formally condemned, and Roman Catholicism modified its general anti-evolutionist stance several times in the twentieth century; this was a process that culminated in the conditional approval of the theory by Pope Pius XII in 1950.[18]

A tentative approach: the initial response to Darwinism, 1875–1895

During the first phase of the Catholic response to the Darwinian revolution, conservatives dominated the debate. Led by Jesuits of the neoscholastic school – a version of Thomism tailored to the apologetical, catechetical, pastoral, and institutional needs of the post-Tridentine Roman church[19] – the antievolutionists of the 1870s and 1880s seized upon the morally objectionable doctrines of Darwin's cultural disciples, Herbert Spencer and Thomas Huxley. Excluding direct creation and design from his scheme of development, Darwin, they charged, had provided the grounds for a solely naturalistic world view. Father F. P. Garesché of St. Louis University was not one to make fine distinctions: he labeled the scientists and philosophers who supported Darwin conspirators against revealed religion heretics who worshipped matter as their God.[20] Camillo Mazella, S. J., a leader in the Italian Thomistic revival who served on the first faculty of Woodstock College, the American Jesuit theological center in Maryland (and later, as Cardinal Mazella, became the prefect of the Congregation of the Index), warned his colleagues in 1877 that the idea of human evolution as developed by Darwin was directly contrary to divine revelation and to the teaching of the Church Fathers.[21]

In this phase of the response, Darwin was clearly the enemy, but there was uncertainty about the general idea of evolution itself. Mazella cited the English Catholic biologist St. George Jackson Mivart against Darwin, despite the fact that Mivart, too, was an evolutionist who sought to reconcile developmentalist thought and Catholic doctrine.[22] Restricting the attack to Darwin was part of an early Catholic strategy, Michael Gannon writes, of "granting the probability of evolution without fully accepting it and rejecting it without fully condemning it."[23]

But this proved difficult, for the issue provoked polarization. Some prominent American Catholic thinkers opposed evolution itself from the outset. When the *Catholic World* endorsed Mivart's book *On the Genesis of Species* and its claim that the theory of evolution, purified of its Darwinian elements, is "consistent with the strictest and most orthodox Christian theology," the ubiquitous Catholic convert and anti-evolutionist Orestes Brownson expressed outrage. Mivart's claim that Adam's body had been formed over a period of time by purely natural causes was heretical, Brownson wrote, because it made science independent of the church's authority.[24]

Liberal Protestantism's acceptance of Darwinism also troubled conservative Catholics; they pointed indignantly to works such as Lyman Abbott's *Evolution of Christianity* and its Darwin-influenced dismissal of traditional doctrines of heaven and hell, the Fall, and the divinity of Christ.[25] Abbott's work, which seemed to be a straightforward attack on traditional Christian faith and a capitulation to the materialistic spirit of the times, moved Jesuit professor Joseph Selinger to condemn any expression of Catholic support for theory of organic evolution.[26] Conservative Catholic journalist Arthur Preuss was also concerned about doctrinal integrity, but he chose to defend religious truth by attacking Darwin's credentials as a scientist, claiming in 1898 (against considerable evidence to the contrary) that Darwinism could not admit of one defender among the world's eminent scientists.[27]

The progressives' early response to Darwin also relied upon religious and theological rather than strictly scientific arguments. Progressive bishops such as Gibbons, Ireland, Keane, and Spalding encouraged Catholics to enter the mainstream of American intellectual life and culture.[28] In this they followed the example of Father Isaac Hecker, the founder of the Missionary Congregation of St. Paul the Apostle (commonly known as the Paulists), the first male religious order of American origin. Hecker urged the accommodation of Catholic life to "the American experiment," especially the constitutional principle of freedom of religion, and taught his followers that the Holy Spirit of God, "the Sanctifier," was at work in a new and dramatic way in nineteenth-century America. In this new dispensation of salvation history, Hecker claimed, American and Catholic ideals merged and reinforced one another: the freedom of the will, individual liberty, moral growth, "the perfection of activity." By celebrating material and spiritual progress, encouraging advances in biblical criticism, and praising the separation of church and state in America, Hecker played a central role in the emergence of progressive Catholicism. Significantly, he endorsed the scientific investigation of human origins and development.[29] After Hecker's death, a French translation of his biography launched the Americanist crisis. Gibbons, the leader of the American hierarchy, came to Hecker's defense, calling him a "true child of the Church."[30]

Gibbons's closest episcopal allies, citing Hecker's example, offered official endorsement for experimentation in Catholic science. Archbishop Ireland of St. Paul assured the Catholic faithful that "between reason and revelation there can never be a contradiction." The "so-called war between faith and science," he said, "is a war between the misrepresentations of faith, or, rather, between ignorance of some scientists and the ignorance of some theologians." Ireland, neither a scientist nor a theologian, expected that "the fruits of all historical research, of all social and moral inquiry, [will] give us Christ rising from the dead." Thus he called the Church "to stimulate the age to deeper researches, to more extensive surveyings, until it has left untouched no particle of matter that may conceal a secret, no incident of history, no act in the life of humanity, that may solve a problem."[31] On the eve of the promulgation of Pope Leo XIII's encyclical condemning the ideas associated with Hecker under the name "Americanism," Ireland was still praising Hecker as a "providential" leader.[32]

John Lancaster Spalding, bishop of Peoria, went even further than Ireland. Refuting accusations by the American Protective Association that Catholics were "papal obscurantists," he declared that American Catholics "are devoted to the Church; they recognize in the Pope Christ's Vicar, and gladly receive from him the doctrines of faith and morals; but for the rest, they ask him to interfere as little as may be."[33] Influenced by the idea of evolution, Spalding applied it to church history. The "indefectible power" of the church was its ability to "survive the destruction of social forms which seemed to be part of her life, and develop new strength in surroundings which had been held to be very fatal to her very existence." Nineteenth-century scholasticism would have to adapt itself to new philosophical insights. "Aristotle is a great mind, but his learning is crude and his ideas of nature are frequently grotesque," he told the audience at the ceremony marking the laying of the cornerstone of Catholic University in 1888. "Saint Thomas is a powerful intellect but his point of view in all that concerns natural knowledge has long since vanished from sight."[34] The United States, receptive to the latest advances in scientific method, was the perfect setting for the passing of the torch of knowledge and truth to a new generation. Catholic scientists and theologians must be willing, if necessary, Spalding said, "to abandon positions which are no longer defensible, to assume new attitudes in the face of new conditions."[35]

The priests who took Spalding's admonitions to heart were not satisfied that their conservative colleagues had really understood either Darwin or his interpreters. Nor were they assured that Thomism provided the best framework for doing so. Is Darwinism or evolutionism tenable on strictly scientific grounds? they asked. If so, should the theory of the evolution of species be restricted to nonhuman species and forms of life? Could this restricted version of evolution pass dogmatic tests? Eventually, progressive Catholic thinkers posed the most dramatic question: Could the theory of human evolution be held scientifically? If so, would it necessarily contradict scriptural and traditional understandings of human creation? Some progressive priests, in fact, were willing to consider a modified form of Darwinism that allowed Adam's body to be a product of evolution, while claiming that his soul had been created immediately by God. Conservatives found this to be scientific nonsense and Christian heresy; no reading of Genesis, to their minds, could justify it. To suggest that humans had evolved from lower forms of animal life was to deny the doctrine that God had created man in His image.[36]

The progressives shared a respect for the legitimacy of scientific data on its own terms, as well as a disdain for what they saw as the facile syllogisms of traditional scholasticism. "Medieval armor will not turn a bullet from a modern rifle," one priest wrote, "nor will the authority of a Medieval philosopher be secure behind which to fight a modern evolutionist."[37] The general idea of evolution would not fade away in the face of scholastic condemnations; it must be understood, accommodated, and interpreted according to the "expansive" mind of the church.

Between the neoscholastics and those who would become known as theistic evolutionists (and later tainted with the label "modernists") a middle ground did exist, if only for a fleeting moment. The Paulist missionary Augustine F. Hewit, a disciple of Hecker, occupied it. He devoted a series of articles in the *Catholic World* to his conviction that a qualified general theory of evolution posed no problems for faith and indeed bespoke God's omnipotence in a way that special creation did not. Hewit denied an inherent connection between evolution and materialism or atheism, and insisted on maintaining the important distinction between the scientific theory and the various dangerous philosophies that had grown up around it. Like the progressives, Hewit stressed the distinction between matters of truth revealed by God and matters of science. But unlike the theistic evolutionists, he maintained that Scripture said nothing whatsoever (to be taken literally) about the means by which God created the world; nor had the church fathers entertained an opinion one way or another about evolution as a means of creation. The question, in short, was theologically neutral. Thus science was free to proceed in its hypotheses concerning the ways in which creation unfolded.[38]

Others also tried, in vain as it turned out, to avoid intra-ecclesial polemics by focusing on the scientific merits of the unfolding evolutionary theory. Paulist priest George M. Searle, an astronomer, argued that evolution was a theory based on facts, not speculation. The church would welcome the valid conclusions of scientists; even if human evolution were one day unanimously accepted, it would not endanger the principles of the Catholic faith.[39] In 1887, the same year Hewit advanced his argument for evolution, Thomas Dwight, the Roman Catholic anatomist at Harvard, reported that he and "a large proportion of Catholic scientists believe in [the theory of organic evolution] to a greater or less extent."[40]

Although Hewit, Searle, and Dwight attempted to lift the evolution controversy out of the realm of church politics, other American Catholic thinkers recognized the futility of this strategy. Scientists addressing the question of life and its development could not realistically hope to skirt the theological issues. Thus, as early as 1884 John Gmeiner, a professor of theology at St. Francis de Sales Seminary in Milwaukee, attempted to provide theological justification for the acceptance of evolution. Noting the church's constant teaching that *"no truth of Science does, or ever can, contradict any truth of Divine Revelation,"* he devoted his book on evolution, *Modern Scientific Views and Christian Doctrines Compared,* to a demonstration of the compatibility between a modified form of Darwinism and traditional Catholic teaching. The Mosaic account of creation, Gmeiner wrote, "teaches that the existing kinds of plants and animals *were brought forth by the Earth and water at the command of God,*" but whether this happened at once, at the command of God, or gradually, "on these points the Mosaic account gives us no information."

On this and similar issues "there may be permitted, without danger of contradicting Divine Revelation, a great latitude of opinion among Christians."[41]

In Gmeiner's work can be seen the first outlines of a Catholic understanding of biology based on the concept of theistic evolution. His call for "a believer's interpretation of Darwinism" borrowed from the "pioneering ideas" of liberal Protestant theologians and scientists, especially Joseph LeConte, upon whose book *Religion and Science* (1873) Gmeiner based his own, and his brother John:

> The guiding providence of the Creator in bringing forth, from quite simple forms of life, the countless now existing species, has not been made superfluous by Darwin's theory. On the contrary, assuming the Darwinian theory to be correct, one will have to exclaim, as Prof. John LeConte did, "How simple the means – how multiform the effects – how far-reaching and grand the design! How deeply they impress us with the wisdom, power, and glory of the Creator and Governor of the Universe!"[42]

Whereas Mivart claimed that Church Fathers like Augustine had articulated a view of nature expansive enough to allow for evolution, Gmeiner insisted that Augustine had actually *taught* evolutionary theory. Darwin, according to this view, had simply lent the weight of empirical evidence to the teachings of the great theologian.[43] Gmeiner also cited St. Thomas Aquinas's commentary on Genesis in support of Augustine's "evolutionist" interpretation of Moses: "In the day, when God created Heaven and Earth, he created also every herb of the field, not actually, but before it grew upon Earth, that is, potentially."[44]

In 1887 Archbishop Ireland appointed Gmeiner to the diocesan seminary in St. Paul. There Gmeiner introduced seminarians not only to theology but also to the scientific evidence in favor of the theory of evolution (e.g., the similarity in human and anthropoid skeletal structures and the paleontological evidence linking the two) and the scientific objections to it (e.g., the absence of a missing link between the two species). No evidence exists to determine whether the seminarians were assigned *Modern Scientific Views and Christian Doctrines Compared*, but it seems likely that Gmeiner shared its conclusions with his students. The Darwinian theory "contains valuable grains of truth," he wrote, "yet, as commonly understood, it is far from being a well-founded theory; it is no more than a bold hypothesis."[45] In a later article, published in 1888, Gmeiner predicted that ecclesiastical opposition to evolution would nonetheless fall away as Darwinism gained widespread acceptance in the scientific community. "Perhaps (due to indisputable proofs establishing their contradictories) some other views widely held among theologians and educated men generally," he wrote, "will gradually be given up, and that even before this century closes."[46]

In 1903 James Driscoll, the Sulpician president of St. Joseph's Seminary, Dunwoodie, New York, and a disciple of Spalding, attempted to give a bit of life to Gmeiner's optimistic vision of a Catholic future more accepting of Darwinism. Among the curricular reforms Father Driscoll introduced in the seminary was a course, which surveyed the developments in the general theory of evolution. He also set about to promote theistic evolution in the pages of the *New York Review*, the scholarly journal published by the seminary faculty.[47] The openings to evolutionary thought at Dunwoodie were part of what one historian has described as "the fresh wave of intellectualism that surged through the 'elite' of the American clergy in the early 1900s [which] was, at least partially, due to the encouragement given progressive thought by the Americanists in the previous decade."[48] Although Driscoll did not know it, or did not want to believe it, however, the church had already decided that theistic evolution, among other progressive ideas, would not be countenanced.

The defining episode: the Zahm Affair, 1892–1899[49]

Toward the end of his teaching career, Gmeiner recalled: "Years ago I already held and still do hold the views [on evolution] now advocated by Dr. Zahm."[50] Born in Ohio in 1851 to immigrant parents, John Zahm entered the seminary at Notre Dame at the age of fifteen. Gifted with a natural aptitude for science, he rose rapidly in the ranks of the Congregation of Holy Cross, and by the age of twenty-three was a professor of chemistry and physics, codirector of the science department, director of the library, curator of the museum, and a member of the board of trustees of the fledgling university. Recognizing that Catholics lagged far behind American Protestants in experimental capabilities, Zahm concentrated on building the science department at Notre Dame into a first-rate facility.[51] In 1883 Zahm, secure in his position at Notre Dame, entered the public controversy over evolution. In doing so he used every means of publicizing his ideas: he published articles and books, preached guest sermons, delivered public lectures, and debated prominent Protestant scientists and theologians.

In the 1880s, as nativism heated up and anti-Catholic publicists depicted "the Irish-Catholic Baboon," Zahm attacked *The Descent of Man* as the source of social discrimination and claimed that Darwinism in general was an insubstantial and unprovable hypothesis. Like other Catholic priests and educated laymen, Zahm was disturbed by "those agnostics, materialists, and atheists" who transformed a "shaky" scientific theory into a "full-blown" philosophical system. Zahm asserted the priority of dogma over the claims of science, for "no liberalism in matters of doctrine can be tolerated.... What the Church teaches must be accepted as divine truth." Yet one must distinguish between official teaching of the magisterium and the private opinions of theologians and commentators. The "official church" had never and would never define the exact age of the world or of the human species, for such matters "have nothing to do" with faith and morals. Similarly, the church permitted an understanding of biblical inspiration that allowed for errors on the part of the authors in matters of history, biology, geology, and other scientific disciplines. (Zahm made this assertion in 1886; twenty-one years later, in 1907, the Pontifical Biblical Commission ruled otherwise.)[52]

On the question of evolution in general, Zahm was ambivalent. On one hand, he rejected Darwinism because it was based, he believed, on a number of highly questionable assumptions: the spontaneous generation of life from inorganic matter, the "nebular hypothesis" that the earth and all heavenly bodies were originally in a state of incandescent vapor, and the "unprovable" notion that one species evolved from another by a process of transmutation. "Not a single fact in the whole range of natural science can be adduced favoring the truth of the transmutation of Species," he wrote.[53]

On the other hand, Zahm argued that Catholics could accept a form of theistic evolution. Nothing precluded the Catholic definition of creation from being broadened to include derivative creation "when God, after having created matter directly, gives it the power of evolving under certain conditions all the various forms it may subsequently assume." Neither church teaching nor Scripture stood in the way of this understanding of creation; indeed, "according to the words of Genesis, God did not create animals and plants in the primary sense of the word, but caused them to be produced from preexisting material." The evolutionist simply maintained that "God did potentially, what the ordinary Scripture interpreter believes he did by a distinct, immediate exercise of infinite power."[54] Like Gmeiner, Zahm plumbed traditional teachings for opinions on evolution and concluded that St. Augustine had taught that animals and plants were brought into existence by the operation of natural causes. On the question of human evolution Zahm followed Mivart, who taught that "theistic evolution may embrace man's body, considered as separate from, and independent of, the soul, which was, Catholics must affirm, created immediately by God." Zahm concluded his careful discussion by stating that "as matters now stand, evolution is not contrary to Catholic faith; and anyone is at liberty to hold the theory, if he is satisfied with the evidence adduced in its support."[55]

Zahm's early rejection of Darwin stemmed in part from an identification of scientific inductionism with atheism. In *Catholic Science and Scientists,* he reflected upon the "proper" relationship between metaphysical truths and scientific, empirically based truths. Philosophers such as Spencer and Huxley err, he decided, when they subject statements of faith to empirical tests, just as religionists err when they seek to validate faith claims by empirical induction. Darwin had "sinned" in presuming that science might dispense with metaphysics in drawing a picture of human design and purpose. "What has been said of Tyndall and Huxley can, in great measure, be repeated respecting Darwin," he charged. "As a close, patient observer of facts and phenomena in the various forms of animal and vegetable life he has had few, if any, superiors. But here his merit ends [for] he is all along directing his energies not so much to increase our knowledge of nature as to establish and corroborate a pet theory. Facts are presented, assumptions made, and conclusions drawn with a recklessness and a disregard for the simplest rules of dialectics .... This is what is called 'science'!"[56]

Science and religion each proceeded from its own principles, operated by its own distinctive methods, and arrived at its own conclusions. Yet these conclusions were not, *could not be,* mutually exclusive nor ultimately antagonistic, Zahm wrote, for both science and religion sought to explain and describe the wonders of God's creation. The priest-scientist was, therefore, the perfect person to demonstrate the harmony between the conclusions of science and of theology, for he respected each method in its integrity while appreciating the implications of one for the other.[57]

Having settled the methodological question to his satisfaction, Zahm entered a new phase in his career, marked by two themes. First, he came to respect the general idea of evolution, apart from its specific post-Darwinian formulations, as a viable explanation of creation and of the development of plant, animal, and human life. Following upon this reconsideration of evolution, second, Zahm recognized the need for a new narrative of the unfolding of God's providence in human history, one that would enliven the traditional Roman Catholic account with insights into creation offered by the evolutionists. These two themes informed Zahm's work from 1893 to 1896, the period when he formulated his version of theistic evolution as a way to modify Darwinism. As he embarked on this phase of his career, Zahm harbored great expectations. "I am beginning to feel that I have a great mission before me," he confided to Edward Sorin, C.S.C., the founder of Notre Dame in 1892, "in making known to the Protestant world the true relation of Catholic dogma towards modern science."[58] During the next four years he produced five books and several articles assessing evolutionary theory vis à vis Catholic dogma and popularizing theistic evolution as a viable and desirable intellectual position for American Catholics.

Zahm's great period of creativity coincided with developments in the Catholic scientific community. By 1894 the small group of European Catholic scientists with whom Zahm corresponded had given their blessing to the general theory of evolution. Mivart himself thanked Zahm for "carrying on my work" in the United States.[59] Zahm, increasingly more comfortable with the scientific method of inquiry, became correspondingly less patient with fellow priests who retained a vague apprehension of modern science.[60] Perhaps equally important to Zahm's development was the warm response he received as he toured the country on the lecture circuit for the Catholic Summer and Winter Schools beginning in 1893. The topic of evolution seemed of great concern to the middle-class Catholics who attended the lectures.[61]

At the Catholic Summer School in Madison, Wisconsin, and the Catholic Winter School in New Orleans, Zahm announced that he was neither a Darwinist nor a Huxleyan, yet he did not wish to imply that he found "nothing good" in their work. The transmutation of species Zahm now identified with Augustine's belief in a successive creation governed by divine law; likewise, the theory of spontaneous generation could be understood as a tentative explanation of the performance of God's will for the world through a series of secondary causes.[62] Furthermore, Zahm contended, there was nothing in Catholic dogma to preclude the possibility that man descended from an anthropoid ape or some other animal. Mivart's doctrine that man's body had evolved with the subsequent infusion of a rational soul by God had not been condemned; indeed, Zahm contended, it was congenial to the teaching of St. Thomas Aquinas. The account of creation in Genesis, moreover, is in strict accord with that of modern science, Zahm taught, for both accounts allowed the possibility that God had created primitive matter and, after an indefinite period of time and proceeding from the simpler to the more complex, had supervised the development of the innumerable forms of the organic and inorganic world.[63] The enthusiastic public response to the lectures and to their publication was deeply encouraging to Zahm.[64]

Zahm's most mature work, *Evolution and Dogma*, appeared in February 1896. He was not an original theorist, nor did he aspire to be one. Rather, his purpose was to demonstrate the advantages and disadvantages of the various schools of thought regarding evolution and to preserve the concept of evolution itself. Zahm described this general theory of evolution, shared by Darwin and Lamarck, as "ennobling" and "uplifting."[65]

Although Zahm proposed no new theories, he was the first American Catholic scientist to accept and apply Darwin's own innovative mode of scientific reasoning. Before Darwin, the prevailing philosophy of science had promoted a type of Baconian inductivism that held that the systematic collection and classification of data, done without preconceptions on the part of the classifier, would eventually lead to "those laws and determinations of absolute actuality" that can be known to be certainly true. Darwin rejected this quest for ultimate certainty in scientific inferences; instead, he formulated hypotheses that governed the choice of facts and accounted for those chosen for every investigation. He set forth natural selection, for example, not as a theory for which absolute proof had been obtained, but merely as the most probable explanation for the largest number of facts pertaining to the origin of species. "The line of argument pursued throughout my theory," he explained, "is to establish a point as a probability by induction, and to apply it as a hypothesis to other points, and see whether it will solve them."[66]

In *Evolution and Dogma* Zahm adopted this form of reasoning as a *via media* between an "inconceivable" empiricism that guaranteed final certitude and an "ultraconservatism" based exclusively on deduction from a priori principles, which leads to "a fanatical obstinacy in the assertion of traditional views which are demonstrably untenable."[67] Zahm did allow himself one a priori principle: science would not, could not, overturn truths revealed by God in scripture and tradition, for God is one and truth is one. In rejecting dualism, Catholicism allowed its scientists to proceed in their own realms of investigation, accepting the explanation of the origin of species most credible on scientific grounds, confident that the procedure would bolster theism.[68] This approach did not hide fallible human science from the light of revelation, Zahm insisted, but it did restrain metaphysics from imposing prematurely upon the course of rational inquiry. "That the theory of evolution should be obliged to pass through the same ordeal [as Copernicus and Newton] is not surprising to those familiar with the history of science," he wrote, "but there are yet those among us who derive such little profit from the lessons of the past, and who still persist in their futile attempt to solve by metaphysics problems which, by their very nature, can be worked out only by methods of induction."[69] In a misbegotten attempt to safeguard divine providence, Zahm charged, opponents of evolution accepted the special creation of immutable species despite overwhelming evidence of transmutation. To preserve belief in human creation in the image of God, they excluded a priori the possibility of human descent from lower forms of animal life.

Zahm concluded, by contrast, that "everything seems to point conclusively to a development from the simple to the complex, and to disclose, in Spencer's words, 'change from the homogeneous to the heterogeneous through continuous differentiations and integrations.'"

The changes and developments are the result "not of so many separate creative acts, but rather of a single creation and of a subsequent uniform process of Evolution, according to certain definite and immutable laws."[70] Consequently, the older views regarding creation must be materially modified to harmonize with modern science: "Between the two theories, that of creation and that of Evolution, the lines are drawn tautly, and one or the other theory must be accepted.... No compromise, no *via media*, is possible. We must needs be either creationists or evolutionists. We cannot be both."[71]

For Zahm, evolution was a question "of natural science, not of metaphysics, and hence one of evidence which is more or less tangible." The general theory of evolution, he was convinced, eventually would be borne out by the evidence. Zahm thus refrained from endorsing in toto any particular system, lest its imperfections detract from the general theory; Darwin himself, Zahm recalled, had modified his theory as new data demanded.[72] Eventually, a modified general theory of evolution would admit "of a preconceived progress 'towards a foreseen goal' and disclose the unmistakable evidence and the certain impress of a Divine Intelligence and purpose." Zahm hastened to add that "the lack of this perfected theory, however, does not imply that we have not already an adequate basis for a rational assent to the theory of Organic Evolution.... Whatever, then, may be said of Lamarckism, Darwinism, and other theories of Evolution, the fact of Evolution, as the evidence now stands, is scarcely any longer a matter for controversy."[73]

In spite of Zahm's zeal for the independence of scientific inquiry, dogmatic considerations did play a role in his treatment of the origin and development of the human race. On this point Zahm sided with the neo-Lamarckians. Lamarck's theory of transmutation embraced two causal factors: an innate power conferred on nature by God that tends to produce a series of plants and animals of increasing complexity and perfection; and an inner, adapting disposition peculiar to living bodies that assures the performance of actions sufficient to the needs created by a changing environment, those actions becoming instinctive and heritable. Zahm integrated the Lamarckian "powers" into a Christian understanding of the ordered progression of species under God. The addition of the Lamarckian "powers" helped to make credible a scientific worldview that featured design and intention in nature and purposeful, teleological variations in organisms.[74]

Zahm drew on this tradition when he endorsed the theory of human evolution explicated in Mivart's 1871 work, *On the Genesis of Species*, which subordinated natural selection to the role played by "special powers and tendencies existing in each organism." According to Mivart, these special powers were the divine instrument employed in directing organisms to produce those forms that God had preconceived. The human body was derived by this evolutionary process, while the soul, source of humanity's ethical and rational nature, appeared in each case by divine fiat.[75] Zahm came to be known as "the American Mivart" for his endorsement of this theory.[76]

In adopting this position Zahm attempted "a perfect synthesis between the inductions of science on the one hand and the deductions of metaphysics on the other."[77] As a naturalist, he surrendered strict adherence to inductive method by positing a supernatural act of God in infusing the rational soul; as a priest, he surrendered the traditional view of the direct creation of the body of Adam. As it turned out, however, Zahm ended up satisfying neither scientific nor religious purists.

His greatest offense in the eyes of the latter was his treatment of St. Thomas. To lend additional support to Mivart, Zahm provided a lengthy exegesis of texts of Aquinas that seemed to affirm that the rational soul is specially created and infused into the human body by God. Moreover, he argued that acceptance of evolution would enhance rather than imperil authentic Catholic teaching on divine providence and human nature:

> And from the theistic point of view [evolution] exhibits the Deity creating matter and force, and putting them under the dominion of law. It tells of a God who inaugurates the era of terrestrial life by the creation of one or more simple organisms . . . and causing them, under the action of His Providence, to evolve in the course of time into all the myriad, complicated, specialized, and perfect forms which now people the earth. Surely this is a nobler conception than that which represents him as experimenting, as it were, with crude materials and succeeding, only after numerous attempts, in producing the organism which He is supposed to have had in view from the beginning. To picture the Deity thus working tentatively, is an anthropomorphic view of the Creator, which is as little warranted by Catholic dogma as it is by genuine science.[78]

The fact that scientific discoveries called into question the scriptural accounts of the universality of Noah's flood, or the creation of Adam and Eve, did not trouble Zahm; even Pope Leo XIII, he noted, had written that teaching science was not the purpose of sacred Scripture.[79] Zahm claimed a freedom of interpretation in matters not defined dogmatically by the church, including the question of human origins: "Hence as a Catholic I am bound to no theory of Evolution or special creation, except in so far as there may be positive evidence on behalf of such theory."[80]

Evolution and Dogma nonetheless became embroiled in intra-ecclesial politics as soon as it appeared. In the years immediately preceding its publication, the anti-modernists serving in the Curia, the administrative bureaucracy of the church, had added evolution to their growing list of pernicious modern ideas. While in Europe in 1894, Zahm had learned that Mivart had come under suspicion, and the rumor circulated in ecclesiastical circles that the theory of evolution itself was to be condemned by the Vatican.[81] Fueled by the debate over Leo XIII's encyclical on biblical criticism, *Providentissimus Deus*, progressive and conservative Catholics were at odds over the meaning of the book of Genesis and the authorship of the Pentateuch.[82] When plans to translate *Evolution and Dogma* into Italian (1896), French, and Spanish were announced by its publisher, D. H. McBride (whose advertising campaign for the book played up its controversial aspects), it did not take much prodding for the traditionalists and anti-evolutionists in Rome to lump together Zahm's advocacy of theistic evolution with the application of the higher criticism to sacred texts, the separation of the church from the academy, and other "misguided" attempts to adapt Catholicism to local and national cultures.

Furthermore, Zahm's opponents counted him as a member of two international networks that were in bad odor at the Vatican. The first was the circle of Catholic thinkers who sought to harmonize Catholic doctrine and evolution – Mivart, Zeferino Cardinal González of Spain, Edouard Léroi in France, and Henri Breuil – all of whom were eventually silenced. Second, Zahm was an Americanist; indeed, Zahm had admitted that his writings and public lectures on science and theistic evolution were his way of promoting "the movement" – by shock therapy, if necessary.[83]

In 1896, Zahm was transferred to Rome by Gilbert Francais, superior-general of Holy Cross, to take the post of procurator-general for the congregation. Critics in the Catholic press speculated that the transfer was designed to deter Zahm from further publication. "The evolution bacillus is a dangerous thing," chirped Arthur Preuss, editor of the conservative *Review*. He suggested that the pure air of Catholic orthodoxy at the Vatican would help Zahm recuperate. "I have never been 'disciplined,' as they put it, and it is not likely that I shall be," Zahm replied bravely. "My views may be not looked upon with favor by all in Rome," he admitted, "but I know that every eminent man of science throughout Europe is in perfect sympathy with my opinions." Nonetheless, he rushed into print a slender volume, *Scientific Theory and Catholic Doctrine*, in which he repudiated Darwin and Huxley unequivocally.[84]

When Zahm arrived in Rome on 1 April, 1896, Salvatore Brandi, S. J., editor of the influential journal *La Civiltà Cattolica*, took note. In July

Brandi wrote to Archbishop Corrigan of New York to inform him that Zahm's "recent utterances on transformism, and his relations with the liberal party, well known in the Vatican and Propaganda, will interfere with his work as Procurator of Holy Cross."[85] The Italian translation of *Evolution and Dogma*, which appeared that fall, exacerbated Zahm's problems. He reported to his brother in December that "the Jesuits are already training their biggest guns on me.... The die is cast."[86] In January 1897 the first in a series of negative reviews of *Evolution and Dogma* appeared in *La Civiltà Cattolica*. Zahm delivered his last formal paper on evolution five months later at the fourth International Catholic Scientific Congress.[87] In December Zahm returned reluctantly to Notre Dame to serve as provincial of the congregation in America. On 10 September, 1898, he received word from Francais that an edict by the Congregation of the Index had banned *Evolution and Dogma*.[88]

Curial officials were displeased that Zahm had portrayed Augustine and Aquinas as evolutionists; they felt that Zahm and Mivart had jeopardized the integrity of Scripture by threatening to reduce the story of Adam and Eve to a myth; they resented Zahm's support of the Americanists.[89] The decree banning *Evolution and Dogma* was followed four months later by the encyclical condemning Americanism. By the turn of the century bureaucrats in the Vatican had begun to perceive Americanism and evolutionism as aspects of a larger intellectual movement that threatened to undermine the philosophical and theological principles upon which the institutional church of their era was based.[90] In the view of the editor of *La Civiltà Cattolica*, Zahm embodied the link between Americanism and evolutionism.[91]

Zahm's methodology troubled Brandi most about *Evolution and Dogma*. The American priest took as a starting point not deductions from revealed truth, but "unbiased" inductions from empirical data. He promised a "synthesis" of these inductions and the "authentic" teaching of the Catholic tradition. He interpreted Scripture critically, assigning different levels of authority to different passages and scientific competence to very few. And he claimed that the defined teachings of the church on these matters were few in number, which allowed him to proceed liberally in most questions.[92]

Zahm's grounding of his positions in Pope Leo's teaching on Scripture and in the authority of the church fathers and medieval scholastics was particularly troubling to Brandi. At stake was the interpretation of Aquinas, upon whose authority the neoscholastic world view was based. Brandi charged that Zahm's mistakes reflected an insufficient training in neoscholasticism: "Because he does not seem to be familiar with Thomistic philosophy, he has misinterpreted these principles and for this reason he cites and makes application of them incorrectly." Most disturbingly, Brandi concluded, Zahm assured the victory of evolution over other explanations of creation and recommended it "not only for Christian philosophy but also for Catholic apologetics."[93]

To Zahm's ecclesial opponents, Darwinism embodied scientific method in its most irreligious form. And, as Father Zahm's biographer concluded, "More than any other American Catholic, [Zahm] translated Darwin's theory into terms understandable and at least partially acceptable to his American and European Catholic audiences."[94]

Battling the social consequences of Darwinism, 1900–1925

The Vatican's action against Zahm alarmed progressives and encouraged the conservatives, especially the Jesuits, who made of Zahm an object lesson. His advocacy of Darwin's "hidden agenda" had derailed the Notre Dame scientist, the anti-evolutionists offered; by giving his blessing to evolution, whether he knew it or not, Zahm had opened the door to materialism, pragmatism, and liberalism – different faces of the same enemy. This integralist tactic should have been no surprise to Zahm. Even while he had been basking in his short-lived popularity, American Catholic conservatives were writing about the ineradicable connections between Darwinism and Spencerian survival of the fittest, on the one hand, and Marxist atheism, on the other. The type of morality fostered by the "uncritical" evolutionists would lead to Marxist-inspired anarchy and the decay of civilization, they warned.[95] William Poland, S. J., writing in the wake of the Zahm affair, resounded the theme, identifying the enemy as "plain materialism disguised under a hundred learned names." The Church had once vanquished materialism, revealing it to beget "so foul and lawless an ethic for the individual, for the family, and for society at large that no man was willing to be called a materialist." Yet, he warned, "the spirit of the evil" had been "nourished in the shadows of the universities," biding its time for "the occasion to renew its entrance into popular favor." "It found that occasion," Poland recalled, "in the appearance of Mr. Darwin's book upon *The Origin of Species*, in 1859."[96]

The enemy, once again, had become Darwin rather than evolution in general. Following the condemnation of modernism in 1907, the focus of conservative Catholic concern shifted from theology to social theory. Thus, even while American Catholic commentators were promoting "scientific updating," they were opposing the insidious social consequences of Darwinism with a renewed vigor. With the establishment of the IRL and similar organizations in the 1890s, diatribes against the high birth rate of the (inferior) immigrant population had begun to appear, and the public campaign to restrict population growth, though not restricted to nativists, found respectability in the social Darwinian axiom that morality itself had evolved over time along with other human traits. In the early twentieth century social scientists who embraced eugenics proposed public policies (or supported others' proposals) to prevent the reproduction of mentally defective, feeble-minded or criminally inclined people. The most common method resorted to was sterilization. Bills authorizing the sterilization of the so-called unfit came before the Michigan legislature in 1897 and Pennsylvania assembly in 1905; the first bill signed and written into law was enacted in Indiana in 1907.

In this trend the battle lines were joined for an early-twentieth-century version of culture wars. "That the socially conservative Roman Catholic Church invariably opposed all sterilization laws wherever they might be proffered," historian Carl Degler notes, "further convinced reform-minded or Progressive Americans that permanently preventing the unfit from procreating was forward-looking as well as socially necessary."[97] During the 1920s the use of I.Q. tests spread beyond the identification of the mentally deficient; increasingly they were drawn upon to compare the intelligence of the various ethnic and racial groups in the United States, a practice "sparked by the enormous influx of immigrants into the United States during the previous fifteen years" and defended by nativists like Madison Grant, author of *The Passing of the Great Race* (1916). This wealthy New York socialist and amateur zoologist contended that southern and eastern European immigrants – particularly Polish and Italian Catholics and Russian Jews – were decidedly inferior mentally, morally, and physically to the Irish, English, Germans, and Scandinavians who had entered the country in earlier years. The movement to limit immigration culminated in the Immigrant Act of 1924.[98]

Sterilization and birth-control initiatives went hand-in-hand in the Catholic perception. As the new century and a new era of progress dawned, a generation of birth-control advocates argued it was time once again for humanity to move to a higher plane of moral existence.[99] Such arguments obviously troubled and offended many Christians and Jews, against whom the xenophobic diatribes were also aimed. Catholics in particular fought vehemently against birth control and associated its advocacy with a growing acceptance of evolutionism and materialist philosophy. In the second and third decades of the twentieth century, when birth-control activists such as Margaret Sanger came to prominence, American Catholics warned that a host of errors would follow from the American embrace of this scientific-cum-materialist trend.

Again, the *Catholic Encyclopaedic Dictionary* summarized the cumulative teaching of the church regarding the appropriate context – eugenics – in which the advocacy of birth control was to be understood:

> EUGENICS. The science which aims at improving the well-being of the race by studying the factors which affect bodily and mental health, with a view to the encouragement of the beneficial and the elimination of the harmful. Statistics are adduced to show that the chief obstacle is the marriage of the unfit, leading to an increase of hereditable evils, such as insanity, addiction to drink, consumption, venereal disease. The Church has nothing but praise for the aim of eugenics and has no objection to the positive methods proposed as a remedy of the evil, e.g., granting diplomas to the fit, endowing them to encourage the rearing of a large family, providing healthy homes, educating public opinion; but she cannot approve of the negative methods suggested by some eugenists, *viz.*, "birth control" or the compulsory sterilization of degenerates. . . . When eugenists go astray, it is because they forget or deny that spiritual well-being is of far greater importance than material, and that even a tainted existence is better than no existence at all.[100]

With this social and cultural battle in the foreground, Catholic antievolutionism in the early twentieth century took the form of warnings, like the one presented by the editors of the conservative *Ecclesiastical Review*, about "the constant diminution of the birth rate over almost the entire civilized world, [which] is one of the most appalling signs of degeneration in our time, and a subject for the most earnest consideration on the part of our pastoral clergy, since our own country is fast taking a conspicuous part in this triumph of the modern paganism." The culprit was "materialistic education and the overwhelming deluge of Socialistic literature" celebrating "the rapidly decreasing birth rate and the no less rapidly ascending proportion of divorces." In the socialist-materialist scheme of things, "continence, when demanded by the Church, is spoken of as immoral because opposed to nature; the laws of civilized nations in matters of sex are proclaimed as unjust because hindering the free development of normal instincts; the bounds set to the full satisfaction of sexual inclinations are stigmatized as degrading because destructive to character and detrimental to the harmonious expansion of the human faculties." Socialism, the editors continued, "is popularizing among the masses the teachings it has gathered from Darwin . . . and Spencer," among others. Socialist doctrine, which should be made "the center of attack on the part of the clergy from pulpit and platform as well as in our schools of ethics," is quite specific in its prescriptions:

> Woman, in the first place, is to be made economically independent of man.... Every additional child born into the midst of the great class struggle and not demanded by economic necessity, we are told, is only a burden and an encumbrance to render the winning of a strike less possible, to swell the ranks of competition in the search for employment, and to delay by so many hours, days or weeks the coming victory of an inevitably conquering Socialism.[101]

Thomas Dwight, the lay Catholic anatomist at Harvard Medical School, added his voice to the debate at this point. Having earlier seen "no contradiction between evolution on the one hand and design and teleology on the other,"[102] Dwight noted in 1911 that "Catholics have [accepted evolution] upon the understanding that the question is an open one" and that "God breathed into it the breath of life of an immortal soul." He also voiced sympathy for the opposition of conservative Catholic clergy to the social and philosophical applications of Darwinism. Indeed, the trends in the Catholic literature for the ensuing decade would mirror Dwight's concern over "the misuse of Darwin and modern science."[103]

In an attempt to reach partial accommodation, some Catholic commentators went out of their way to distinguish between Darwin, "the scientist," and Spencer, Huxley, and Ernst Haeckel, "the philosophers," depicting the former as "neutral" and "objective." (Materialist philosophers, by contrast, had long been held in disrepute, and no Catholic apologists were claiming about secular philosophy, as they were about secular science, that ultimately it presented no serious challenges to the faith.) Other Catholic critics of evolution, however, while preserving the distinction between science and its misuse by materialist philosophers, blamed Darwin himself for practicing second-rate, or "atheistic," science.[104]

The distinction between science and its philosophical expressions allowed the commentators to focus their fire on the real enemy, the emerging materialist worldview of secular modernity, with its disastrous social and religious implications. Darwinism as a scientific theory reflected the epistemological assumptions of this world view and advanced it in the realm of science; Darwinism as a philosophy, that is, as "a more general and universal theory which is applied to the physical world, to the realm of ethics, to man and to society," as one Catholic commentator described it, lifted materialism to the status of a scientific worldview. Hence Darwinism fit quite snugly into a conservative-turning, anti-modernist Catholic horizon in which "modern thought" evoked adjectives like "materialist," "atheistic," "naturalized" (i.e., de-supernaturalized), "irreligious," and "Protestant." And Darwinism, in all its forms, loomed threateningly as one of the unambigious products of modern thought, the culmination of its irreligious tendencies.[105]

Conclusion

The Roman Catholic reception of Darwinism in the United States centered on the nontraditional, encompassing systems of philosophy and culture – rival worldviews – that accompanied it. These worldviews were judged according to the norms of the neoscholasticism established in Catholic seminaries and institutes of higher learning in the early nineteenth century. The battle for control of both science and philosophy, in short, shaped up as a conflict between "scholastic" (i.e., Aristotelian) and "modern" (e.g., Newtonian) methods and ways of conceptualizing nature and human development.

Brandi in Rome and Poland and his allies in the U.S., it is important to recognize, did not reject the theory of organic evolution as a general scientific explanation of the development of species; they were not particularly *concerned* with the scientific theory, per se, and seemed to accept the possibility of a distinction between evolution as scientific theory and evolution as an encompassing materialist philosophy. It is true that the Roman Jesuits at the turn of the century *were* concerned with the ways in which evolution, in any formulation, contradicted the biblical accounts of creation, insofar as those accounts were to be read uncritically. Nonetheless, the opposition to evolution on grounds of biblical literalism, if not exactly irrelevant in the Roman Catholic case, was subordinated to a prior concern over the status and character of the traditional teachings of the Roman magisterium.[106]

What the Roman Catholic anti-evolutionists found most objectionable, therefore, were the undisciplined interpretations of the theory, that is, interpretations ungoverned by neoscholastic norms and conducted outside a Christian-Aristotelian framework. Such unorthodox interpretations inevitably led to objectionable or "heretical" applications of the theory, including its extension to the human soul and its inflation to the status of a competing "metaphysical" explanation for human spiritual, moral, and intellectual development. Here, as we have seen, is precisely where Zahm went wrong in the eyes of the conservative Catholic clergy and hierarchy: he had challenged the supremacy of the neoscholastic paradigm as an encompassing worldview within which scientific research as well as doctrinal development must take place.

Of course Zahm had not provided an alternative to neoscholasticism, nor had he even departed in significant ways from its metaphysics; he had, after all, turned to Aquinas time and again for justification of theistic evolution. Yet he had intepreted Aquinas, Augustine, and other Church Fathers "loosely" – that is, without sufficient attention to the neoscholastic commentaries. This violated Zahm's ecclesial opponents' understanding of Thomism's special mission in "the age of evolution." The official Roman Catholic philosophy would preserve the primary achievement of civilization: the establishment of moral and social order. Outside of the neoscholastic framework, even "theistic evolution" lent itself readily to the forces of materialism and dissolution – to "social Darwinism," for example – that were eroding traditional order and civility.

Yet evolutionary theory itself was neither rejected or accepted as much as it was seen as a prize in this competition of worldviews. In the early-twentieth century, American Catholic conservatives muted the criticism of science and focused on undisciplined and ideologically inspired applications of the theory to other realms. Despite the fate of *Evolution and Dogma*, Zahm made an impact on the Catholic community by publicizing and interpreting the theory of organic evolution for significant segments of the American Catholic community. Even his conservative opponents accepted Zahm's basic premise, that true religion had nothing to fear from true science. Accordingly, they reasoned, Catholics should join in the efforts to purify science. "Scientific updating" thus became the goal of early-twentieth-century Catholic apologists.

Zahm's error had been to attempt such updating outside the sacred canopy of Thomism. This is perhaps the most significant sense in which the Roman Catholic response to Darwin differed from the Protestant responses discussed elsewhere in this volume. While Roman Catholicism defines itself in every age according to the way its papal-episcopal magisterium, or teaching office, interprets the Christian Tradition (the apostolic faith as it has been handed down, including but not limited to the biblical witness), Protestants in general, and evangelical or fundamentalist Protestants in particular, are bound to the verses of the Bible alone. On a controverted scientific question such as evolution, Catholic tradition, rather than biblical literalism, was the contested arena of meaning for Catholics; in short, neoscholasticism was a bigger obstacle to turn-of-the-century Catholic evolutionists than the general theory of evolution itself. When neoscholasticism eventually fell out of favor, the Church was in a position to reconsider evolution through different theological-anthropological lenses, and it did so. In this regard, the Catholic response to evolution was perhaps closer in spirit and form to the Jewish response. As Marc Swetlitz argues in this volume, the Jewish debates precipitated by evolutionary thought were ultimately about the nature of modern Judaism rather than about evolution per se.

Indeed, the Roman Catholic response to evolution continued to reflect a deeply rooted ambivalence in the years after the Zahm affair. A

Catholic apologist writing in 1928 summarized the Church's encounter with evolution as follows: "Despite the growing caution over man's bodily origins, it did not lead to any public censure by a Roman congregation on the general theory of evolution."[107] The survival of evolution in Catholic scientific circles reflected the judgment of a Jesuit scientist writing in 1911, not long after the condemnation of modernism: "The theory of evolution is not in itself opposed to theism. It has unfortunately been misused and still lies under reproach and suspicion."[108]

Notes

1 Donald Attwater, ed., *The Catholic Encyclopaedic Dictionary* (New York: Macmillan, 1931), p. 187.

2 R. Scott Appleby, *Church and Age Unite! The Modernist Impulse in American Catholicism* (Notre Dame, Ind.: University of Notre Dame Press, 1992), pp. 7–8.

3 Gerald P. Fogarty, S. J., *American Catholic Biblical Scholarship: A History from the Early Republic to Vatican II* (San Francisco: Harper and Row, 1989), p. 174. On O'Connell, see James M. O'Toole, *Militant and Triumphant: William Henry O'Connell and the Catholic Church in Boston, 1859–1944* (Notre Dame, Ind.: University of Notre Dame Press, 1992). After serving for six years as the rector of the American College in Rome, where he had been sent to restore Thomistic orthodoxy after the tenure of the "Americanist agent in Rome," Denis O'Connell, William O'Connell (no relation) was named Bishop of Portland, Maine; in 1905 he was named Co-Adjutor Archbishop of Boston. He became Archbishop of Boston in 1907 and was elevated to cardinal in 1911. O'Toole and Fogarty note that he owed his rise to his friendship with Cardinal Raffaele Merry del Val, the Vatican Secretary of State. Merry del Val was a leading anti-modernist and a force behind the Romanization of the American hierarchy.

4 John L. Morrison, "A History of American Catholic Opinion on the Theory of Evolution, 1859–1950," Ph.D. diss., University of Missouri, 1951.

5 James Turner, *Without God, Without Creed: The Origins of Unbelief in America* (Baltimore: Johns Hopkins University Press, 1985), pp. 150–57.

6 Ibid., p. 240. Turner notes that Charles Eliot Norton, later of Harvard, valued Darwin's honesty and plain-speaking, and believed that "the progress of science" had left Protestantism "vacant of spiritual significance," a turn of events he did not lament. Yet he doubted that science offered very little if any moral foundation to "the ignorant and dependent masses" (quoted in Turner, p. 241).

7 Adrian Desmond and James R. Moore, *Darwin* (New York: Warner Books, 1992), pp. 15–16.

8 From 1820 to 1920, the United States attracted 33.6 million immigrants. See Philip Taylor, *The Distant Magnet: European Emigration to the U.S.A.* (New York: Harper & Row, 1971). Between 1851 and 1920, 3.3 million Irish

immigrants, the vast majority of them Roman Catholic, settled in the United States, bringing the total Irish migration to the United States during the century of immigration, 1820–1920, to 4.3 million people. See Patrick J. Blessing, "Irish," in *Harvard Encyclopedia of American Ethnic Groups*, ed. Stephan Thernstrom (Cambridge: Harvard University Press, 1980), pp. 524–45. Jay P. Dolan notes that the typical Irish emigrant was young, unmarried, and poor. See Dolan, *The American Catholic Experience: A History from Colonial Times to the Present* (Notre Dame, Ind.: University of Notre Dame Press, 1992), p. 129.

9 Carl N. Degler, *In Search of Human Nature: The Decline and Revival of Darwinism in American Social Thought* (New York: Oxford University Press, 1991), pp. 25–40. For a popular and influential presentation of the mentality behind these social policies, see the writings of A.P.A. activist Madison Grant, *The Passing of the Great Race* (New York: Scribner, 1916).

10 *The New York Evening Telegram*, December 10, 1886; reported also in *The Catholic News*, December 15, 1886.

11 The classic study of anti-Catholicism is Ray Allen Billington, *The Protestant Crusade: A Study of the Origins of American Nativism* (New York: Macmillan, 1938); see also Jenny Franchot, *Roads to Rome: The Antebellum Protestant Encounter with Catholicism* (Berkeley: University of California Press, 1994). Both studies focus on the antebellum period.

12 Jeanne Petit, "'The Watering of the Nation's Life-Blood': Manliness, Race, Citizenship and the Immigration Restriction League," unpublished paper, p. 1.

13 Henry Pratt Fairchild, *Immigration: A World Movement and Its American Significance* (New York: Macmillan Company, 1913), p. 397. On the racist cast of immigration restriction movements, see John Higham, *Strangers in the Land: Patterns of American Nativism, 1860–1925* (New Brunswick: Rutgers University Press, [1955] 1994).

14 Charles Darwin, *The Descent of Man*, is quoted to this effect in Degler, *In Search of Human Nature*, p. 15.

15 Joseph LeConte, "The Factors of Evolution," *The Monist*, 1 (1890–91):334.

16 Francis Clement Kelley, *The Bishop Jots It Down* (New York and London: Harper & Brothers, 1939), pp. 79–80; also see W. R. Thompson, "Evolution," *Catholic World*, 34 (1882):683–92.

17 John Rickard Betts, "Darwinism, Evolution, and American Catholic Thought," *Catholic Historical Review*, 45 (1959):174.

18 Pope Pius XII, "The Encyclical Letter *Humani Generis*," in *The Teaching of the Catholic Church*, eds. Joseph Neuner and Heinrich Roos (New York: Alba, 1978). This papal teaching "settled the matter" of evolution by claiming that it is acceptable within the framework of Christian doctrine as long as Catholics confess the individual creation of each human soul, the authority of revelation in speaking to us of the source of our being, and the unity of the human race.

19 On Thomism and neoscholasticism, see Gerald McCool, *Catholic Theology in the Nineteenth Century: The Quest for a Unitary Method* (New York: Crossroad, 1977); and Gabriel Daly, *Transcendence and Immanence* (Oxford: Clarendon Press, 1980).

20 F. P. Garesché, *Science and Religion: The Modern Controversy* (St. Louis: n.p., 1876), pp. 8–9; "Draper's Conflict between Religion and Science," *Catholic World*, 21 (1875):179.

21 Camillo Mazella, *De Deo Creante: Praelectiones scholastica dogmaticae* (Woodstock, Md.: n.p., 1877), p. 344. For a similar comment, see "The Descent of Man," *Catholic World*, 26 (January 1878):508, 511.

22 Mazella, *De Deo Creante*, p. 307.

23 Michael Gannon, "Before and after Modernism: The Intellectual Isolation of the American Priest," in *The Catholic Priest in the United States: Historical Investigations*, ed. John Tracy Ellis (Collegeville, Minn.: St. John's University Press, 1971), p. 314.

24 Henry F. Brownson, ed., *The Works of Orestes A. Brownson*, 20 vols. (Detroit: Thorndike House, 1884), 9:520–25.

25 See *Freeman's Journal*, 28 May 1898, p. 4; "A Protestant View of Christianity," *Catholic World*, 55 (1892):770. See also H. H. Wyman, "Science and Faith," *Catholic World*, 71 (1900):5, 8; George McDermot, "Spencer's Philosophy," *American Catholic Quarterly Review*, 26 (1901):658; and William Poland, S. J., "Modern Materialism and Its Methods in Psychology," *Ecclesiastical Review*, 17 (1897):150.

26 Morrison, "A History of American Catholic Opinion," pp. 181–82.

27 Arthur Preuss, "Darwin's Unprovable Theory," *Fortnightly Review*, 29 September, 1898, p. 3.

28 The best account of the Americanist episode remains Thomas T. McAvoy, C.S.C., *The Great Crisis in American Catholic History, 1895–1900* (Chicago: Henry Regnery Co., 1957). See also Margaret Mary Reher, "Americanism and Modernism – Continuity or Discontinuity?" *U.S. Catholic Historian*, 1 (1981):86–100.

29 David O'Brien, *Isaac Hecker: An American Catholic* (Mahwah, NJ: Paulist Press, 1992), p. 336.

30 John Tracy Ellis, *The Life of James Cardinal Gibbons, Archbishop of Baltimore, 1834–1921*, 2 vols. (Milwaukee: Bruce Publishing Co., 1952), 2:101.

31 "The Church and the Age," in John Ireland, *The Church and Modern Society* (Chicago: D. H. McBride and Co., 1896), pp. 97–98.

32 Ireland's introduction to Elliott's *Life of Father Hecker* was reprinted in Felix Klein, *Americanism: A Phantom Heresy* (Crawford, N.Y.: Aquin Book Shop, 1951), pp. xiii–xxi.

33 John Lancaster Spalding, "Catholicism and A.P.A.ism," *North American Review*, 154 (1894):284.

34 Spalding, "University Education," quoted in David Francis Sweeney, *The Life of John Lancaster Spalding* (New York, N.Y.: Herder and Herder, 1965), p. 186.

35 Quoted in John Tracy Ellis, *The Formative Years of the Catholic University of America* (Washington, DC: American Catholic Historical Association, 1946), p. 113.

36 Selinger, quoted in Morrison, "A History of American Catholic Opinion," p. 183.

37 Quoted in Morrison, "A History of American Catholic Opinion," p. 70. There was at least one unrepentant American Catholic Darwinist, William Seton, grandson of Elizabeth Bayley Seton, novelist, philanthropist, and paleontologist. See William Seton, "The Hypothesis of Evolution," *Catholic World*, 66 (1897):201; and "How to Solve One of the Highest Problems of Science," *Catholic World*, 58 (1894):788.

38 Augustine F. Hewit, "Scriptural Questions," *Catholic World*, 44 (1887):660, 677. Hewit, it should be noted, accepted the notion of evolution only for nonhuman species. He thought that evolutionists, such as Mivart, who extended evolution to humanity, gave support to atheists who would use science to subvert religion; they replaced the creature sublime in the image of God with "a stupid and vicious beast."

39 George M. Searle, "Evolution and Darwinism," *Catholic World*, 56 (1897): 227.

40 Thomas Dwight, "Science or Bumblepuppy?" *American Catholic Quarterly Review*, 12 (1887):636–46.

41 Rev. John Gmeiner, *Modern Scientific Views and Christian Doctrines Compared* (Milwaukee: J. H. Yewdale and Sons, 1884), pp. 3–4, 157. Italics in original.

42 Ibid., pp. 51–52, 159.

43 "If one compares the views of St. Augustine with the speculations of Darwin," he wrote, "one might be tempted to look upon St. Augustine as the venerable teacher who advanced some grand comprehensive ideas, which his disciple Darwin has explained more in detail." Ibid., p. 157.

44 Ibid., p. 158.

45 Ibid., pp. 165–70.

46 Gmeiner, "The Liberty of Science, "*Catholic World* 48 (1888):149.

47 On the Dunwoodie curriculum, see Appleby, *Church and Age Unite!*, pp. 110–116.

48 Reher, "Americanism and Modernism," p. 99.

49 The discussion of John Zahm is adapted from my book, *Church and Age, Unite!*, pp. 27–52.

50 Quoted in the *Catholic Citizen*, 24 August, 1896, Holy Cross Archives, Notre Dame, Ind. (hereafter CSCA-ND).

51 Ralph E. Weber, *Notre Dame's John Zahm: American Catholic Apologist and Educator* (Notre Dame, Ind.: University of Notre Dame Press, 1961), pp. 10–19. Largely through Zahm's efforts, Notre Dame became the first American campus lighted by electricity.

52 John Augustine Zahm, *The Catholic Church and Modern Science: A Lecture* (Notre Dame, Ind.: Ave Maria, 1886), pp. 6–7.

53 Ibid., p. 19.

54 Ibid., p. 44.

55 Ibid., pp. 30–31ff. Zahm was in touch with Mivart; excerpts from their correspondence are available, CSCA-ND.

56 John A. Zahm, *Catholic Science and Scientists* (Philadelphia: H. L. Kilmer and Co., 1893), pp. 133–35, passim.

57 Ibid., pp. 216–17.

58 Quoted in John A. Zahm, *Evolution and Dogma* (reprinted, New York: Arno, 1978), "Introduction," by Thomas J. Schlereth.
59 Mivart to Zahm, postcard, 18 May, 1896, CSCA-ND.
60 Weber, *Notre Dame's John Zahm*, pp. 16–28.
61 Zahm received the most intensive press coverage of his career during the Chautauqua lectures. See, for example, "Dr. Zahm Denies the Universality of Noah's Flood – An Interesting Lecture Delivered before the Catholic Summer School at Plattsburgh," *The Sun*, July 1893, pp. 13–14, CSCA-ND.
62 John A. Zahm, *Bible, Science and Faith* (Baltimore: John Murphy and Co., 1894), pp. 78, 80.
63 Ibid., pp. 84–91, 313–16.
64 *Catholic World* termed it "a veritable Godsend" and recommended it for Catholics and Protestants alike "who wish accurate information concerning current controversies regarding Science and Faith." *Ave Maria* judged it to be one of the most important Catholic books published in English within the last decade. For the *American Catholic Quarterly Review* it was "the most valuable contribution made by an American talent and industry to the cause of Christian apologetics." "To timid and troubled souls," the *Sacred Heart Review* assured, "Father Zahm's book will be a great comfort and support." These excerpts were reprinted in "Opinions of the Press," the *Tablet*, December 1894, p. 28, copy in CSCA-ND. The *Baltimore Sun* praised it as a "most important and learned contribution toward the adjustment of some divergent views of the present day as to the relation of the Bible and modern investigations in the field of the natural sciences." Newspapers in Boston, Buffalo, and London applauded Zahm as "a man of the age, cognizant of all the movements of modern thought and able to give a reason for the faith that is in him." Reprinted in "Review of *Bible, Science and Faith*," *The New World*, 12 January 1895, pp. 30–31, CSCA-ND. The *Independent* perceived Zahm's defense of science as a crusade in which learned Catholics and Protestants might make common cause. One of the most comprehensive reviews, published in the *Tablet*, noted that Zahm's was the first complete work on this theme to appear in English. When European evolutionists' conclusions seemed to depart from the outlines of "true Catholic conservatism," Zahm demonstrated that the church left open to discussion the particular point in question, or he defended the disputed conclusions by citing respected Catholic authorities of the past. "Combining the indications afforded by physical science, by historical archaeology, and by Holy Scripture, [Zahm] reaches the provisional conclusion that man has probably been on the earth not less than 10,000 years; and that there is no evidence to prove a higher antiquity than this," the *Tablet* reported. "The Bible and Science: Review of *Bible, Science and Faith*," *Tablet*, December 1824, p. 27, CSCA-ND.
65 Zahm, *Evolution and Dogma*, pp. 435–38.
66 Quoted in James R. Moore, *The Post-Darwinian Controversies* (Cambridge: Cambridge University Press, 1979), p. 194.
67 Zahm, *Evolution and Dogma*, p. vii.

68 Ibid., p. v.
69 Ibid., pp. xvii, 69–70, 433.
70 Ibid., pp. 50–53.
71 Ibid., p. 75.
72 In the second edition of *The Origin of Species*, for example, he had revised a previous estimate that all animal and plant life derived from four or five progenitors, acknowledging that "all organic beings which have ever lived on the earth have descended from some one primordial form, into which life was first breathed by the Creator." Zahm found this adjustment to be in keeping with scientific fact and thus in close conformity to revealed truth. At the same time, Zahm lashed out at neo-Darwinists for regarding natural selection as the sole and sufficient cause for all organic development even as Darwin was reducing its role by allowing for environmental factors. Ibid., p. 83, 370.
73 Ibid., pp. 200–01.
74 This was a strategy of theistic evolutionists in general. Their softening of Darwinian theory reduced the level of intellectual tension or "cognitive dissonance" between the rival ways of knowing the world presented by modern science, on the one hand, and traditional theism, on the other. See Neal C. Gillespie, *Charles Darwin and the Problem of Creation* (Chicago: University of Chicago Press, 1979), pp. 1–18, 85–108. See especially John C. Greene, *Darwin and the Modern World View* (Baton Rouge: Louisiana State University Press, 1961); and *The Death of Adam: Evolution and Its Impact on Western Thought* (Ames, Iowa: Iowa State University Press, 1959).
75 See John D. Root, "The Final Apostasy of St. George Jackson Mivart," *Catholic Historical Review*, 71 (1985):1–25, for a statement of Mivart's development, and especially for his reaction to Leo XIII's encyclical on biblical interpretation, which, Root argues, "provided a sharp impetus" to Mivart's final break from the church (pp. 7–8).
76 Zahm was in correspondence with Mivart, who thanked him for "carrying on my work in the United States"; Mivart to Zahm, 18 May, 1896, CSCA-ND.
77 Zahm, *Evolution and Dogma*, p. 70.
78 Ibid., p. 122.
79 Zahm, *Bible, Science and Faith*, pp. 121–22; Zahm, *Evolution and Dogma*, pp. 388–90.
80 Zahm, *Evolution and Dogma*, p. xiv.
81 On this point, see George M. Searle, "Dr. Mivart's Last Utterance," *Catholic World*, 71 (1900):353–65; and Searle, "Evolution and Darwinism," *Catholic World*, 66 (1897):227–38. Also compare Weber, *Notre Dame's John Zahm*, pp. 109–12.
82 On responses to *Providentissimus Deus*, see Roger Aubert, ed., *The Christian Centuries*, vol. 5: *The Church in a Secularized Society* (New York, 1978), pp. 164–203.
83 A periodical that otherwise supported Zahm, *The Colorado Catholic*, admitted as much: "when Fr. Zahm tells us that St. Augustine was an incipient

evolutionist ... he doubtless shocks some good people ... who resent as impious any attempt to connect sacred names with so irreligious a thing," 18 February, 1896, pp. 16–17, CSCA-ND.

84 "Dr. Zahm," *The Review*, 23 April 1896, p. 2.

85 Salvatore Brandi, S. J., to Archbishop Corrigan, 13 June, 1896, John Zahm Collection, University of Notre Dame Archives [UNDA].

86 John Zahm to Albert Zahm, 6 December, 1896, John Zahm Collection, UNDA.

87 John Zahm to James Cardinal Gibbons, 9 March, 1897, copy in UNDA.

88 The edict is quoted in Weber, *Notre Dame's John Zahm, p.* 107.

89 Francais to Zahm, 10 November, 1898, John Zahm Collection, UNDA.

90 Lester R. Kurtz, *The Politics of Heresy: The Modernist Crisis in Roman Catholicism* (Berkeley: University of California Press, 1986), p. 13. For a thorough discussion of the neoscholastic world view and the ways in which it was threatened by reliance on secondary causes in the manner described by Zahm and others, see Gabriel Daly, *Transcendence and Immanence: A Study in Catholic Modernism and Integralism* (Oxford: Clarendon Press, 1980).

91 "Leone XIII e L'Americanismo," *La Civiltá Cattolica*, ser. 17, vol. 5, 18 March, 1899, pp. 641–43.

92 "Evoluzione e Domma," *La Civiltá Cattolica*, ser. 17, vol. 5, 7 January, 1899, pp. 34–49.

93 Ibid., pp. 40–41.

94 Weber, *Notre Dame's John Zahm*, p. vii.

95 Francis Howard, "Catholicism, Protestantism, and Progress," *Catholic World*, 62 (1895):145–53. Also see S. Fitzsimmons, "The Rise and Fall of Evolution by Means of Natural Selection," *American Catholic Quarterly Review*, 26 (1901):87–107.

96 William Poland, S. J., "Modern Materialism and Its Methods in Psychology," *American Ecclesiastical Review*, 17 (1897):150–1.

97 Degler, *In Search of Human Nature*, p. 48.

98 Ibid., p. 53.

99 Ibid., p. 55.

100 Attwater, ed., *The Catholic Encyclopedic Dictionary*, p. 190

101 "Editorial," *The Ecclesiastical Review*, 15 September, 1911:276–279.

102 Thomas Dwight, "Description of the Human Spines Showing Numerical Variation in the Warren Museum of the Harvard Medical School," *Memoirs of the Boston Society of Natural History*, 5 (1901):242–43.

103 Thomas Dwight, *Thoughts of a Catholic Anatomist* (New York: Longmans, Green, 1911), p. 12.

104 J. B. Ceulemans, "The Modern Schools: Evolutionism," *The Ecclesiastical Review*, 7 (1912):257.

105 Roman Catholics were not alone, of course, in worrying about the negative consequences of modern materialism; Protestant and Jewish (and some secular) intellectuals shared many of the same general concerns. Among many sources, see Frederick Gregory, "The Impact of Darwinian Evolution on Protestant Theology in the Nineteenth Century," in *God and Nature: Historical Essays on the Encounter between Christianity and Science*, eds. David C. Lindberg and Ronald L. Numbers (Berkeley and Los Angeles: University

of California Press, 1986), pp. 369–90; Jon H. Roberts, *Darwinism and the Divine in America: Protestant Intellectuals and Organic Evolution, 1859–1900* (Madison: University of Wisconsin Press, 1988); and Marc Swetlitz, "American Jewish Responses to Darwin," in this volume.

106 The gradual and measured incorporation of the higher criticism of the Bible into Catholic teaching began in the 1930s and 1940s, buoyed by Pius XII's encyclical *Divino Afflante Spiritu*, a cautious endorsement of scientific criticism of the Bible.

107 James J. Walsh, *A Catholic Looks at Life* (Boston: Stratford, 1928), p. 29.

108 Karl Aloisneller, S. J., *Christianity and the Leaders of Modern Science*, trans. T. M. Kettle (Freiburg: B. Herder, 1911), p. 366.

8

American Jewish responses to Darwin and evolutionary theory, 1860–1890

MARC SWETLITZ

> Darwinism declares ... that creation is not to be explained through a miracle but through the natural law of progressive development of life under favorable circumstances.... Does not this view of life ... harmonize perfectly with our comprehension of religion, which we do not recognize in form, but in reform, which has its living power in the internal remodeling of Judaism and its Messianic mission as progress toward a completed humanity?[1]
>
> —Kaufmann Kohler, 1874

During the 1870s and 1880s, American Jews discussed Darwin and scientific theories of evolution in the context of debates within the Jewish community on the future of American Judaism. Kaufmann Kohler, a prominent American Reform rabbi, appealed to Darwin and evolutionary theory as support for radical Reform Judaism, which emphasized the idea of progressive revelation. The legitimacy of radical Reform, however, was subject to much debate. More moderate Reform rabbis as well as traditionalist Jewish leaders criticized Kohler's support for evolution and the links he made between evolution and radical Reform. Moreover, traditionalists who accepted evolution also attacked Kohler

My research and writing have benefited from the help of several individuals and institutions. I am especially indebted to Jonathan Sarna for helping me into the world of nineteenth-century American Jewry and for his comments on earlier drafts. Robert Singerman's vast knowledge of nineteenth-century American Jewish literature saved me much time and wasted effort, and I thank him for that. The assistance of the staff at the Goldfarb Library at Brandeis, the Hebrew Union College Library, and the American Jewish Archives was invaluable. Comments and questions by those who heard various presentations of this material at the Dibner Institute for the History of Science and Technology, Brandeis University, the American Jewish Archives, and the "Responding to Darwin" conference in New Zealand helped to stimulate my thinking and refine my presentation. Finally, this project was made possible only through a fellowship from the Dibner Institute for the History of Science and Technology and a Bernard and Audre Rapoport Fellowship from the American Jewish Archives.

and his fellow radicals, claiming that evolutionary theory actually gave more support to traditional Judaism!

Intracommunity rivalry was certainly not alone in shaping the responses of Jews to evolution. The American Jewish community faced many challenges from without that threatened the integrity and survival of Judaism. Religious leaders pointed to assimilation as the community's most urgent problem, whether it resulted from conversions to Christianity, the appeal of the Ethical Culture Society (founded in 1876 by Felix Adler, son of a prominent Reform rabbi), or the antagonism to religion generated by the growth of agnostic and materialistic thought.[2] Each of these factors had been linked to evolutionary theory – Christians used evolutionary theory to argue for the superiority of Christianity over Judaism; Adler had invoked modern science and evolution to support Ethical Culture; and public scientists, such as Thomas Henry Huxley, John Tyndall, and Ludwig Büchner, connected evolution with agnosticism or materialism.[3] Therefore, when rabbis and lay leaders discussed evolution, they were typically more concerned with the specter of materialism and its supposed effects on lowering synagogue attendance than with the details of evolutionary theory. Jewish leaders, in the words of historian Naomi Cohen, aimed "to refute the skeptic" both inside and outside the Jewish community.[4]

The relationship of responses to evolution and debates about Judaism *within* the Jewish community, however, deserves more attention. Previous scholars have examined only Reform responses, focusing on the few rabbis who enthusiastically supported evolution.[5] The opposition to evolution among Reform Jews, in contrast, is virtually unexplored territory.[6] Moreover, Reform Jews were not the only Jews who had the motivation, leisure, or intellectual resources to address evolutionary theory. Almost every Jewish newspaper published between 1870 and 1890 that I examined contained editorials, sermons, or reprints from the non-Jewish press on Darwin and evolution.[7] In this chapter, I chart the chronology of Jewish responses and pay special attention to the relationship of their quantity and substance to debates about the future of Judaism in America. The first section examines the paucity of responses to evolution in the 1860s and the widespread opposition that emerged after the publication of Charles Darwin's *Descent of Man*. In the second section, I explore the initial support for evolutionary theory by Kohler and other Reform rabbis in the mid-1870s and the strong reaction of Rabbi Isaac Mayer Wise. I then turn to a broader sketch of the spectrum of American Jewish responses to evolution from the mid-1870s through the mid-1880s. The fourth section examines two episodes in the 1880s that illustrate the links between discussions about evolution and growing tensions between Reform Jews and traditionalists. In the penultimate section, I briefly assess the fate of evolutionary ideas among Reform and traditionalist Jews in the 1890s. Finally, I summarize the chronology and pattern of Jewish responses and conclude with a brief comparison with responses to evolution by American Protestants.

Before beginning, I must provide some introductory remarks about the American Jewish community and Judaism in the last four decades of the nineteenth century and indicate the scope and limits of my study.[8] My research focuses on the rabbinic and lay leaders of two groups of Jews, that is, Reform and traditionalist. The Jewish community was largely an immigrant population, most of whom, although not all, came from the German-speaking countries of central Europe. While the majority of Jews were not well educated, some Reform and traditionalist rabbis had received a Ph.D. at a European university, typically involving research in languages, literature, or history. All had some interest in a modern, critical study of texts and Jewish history, but, at the same time, were wary of the way in which such scholarship had been used to undermine religious belief. Nevertheless, Reform and traditionalist Jews alike were accommodationists. They believed that new forms of Jewish life and thought were necessary to mediate between their desire for acculturation and their commitment to preserve a distinct Jewish identity and community in America.[9]

Reform and traditionalist Jews differed regarding the precise form that this accommodation should take. In both groups, a spectrum of beliefs and practices existed, resulting in some overlap. Reform rabbis considered belief in one God and the moral law as the essential spirit of Judaism; however, they also had much to disagree about, e.g., the nature of revelation, the validity of biblical criticism, and the need to preserve traditional ceremonies even if they were not binding. Radicals tended to be more innovative than moderates in theology and practice, with a few even switching observance of the Sabbath to Sunday since many Jews had to work on Saturday. Traditionalist Jews differed from Reform Jews most essentially in their view that *all* commandments – ritual and ceremonial as well as moral – were equally binding on the Jewish people. While this did not entail a rejection of change, preservation was as much a concern as innovation. At the same time, a spectrum of belief and practice existed among traditionalists, which became the historical antecedent for the institutional division between Conservative and modern Orthodox in the twentieth century.[10]

In the second half of the nineteenth century, however, the institutional structure of Jewish religious life was quite different. While Reform and traditionalist leaders had long engaged in polemics regarding the proper direction of American Judaism, no effort to institutionalize such differences occurred prior to the 1880s. Indeed, Rabbi Wise, the leading advocate of moderate Reform, organized both the Union of American Hebrew Congregations (UAHC) in 1873 and the Hebrew Union College (HUC) in 1875 with the hope that they would unite and serve all American Jews, a hope initially shared by most traditionalist leaders. Such hopes soon faded. Radical Reform rabbis, less willing to compromise with traditionalists, increased in numbers and strength during the 1870s. At the decade's end, a group of young, articulate, and vocal traditionalist Jews emerged, who aimed to reinvigorate Judaism and who attacked Wise as well as other HUC faculty for denying the binding nature of the full corpus of Mosaic legislation. Eventually, in 1886, after five years of strong polemics and organizational efforts, traditionalists announced the opening of the Jewish Theological Seminary of New York, which in the words of its first president, Rabbi Sabato Morais, would preserve what he called "Historical and Traditional Judaism."[11]

This article focuses on Reform and traditionalist Jews, who together constituted a large majority of American religious Jews prior to the early 1880s.[12] At the same time, there existed another group of religious Jews who had emigrated from Eastern Europe and Russia and remained strongly attached to the lifestyle, language, and religious observance of their homeland. While a small minority in the 1870s, their numbers dramatically increased due to the massive immigration of Eastern European and Russian Jews to the United States beginning in 1881. Very resistant to any sort of accomodation with American life and modern scholarship, they strongly opposed the founding of the Jewish Theological Seminary, which they believed would only hasten the assimilation of Jews and the disappearance of true Judaism.[13] Responses of these Jews to evolution, however, have yet to be explored by historians and are not included in this paper.

From no response to vocal opposition: 1860 to 1873

Prior to 1874, American Jewish opinion predominantly opposed the transmutation of species. However, during the 1860s there was virtually no direct response to Darwinism at all. No mention of evolution appeared in the traditionalist newspaper *The Occident*, which stopped circulation at the decade's end. *The Jewish Messenger*, edited by the traditionalist Rabbi Samuel Myer Isaacs, first mentioned the topic in 1869, when it carried a supportive review of *Man in Genesis and Geology*, written by the Protestant cleric Joseph P. Thompson, who opposed the evolution theory.[14]

Discussions of evolution appeared only slightly more frequently in Reform newspapers. Rabbi Isaac Mayer Wise, editor of *The American*

Israelite and leader among moderate Reform Jews, published four articles or notices on science and religion during the 1860s that mentioned evolution. In 1868 Wise himself penned an editorial note, pointing to Christoph Aeby's *Die Schadelformen des Menschen und der Affen* (which discussed the differences between human and ape skulls) as providing empirical evidence against "all gorilla theories of modern naturalists." The next year, a new voice for radical Reform appeared, *The Jewish Times*. The newspaper reprinted a sermon by America's leading radical Reform rabbi and theologian, David Einhorn. While not explicitly referring to organic evolution, he spoke at length about the special creation of the human body and soul. At stake for Einhorn was a central theological doctrine: the creation of human beings in the image of God.[15]

The relative paucity of responses to Darwin's theory of evolution in the 1860s compared to the number of responses in the 1870s and afterward was not limited to the Jewish community. The historian Jon Roberts has found the same phenomenon among American Protestants, attributing it to their confidence that the scientific community would, after brief consideration, detect the errors in Darwin's theory and repudiate it.[16] A similar reason might have also applied to Jews. Some opponents of evolution, when writing in the 1870s and 1880s, maintained that the acceptance of evolution was only temporary and that investigation would eventually prove it a false theory.[17]

The American Jewish community, however, was particularly indifferent to the subject of organic evolution in the 1860s. No single article or sermon devoted primarily to evolutionary theory appeared in print before 1871. Offering explanations for such nonevents is fraught with difficulties, especially given the absence of systematic information about Jewish responses to evolutionary theory in other countries. Nevertheless, several characteristics of mid-nineteenth-century American Jewry are worth mentioning as possible factors to explain the lack of Jewish responses during the 1860s. These characteristics will also help provide a context in which to understand Jewish responses to Darwin when they eventually appeared.

American Jews did not possess a high degree of literacy in scientific subjects related to organic evolution – paleontology, geology, morphology, and embryology. While some rabbis had received a Ph.D. degree, they did not write their dissertations on scientific topics. Moreover, there were no institutional motivations to acquire such learning. Jewish seminaries in America as well as in Germany did not teach in the natural sciences. In addition, American Jews did not engage in research in areas related to evolutionary theory in the 1860s, and those that eventually did were not very involved in Jewish community life. The first such scientist was Angelo Heilprin, who studied at the Royal School of Mines in the 1870s, taking courses with Thomas Henry Huxley. He then returned to become professor of invertebrate paleontology at the Academy of Natural Sciences of Philadelphia in 1880. Heilprin, however, wrote no book or article on issues relating his studies to Judaism or even religion generally.[18]

Therefore, as a community, American Jews approached the subject of evolution very much as outsiders. Perhaps the most scientifically literate among them were physicians; indeed, a physician wrote the first full-length article on evolution to appear in the Jewish press.[19] This general condition provided ammunition for those Jews who accused their co-religionists of lacking the knowledge necessary for judging the strengths and weaknesses of evolutionary theory. It also led to a reliance on Christian authors for articles addressing the topic of evolution, articles that were sometimes written expressly for a Jewish newspaper but more often were reprinted from non-Jewish sources.[20] The lack of knowledge made it difficult for Jews to participate in technical discussions that seemed to dominate the 1860s.[21] In addition, this might have reinforced the position that there was no need to rush and take a position on evolution when the outcome of the debate among scientists was still quite uncertain.

A second characteristic of mid-nineteenth-century American Jewry was the absence of a Paleyite tradition of natural theology that would have been threatened by Darwin's theory of natural selection. When writing about nature and nature's God, Jews more often developed arguments based on nature's beauty, order, and regularity, arguments that could better withstand the challenge of Darwin's evolutionary theory.[22] At the same time, Jewish writings did include reference to the argument from organic design. Interestingly, this type of argument began to appear more frequently in the American Jewish press after 1860. This resulted in large part through the efforts of the London-based Jewish Association for the Diffusion of Religious Knowledge, which printed two pamphlets, in 1861 and 1863, widely reprinted in the United States. The first pamphlet included a selection by the medieval Jewish philosopher Bachya ibn Pakuda, whom the editor compared favorably to William Paley: "it is a singular fact, that in his *Natural Theology*, Paley, who wrote about six centuries after our author, follows a precisely similar line of argument deduced from a like illustration."[23] Ironically, the most immediate response to the publication of the *Origin of Species* appears to have been increased attention to Paleyite design arguments.

American Jewry in the 1860s was largely an immigrant community, with ties to both Britain and German-speaking countries. While the links with Britain worked to encourage a greater emphasis on nature, the connections of American Jews to Germany and its intellectual traditions worked in the opposite direction. The historian Frederick Gregory has examined a group of nineteenth-century German Protestant theologians, influenced by Kantian philosophy, who accepted a radical separation between reason and science, on the one hand, and morality or feeling and religion, on the other hand. For these theologians, the controversy over evolutionary theory did not challenge the foundations of religious faith.[24] Some German Jews too fell under the influence of Kant and his successors in Germany, and this tradition came to America with the migration of German Jews in the 1840s and 1850s.

Kantian philosophy especially influenced the formulation of the theology of Reform Judaism, which emerged in early nineteenth-century Germany. In brief, Kant helped to shape the core convictions of Reform Judaism: a belief in one God and a commitment to His moral law. Moreover, accepting Kant's critique of natural theology, Reform Jews looked to conscience, feeling, or intuition as the ultimate source of belief in God and knowledge of the moral law.[25] The impact of this theology is strikingly illustrated by the case of Rabbi Bernhard Felsenthal, who immigrated to the United States in 1854. In a widely publicized sermon in 1870, he argued that Judaism demanded a commitment to only two claims: the existence of "one supernatural and preternatural God" and of eternally valid "moral laws." So conceived, Judaism gave "the fullest and most unlimited freedom" to scientific inquiry. He maintained that "a [Charles] Darwin and a [Thomas Henry] Huxley, a [Carl] Vogt and a [Jacob] Moleschott," speculators on "the origin of species," could proceed "without meeting any interference or any protest" from Judaism.[26] Indeed, Felsenthal interfered so little that he never discussed scientific issues throughout his long and prolific career.

While few Reform Jews followed Felsenthal in strictly separating science from Judaism, he exemplifies a tendency among nineteenth-century American Jews: to focus on human beings – morality, mind, and history – rather than the rest of the natural world when discussing God and Judaism. Whether due to a lack of knowledge about natural history, the absence of a strong tradition of natural theology, or the more specific influence of Kantian philosophy, American Jews rarely wrote on the topic of evolution in the 1860s. When they did, they focused on human evolution. Most of this occurred in the late 1860s, in the form of a brief comment on the subject or a brief notice about a scientist opposed to evolution. Only after the publication of Darwin's *Descent of Man* in 1871 did American Jews begin to devote time and attention to scientific theories of evolution.

It remains uncertain why it was Darwin's book that stimulated American Jewish responses, rather than Thomas H. Huxley's *Evidence as to Man's Place in Nature* (1863), Ernst Haeckel's *Naturliche Schopfungsgeschichte* (1868), or the many articles by American scientists and

Protestant clerics about human evolution. The few references to evolution that appeared in the 1860s clearly indicate the American Jews were aware of an ongoing debate about human evolution. Perhaps they hoped that Darwin would not take sides on the matter, striking a blow against those in favor of applying evolution to human beings. Jews viewed Darwin as a cautious, empirical, and believing naturalist, in contrast to the speculative, philosophical, and materialist scientists such as Huxley and Haeckel. Indeed, Wise, American Jewry's harshest and most persistent critic of evolutionary theory, suspected that Darwin would not have written the *Descent* if he had not been pressed by his German disciples and admirers, the materialists Vogt, Moleschott, and Haeckel.[27] When Darwin finally entered the controversy in support of human evolution and applied his theory of natural selection to the human mind, the moral sense, and human history, the challenge could no longer be ignored.

Just a few weeks after the publication of Darwin's *Descent of Man*, the radical Reform *Jewish Times* began a twelve-part series entitled "The Origin and Antiquity of Man," written by Adolph Kessler, a New York physician. In the first and second parts, Kessler offered an extended critique of theories of human evolution. He argued that human beings were anatomically, spiritually, and culturally unique and concluded that the gap between humans and animals could not be bridged. In addition, he raised the specter of philosophical materialism, denouncing scientific assertions that humans differed from apes only in degree as an "exuberant outgrowth of the hypermaterialistic tendencies of our age and scientific research."[28] Over the next couple of years, articles, editorials, and letters to the editor criticizing evolution appeared in *The Jewish Times, The American Israelite*, and *The Jewish Messenger*. These publications included reprints from British and German sources as well as contributions by American rabbis and laymen.[29]

While a few writers offered criticisms of Darwin's reasoning and methodology – e.g., he ignored Baconian inductivism and made unwarranted assumptions – most Jews focused on the problem of human evolution, and the bulk of their discussion centered on mind and morality.[30] The human mind, with its rational capacities and its ability to improve the human condition over time, was unique in creation; no animal possessed a mind that shared these qualities. Such critics regarded religious feeling as a distinctively human character and ridiculed Darwin's attempt to link this human attribute with a dog's reverence for its master.[31] References to morality usually focused on Darwin's theory of natural selection, which was universally presented as leading to egoism and an ethic of might makes right. One writer contrasted the Jewish ethic of compassion and aid to the poor and needy with passages in the *Descent* that referred to the injurious effects of civilized society on the process of natural selection, and they took this contrast as ample reason for rejecting Darwin's theory. Another understood the moral conscience as a unique attribute of the human mind and, therefore, as evidence against the fact of human evolution.[32]

Amidst this strong opposition to Darwin and human evolution, an occasional note of reconciliation could be heard. Some who rejected evolution nevertheless asserted that if the doctrine of evolution proved true, it would in no way subvert Jewish belief in God's existence and unity and in the supremacy of human beings over the rest of creation. For example, Rabbi Solomon Sonneschein, editor of *Die Deborah*, the German-language edition of *The American Israelite*, argued in 1873 for distinguishing between Darwinism and philosophical materialism. Jews opposed all forms of materialism, but he thought Darwinism, by which he meant the transmutation of species, could be reconciled with the activity of a divine intelligence. Yet, in the end, he maintained that the evolution of species, and especially of human beings, was not yet proven.[33] By concluding that evolution was not true, Sonneschein and others obviated any need to grapple with the reconciliation of Jewish belief and evolutionary theory.

The only place to find outright acceptance of evolution was the pages of the recently founded radical Reform weekly, *New Era*, whose motto was "The Voice of Reason is the Voice of God." In the first issue, published in 1870, the Jewish chemist Isidor Walz wrote an article, "The Modern View of Nature," which briefly described Darwin's evolutionary theory and proclaimed it true. In subsequent issues, the editor, Rabbi Raphael D'C. Lewin, occasionally included small notices about discoveries or books about evolutionary ideas. Yet no one *argued* for evolution (they simply endorsed it and reported research in its favor) and no one ever explained how evolution served as the "voice of God."[34] Their acceptance of evolution appeared to involve little struggle and to be of little concern.

Disagreement in the American Reform community: 1874

This changed in January, 1874, when Rabbi Kaufmann Kohler delivered a sermon on science and religion at his first Sunday service at Sinai Congregation in Chicago. He had immigrated to America from Germany in 1868 and had quickly become a leader in the radical Reform community. Kohler's sermon, reprinted in *The Jewish Times*, introduced a number of themes that became standard fare in sermons and talks by Reform Jews supportive of evolution. In the first half of the sermon, he expressed his general views on the relationship of science and religion, which were very similar to the views expressed earlier by Felsenthal. Scientific truth derived from reason applied to phenomena; religious truth, from human feeling and conscience. Kant had established the limits to human reason, denying the validity of any philosophical materialism, and, in Kohler's interpretation, had also established the existence of a moral order beyond nature. Reflecting the lessons in critical biblical study that he had learned while writing a Ph.D. thesis, Kohler maintained that the opening chapters of the Bible contained "popular legend and poems." The truth of Genesis lay in the spirit of the Bible rather than in the letter, and this involved faith in one God and the eternal moral law.

Kohler then turned to Darwinism, which he defined as "the natural law of progressive development of life under favorable conditions." Ignoring the theory of natural selection, he rested his acceptance of evolution in part on the fact that "almost all investigators" in related fields supported it. Equally important, he held that the essential feature of evolution was change through natural law as opposed to miraculous creation. For support, he appealed to the authority of medieval Jewish philosophers who had rejected the supernatural and who had taken the order of nature as "sufficient testimony of eternal omnipotence." Kohler retained a unique status for human beings by arguing that they possessed "more of the creative power and the creative spirit" than any other animal. Indeed, evolution in human history had been as much spiritual as material. This led Kohler to his final claim: that evolution supported the Reform understanding of Judaism, which admitted progress in morality and conceptions of God in biblical and postbiblical times. Indeed, the ability of Reform Judaism to remodel ritual and theology in accord with modern science and at the same time to retain the essence of Jewish teaching – God and morality – meant the ultimate triumph of Reform as the religion of humanity.[35]

Kohler's sermon appears to have encouraged others to support evolution. Five days later, Moritz Ellinger, editor of *The Jewish Times*, changed his position. Earlier a cautious critic of evolution, he now wrote an editorial enthusiastically endorsing Kohler's arguments. In March, 1874, Sonneschein also moved toward a more forthright endorsement of evolution theory. He published a short essay in *The American Israelite* and in *The Jewish Messenger* reporting on what he understood as a foreshadowing of evolution theory in the works of the fifteenth-century Jewish philosopher Joseph Albo. Moreover, Sonneschein argued that Judaism, unlike Christianity, tolerated freedom of inquiry, even to the extent of denying the special creation of human beings. Finally, in July, Jacob Mayer, a Reform rabbi in Cleveland, gave a sermon supporting the evolution theory, also mentioning Albo. He stressed that evolutionary theory, rather than the special creation of Adam and Eve and their fall, provided a natural foundation for belief in the progress exhibited in human history and in the history of Judaism. The not-so-subtle implication was that traditional Christianity remained out of step with recent discoveries in modern science.[36]

What precisely moved Kohler, Ellinger, Sonneschein, and Mayer to support the evolution of species, and at this particular time, remains unclear. Kohler's sermon and the responses of his co-religionists reveal a set of assumptions about religion and Judaism that would have made acceptance of evolution less problematic for them than for traditionalists: science and religion as separate realms; the Bible as myth and poetry; revelation as progressive; God as operating through natural law rather than miracles. At the same time, acceptance of evolutionary theory in the hands of these Jews served as a cultural authority for the Reform version of Judaism as well as for the superiority of Judaism over Christianity as the religion best suited to the modern age of science. The acceptance of evolution was thus one part of a larger rhetoric of triumphalism that was current in Reform Jewish apologetics.[37]

Yet Kohler and his colleagues left themselves vulnerable to attack, for they had not addressed any of the issues that opponents of evolution had earlier raised: the relationship of animal and human minds, the origin of the conscience, and the moral and teleological implications of the theory of natural selection. In brief, they had accepted the fact of evolution on the authority of scientists, ignored disputes among scientists about the mechanisms of evolution, and argued for the compatibility of evolution – as the progressive development of life – with Judaism as they understood it.

Isaac Mayer Wise stepped into the breach. Wise became the most vocal critic of evolution among Reform Jews and, indeed, among American Jews as a whole. Before Kohler's sermon, Wise had criticized evolution only in brief editorial notes. Afterward, he aggressively opposed evolution. In late January, a few weeks after Kohler's sermon, Wise delivered a lecture, "Agassiz and Darwin," which remained unpublished. Wise's first extended critique of evolutionary theory appeared six months later: he published Mayer's sermon (Mayer was an editorial consultant for Wise's newspaper) as well as his own extended critique of evolution.[38]

Wise's major response, however, began in October, 1874, when he delivered the first of twenty-two lectures over the course of several months. The scope of the lectures, published in 1876 as *The Cosmic God*, covered much more than evolution. In fact, in an effort to provide a rational foundation for theology, it encompassed all of nineteenth-century philosophy and science.[39] Wise read all he could on modern philosophy and science and visited several laboratories in order to get a sense of scientific practice.[40] This sort of preparation was necessary, for Wise felt that his Reform colleagues had not grappled with the details of scientific knowledge and hence had failed to recognize the faulty logic and erroneous assumptions that plagued the science of evolution.

Wise devoted three lectures to evolutionary theory. He pointed to several arguments against the general proposition that species evolved, including gaps in the fossil record, the lack of observations of species transformation, and embryological development, which revealed only a series of stages and no transitional forms. Darwin's theory of natural selection also came under attack because of its reliance on several unproved assumptions, e.g., unlimited variability and a struggle for existence in nature. Most of Wise's attention, however, focused on the particular problems posed by human evolution. He devoted a considerable amount of space to criticism of the adaptive stories offered by Darwin to explain particular anatomical characters, such as upright posture, the lack of a tail, and the absence of hair on large parts of the body. The most critical and emotionally disturbing issues remained those related to morality and mind. Wise understood Darwin's principle of natural selection to be a natural law, in which survival of the strongest ruled. Nature was "a battle-ground" in which "war to the knife, perpetual warfare of each against all" was the human law. This robbed the moral law as taught by Judaism of all its legitimacy. More generally, Darwin's theory, by placing human conduct under the rubric of natural law, left no room for human freedom and moral responsibility.[41]

However, Wise opposed all theories of species transmutation, not only Darwin's, and his arguments focused on the postulated continuity between animal and human minds. He illustrated his position by pointing to the discussion in the *Descent of Man* that connected the feeling of religious devotion to the love of a dog for its master. Although Darwin recognized that the two states of mind were different, he nevertheless argued that the mind of the dog represented a "distant approach" to the human mind. Wise disagreed, insisting that the "power of abstraction," essential to the religious state of mind, was unique to humans and had no precursor in the animal world. Human and animal minds bore no resemblance to one another. Because the human mind alone possessed abstract rationality, conscience, and self-consciousness, human evolution was impossible.[42]

In addition to criticizing Darwinism, Wise offered his own theory of the history of life, a theory that combined the periodic, progressive creation of the cosmos and life with a metaphysics derived from German *Naturphilosophie*. Like most scientists prior to 1859, he took the results of geology and paleontology as evidence for a series of successive creations, progressing from lower to higher forms and culminating in the creation of human beings. This did not mean that God performed periodic miracles. Wise's panentheistic God was "everywhere, in all space, in all objects of nature, in every attribute of matter and in every thought of the mind." This "Cosmic God" operated through "the continuous chain of cause and effects," and cosmic development occurred by a teleological unfolding of what was originally in God's mind. More specifically, Wise used the terms "evolution" and "differentiation" to refer to the two processes occurring in this teleological unfolding. Evolution signified the progress from lower to higher organic forms; differentiation, the diversity of species, each adapted to its proper time and place.[43]

Wise offered empirical, logical, moral, and metaphysical arguments against the truth of scientific theories of evolution, including Darwin's. However, he did not advance these arguments in a vacuum. Wise was clearly concerned to stem the tide of assimilation, which he and other Jewish leaders thought was due at least in part to promotion of atheism and materialism in the name of science. At the same time, Wise objected to evolution and created his own cosmological theory as a way to provide support for his own version of Reform Judaism.

While not referring to Judaism in *The Cosmic God*, Wise did not hesitate in other articles and sermons to link his theory of the history of life to his understanding of Judaism and the Bible. He maintained that the Jewish religion, like nature, was bound by a law of evolution, meaning progression from lower to higher forms. Wise's Reform Judaism, like that of Kohler, understood Jewish law and modes of worship as continually and progressively changing. But when he talked about principles, Wise shifted from the language of continuous progress to the language of discontinuity and stability. He never adopted the theology of progressive revelation, and he rejected critical biblical study. God had directly revealed fundamental moral principles and ideas about himself to Moses at Sinai, and Moses had composed almost all of the Pentateuch. For Wise, discontinuity and stability marked the history of life and of Judaism. Just as God created each species in a unique act and required that it remained fixed as long as it lived, so, too, did God reveal to Moses a unique message that did not change throughout Jewish history.[44] Wise's opposition to the gradual, continuous transmutation of species, therefore, was more than just a defense of human dignity, freedom, and moral duty. With his own version of the evolution of the cosmos and life, Wise had formulated a defense of his particular version of Reform Judaism.

Opposition and acceptance among American Jews: 1874–1886

Wise's lecture series, reprinted weekly in *The Israelite*, did not generate any debate. No one endorsed his philosophy of cosmic evolution and no one took the time to refute it. From the mid-1870s through the early 1880s, Wise continued to denounced evolution theory in editorials and sermons.[45] However, he was not alone in the Reform community in his opposition to the transmutation of species. Rabbi Adolph Huebsch had strongly opposed evolution in 1873. While his later sermons later emphasized that Judaism would not suffer if scientists proved evolutionary theory to be true, Huebsch never explicitly advanced the truth of evolution.[46]

In 1881, Rabbi Aaron Hahn joined the ranks of Reform opponents of evolution. Hahn, like Huebsch, argued that if scientists proved evolution true, then "Jewish theology will bow its head before the majesty of truth and adopt it without further delay." In addition, in *History of the Arguments for the Existence of God*, he discussed the argument from design at length, advocating the position advanced by Asa Gray and Noah Porter, namely, that natural theology could withstand the challenge of Darwinian natural selection. Nevertheless, Hahn firmly believed that the evolution of species was improbable, drawing his support from Wise's *Cosmic God*. He quoted passages from *The Cosmic God* that argued for a distinction in kind between animal and human minds and concluded that they "contain more common sense and more true philosophy than all the defenses of Darwinism together."[47]

While Reform Rabbis Wise, Huebsch, and Hahn continued to reject evolution, opposition was stronger and more widespread among traditionalist Jews. Samuel M. Isaacs, the antievolution editor of *The Jewish Messenger*, printed a few editorials of his own in the mid-1870s. He was joined by the editor of *The Hebrew Leader*, who also wrote several editorials criticizing evolution. In the early 1880s, the new editor of *The Hebrew Leader*, Abraham S. Cohen, continued the opposition of his newspaper to organic evolution. And in 1886, Rabbi Alexander Kohut, a recent émigré from Hungary who quickly became a leading scholar among traditionalist Jews, delivered a talk on "Science and Judaism," in which he maintained that human beings had originated only 5740 years earlier.[48]

Each of these men tended, like Wise, to emphasize empirical and logical arguments, invoking such scientific authorities as the aging Canadian geologist Dawson and the deceased American naturalist Louis Agassiz, both of whom rejected evolution. They also found evolution degrading to human dignity and threatening to the authority of the Hebrew scripture. Arguments against evolution based on specific references to the Genesis text, e.g., God "created man in his image," occasionally appeared, but this was not typical.[49] More common were theologically based objections to evolution, such as those offered by Rabbi Sabato Morais, a leader among traditional Jews and a founder of the Jewish Theological Seminary in 1886. In the mid-1870s and then again in the 1880s, he delivered two sermons that severely criticized scientific theories of evolution. Evolution, according to Morais, meant that "all things which exist developed themselves accidentally by themselves from an original undefinable speck." As such, he argued that evolutionary theory denied the belief that God created the world according to His free will. Equally important, the materialism and atheism commonly associated with evolution by scientists opposed firm belief in the immortality of the human soul. This, Morais argued, undermined the ultimate accountability of man before God and so encouraged nihilistic and immoral behavior.[50]

The most scientifically detailed criticism of evolution by a traditionalist Jew was offered by Rabbi Abraham de Sola. Born into a British rabbinic family, De Sola immigrated to Montreal, Canada, in 1847, where he became a leading intellectual in the Montreal as well as in the Jewish community. While resident in Montreal, De Sola kept in close contact with traditionalist Jews in New York and published often in *The Occident* and *The Jewish Messenger*. He had a deep interest in medicine and natural history, becoming president of Montreal's Natural History Society in 1868. On the issue of evolution, he was an avid supporter of Dawson, whose books he reviewed for Jewish newspapers.[51] His most explicit and extensive discussion of evolution occurred in a long essay-review of Dawson's creationist book, *The Origin of the World*, published in 1877. De Sola emphasized two points. First, he discussed Dawson's notion of the plenary inspiration of the Scripture, which required that Genesis report the truth about the origin of the world. Second, he emphasized in some detail the empirical arguments against the evolution of life, for example, the lack of observation of species transmutation and the lack of transitional forms in the fossil record. The evolution of species was "unproved," and thus no compelling reason existed to abandon the Genesis creation story. De Sola recommended that Dawson's works "be in every Jewish home, side by side with the sacred Book which [they] so eloquently and effectively defend."[52]

De Sola was unique. Within the North American Jewish community at large, he appears to have been the only religious Jew with a leadership role in activities related to natural history. Among traditionalist Jews, he alone wrote lengthy and detailed articles about evolution and historical geology. Most traditionalist Jews discussed evolutionary theory only briefly, in one or two sermons or editorials.

Morais' two sermons offer some insight into the meagerness of traditionalist responses. While explicitly addressing issues related to evolution, both sermons aimed, in fact, to explain why "philosophy and science" are not topics fit for the pulpit. In the first place, Morais admitted his own ignorance and that of his audience in regard to modern science and philosophy. This made it difficult both for the rabbi to address the logical and empirical problems of evolutionary theory and for listeners to follow such technical details. The preacher's duty to instruct, exhort, and warn would be hampered if the subject matter was inaccessible to his audience. In the second place, Morais maintained, as did some contemporary Protestant opponents of evolution, that evolutionary theory, like many past scientific theories, would be altered or abandoned in the future. Thus, it was not worth the trouble to take it seriously.[53] Moreover, he argued that Jews did not require a temporary reconciliation with evolutionary theory since Judaism possessed an inherent vitality that could sustain faith without the props afforded by the fashionable science or philosophy. Finally, Morais insisted that the materialism and atheism commonly associated with evolutionary theory made it best to ignore the topic for fear of inadvertently encouraging infidelity. "Better would it be," he wrote, "for the peace of your mind, for the advantage of Judaism, and for that of the human race at large, that you remain in ignorance of abstruse studies, than you be godless." Sermons should be designed to fortify belief and virtuous behavior, not to engender unbelief.[54]

Not all traditionalist Jews, however, followed Morais' admonitions. Indeed, not all rejected evolution. For example, the merchant Alfred T. Jones, editor of *The Jewish Record,* who supported Morais in his efforts to establish a Jewish Theological Seminary, looked favorably upon evolutionary theory.[55] However, the bulk of support for evolution by traditional Jews came from the editors of *The American Hebrew.* This weekly, founded in 1879, was run by a younger generation of observant Jews that included, among others, Frederick de Sola Mendes, Solomon Solis-Cohen, and Henry Pereira Mendes. All were in their twenties and all were trained (or were training) in science or medicine: Frederick de Sola Mendes received a Ph.D. at the University of Jena, where he studied with Kuno Fischer and Ernest Haeckel; Solis-Cohen was currently engaged in medical studies at Philadelphia's Jefferson Medical College; and H. P. Mendes soon took up the study of medicine, receiving his M.D. from New York University in 1884. The founders of *The American Hebrew* reflected a desire for Jewish renewal and to reinvigorate traditional Judaism in America, and they had no hesitation about using medicine and science, including evolution, to advance this goal.[56]

The first support for evolution in *The American Hebrew* appeared in 1880, in a laudatory review of Asa Gray's *Natural Science and Religion* in which the reviewer heartily endorsed Gray's theistic version of evolution. That same year, H. P. Mendes delivered a sermon in which he argued that the text of Genesis offered no barrier to the acceptance of the evolution of species. For Mendes, the Bible was not a scientific work but a depository of moral truths. Moreover, evolution from a primordial form reflected God's glory more than separate acts of creation for each species. These remarks are all the more remarkable, retrospectively, in light of the fact that H. P. Mendes became one of the founders of the Orthodox movement in America in the early-twentieth century.[57] During the 1880s, *The American Hebrew* editors continued to publish editorials and articles supportive of evolution. Most importantly, their advocacy developed in the context of a deepening opposition to Reform Judaism, which was being shaped more and more by its radical leaders.

Among Reform supporters of evolution, only occasional references to evolution appeared in sermons and lectures in the mid-to-late 1870s.[58] However, beginning in 1880, support for evolution became more and more visible in the Reform community, especially in sermons and essays by radical Reform Jews. Kohler, who himself returned to the theme of evolution, was joined by several others. The lawyer Morris M. Cohn and Rabbi Emil G. Hirsch spoke in favor of evolution in 1880. Between 1883 and 1885, Rabbis Emanuel Schreiber, David Stern, and Solomon Schindler all voiced support for the transmutation of species. They viewed evolution as an established scientific fact. Schreiber expressed this most directly: "it is no use denying the theory of evolution in our days. We are too far advanced for that. This theory is an established fact, although there are existing among scientists a few unsettled questions." The discussions by these rabbis focused on explaining how evolution demanded an active God or how evolution supported Reform Judaism's view of revelation or of ritual change.[59] Indeed, it was this aggressive appeal to scientific theories of evolution as support for radical Reform that initiated some heated polemics about evolution between Reform and traditionalist Jews in the 1880s.

Evolution and the Reform-traditionalist controversy: 1881–1888

Again, it was Kaufmann Kohler who initiated this more aggressive strategy to defend radical Reform Judaism. In a widely discussed article published in January, 1881, "Old and Modern Judaism," Kohler proclaimed that "evolution is the key offered by our age for the solution of the great problems of growth in the world of matter or mind." His central concern was the concept of progressive revelation, which he cast as a subspecies of a more general law of evolutionary progress.[60] Later in 1881, Emil Hirsch, who was fast becoming a leader in the Reform community, joined Kohler in declaring that the idea of God and the moral law had been revealed gradually and progressively in history. Hirsch appealed to evolution to explain the view that the rituals and ceremonies of Judaism had to adapt continually to the spirit of the times. "Evolution follows an inherent law," he wrote, and "only such reforms will stand as are in keeping with the tendency of this historical channel." For Hirsch, many traditional laws, such as those that regulated diet, priestly purity, and dress, had to be abandoned since they now hindered rather than aided the task of inculcating the moral law and the proper conception of God.[61]

The traditionalist editors of *The American Hebrew* disagreed strongly with the Reform rejection of ceremonial law as essential to Judaism. In the early 1880s, they began to more aggressively challenge the legitimacy of Reform Judaism.[62] Their view of Judaism was similar to the "Positive-Historical" interpretation of Judaism developed in Germany in the 1850s in opposition to the emergence of Reform. This interpretation emphasized the historical development of the practices and beliefs of the Jewish community, itself bound together by ties of consanguinity and by adherence to an all-encompassing set of legal strictures. These traditionalist Jews focused on interconnections, slow development, and practice as the binding forces that provided coherence and meaning to Jewish life. To implement radical change, as some Reform Jews had suggested, would only ensure the death of Judaism and the disappearance of the Jewish people.[63]

Equally disturbing to the editors of *The American Hebrew* was the use by Reform rabbis of evolutionary theory to legitimize Reform Judaism. *The American Hebrew* editors accepted organic evolution, and they considered themselves as much in line with modern science as did contemporary Reform Jews. Their focus on the historical development of the Jewish people and ceremonial law provided the basis for an appeal to evolutionary theory as support for their version of Judaism. They took advantage of this opportunity to assert their own claim that traditional Judaism rested on the authority of the new science of evolution.

In early 1882, *The American Hebrew* editors accused Hirsch of misusing Herbert Spencer to promote the "reckless demolition" of whatever Jewish ceremonies appeared incongruous with modern times. Hirsch had pointed to the history of religions in the writings of contemporary anthropologists as well as in Spencer, e.g., there had been a series of successive stages to the evolution of religion, whereby older ideas and practices were replaced by newer, better ones. For Hirsch, this supported his view that Judaism, too, had to undergo such evolution and not remain forever attached to ancient laws and ceremonies. The editors of *The American Hebrew* responded by arguing that Hirsch had not read Spencer very closely. Quoting passages from Spencer's *Data of Ethics*, the editors alluded to his notion of society as an organism, where all of society's institutions were part of a harmoniously working unity. They specifically mentioned Spencer's discussion of ceremonial institutions and his claim that these were necessary until a stronger moral sense had been evolved. The premature abolition of ceremony would lead to social anarchy. From this, the *American Hebrew* editors argued that the gradual evolution of ritual and ceremony rather than radical rejection or innovation had been and should be the mode of change in Jewish life. They were not opposed to change per se, just piecemeal and uncoordinated change.[64]

During the next few years, the editors of *The American Hebrew* reiterated this argument, emphasizing that both Jews and Judaism were bound by the eternally valid law of gradual evolution. Just as the evolution of organic life occurred gradually and slowly and not in fits and starts, so too the evolution of Judaism had proceeded and should proceed in the same manner. They characterized radical Reform as a mandate for revolution, not evolution, which would only destroy the perfection of Judaism.[65]

The American Hebrew editors offered another argument beginning in 1884, in the aftermath of the "treyfa banquet." At a banquet celebrating the first graduating class from HUC, shellfish, meat, and dairy products were all served. Traditionalist Jews attending the banquet stormed out. This episode, the causes of which are still shrouded in mystery, stimulated the debate about the role of ritual in Jewish life and greatly weakened the ability of Reform and traditionalist leaders to work together despite their differences.[66] In defense of ritual practice, the editors of *The American Hebrew* appealed to the law of "survival of the fittest." Jews had survived thousands of years, constantly struggling against all sorts of hardships and surviving. Their success testified to the fitness and, hence, the necessity and value of the Jewish ceremonial practices that had bound the Jewish community together. Moreover, the separateness that resulted from adherence to Jewish law had biological consequences, i.e., it had produced a superior race. "The law of fittest surviving, aided by the breeding of hereditary qualities in a pure race," wrote the editors of *The American Hebrew*, "has given the Jews a physiological and mental superiority which can be perpetuated only by the perpetuation of the race purity." Reform proposals to discard the law would eventually undermine Jewish separateness and eliminate the superior biological and psychological traits possessed by the Jewish people.[67]

This argument drew upon an established apologetic tradition among traditional Jews, in which Jewish ritual was justified because of its hygienic effects. While precedents existed, this apologetic strategy received a new impetus in early-nineteenth-century Germany. Both Christians and Enlightenment *philosophes* had viewed rituals and laws that distinguished Jews from others – e.g., circumcision, family purity, dietary laws – in a negative light. Nineteenth-century Reform Jews added their voices to this chorus of disapproval, in part reiterating the Enlightenment viewpoint that such rituals could not withstand the critical light of reason. In response, Jews committed to ritual practice turned the tables. Reason, in the form of medical assessments of past and present Jewish communities, proved that these rituals and laws improved and maintained the physiological well-being of Jews.[68] In the wake of Darwin, this medical justification for rituals that separated Jews from others could be reformulated in the context of evolution by survival of the fittest. In the mid-1880s, the editors of *The American Hebrew* offered this new argument in support of Judaism as they understood it.

The polemics between radical Reform rabbis and the editors of *The American Hebrew*, in relation to evolutionary theory, lasted through the spring of 1886. By this time, Reform rabbis had staked out their position in the Pittsburgh Platform (November, 1885) and traditionalist Jews had established The Jewish Theological Seminary (January, 1886). Debate about evolution abated, but only until the winter of 1886–87, when a young radical Reform rabbi provided another occasion for opponents of evolution to voice their objections. Four years after graduating in the first class from HUC, Rabbi Joseph Krauskopf delivered a series of sixteen detailed lectures, entitled "Evolution and Judaism," the longest supportive discussion of evolutionary theory among nineteenth-century American Jews. Wise published the lectures in *The American Israelite*, and they appeared in book form in 1887.[69] While Wise offered no comments on the lectures as he published them, other Jews publicly opposed Krauskopf's defense of evolution and its application to human history and to Judaism.

Krauskopf took seriously Kohler's 1881 statement that evolution provided the key to solving all material and spiritual problems. In his lectures, he applied evolution to various aspects of historical and human development: religion, the Bible, primeval man, the intellect, social behavior, the religious sentiment, morality, God, worship, and, finally, Judaism proper. On the topics of morality and God, he was most provocative. Krauskopf's theory of moral evolution combined some of Spencer and some of Darwin and made moral development dependent on the evolution of society and the intellect. The conscience itself was a product of historical development.[70] Krauskopf's notion of God was also provocative. In a discussion of Darwinism, he argued that the evolution of life manifested design and purpose. This teleological order reflected the operation of "an active, never-changing law which shapes all matter." Theologians called this law by the term God. "With this conception of the nature of God," he concluded, "every difference between science and religion disappears."[71] Indeed, Krauskopf paid great tribute to science, going so far as to say that worship could take place better in nature and in the laboratory than in the synagogue.[72] And he packaged all of this as a support for Reform Judaism, an evolving religion that combined the best of modern science with the essence of religion, namely, the moral law.[73]

For many, this "religion of science" was too much. Krauskopf himself reported in his penultimate lecture that the "Orthodox press" had lambasted him for the series, recoiling in horror that such teachings should be expounded by a Jewish clergyman. While I have been unable to identify these specific criticisms, other responses reveal the continued existence of opposition to evolution by a variety of American Jews. One response appeared in a letter by "Justitia," a member of Krauskopf's congregation in Kansas City, which was published in *The Jewish Free Press*, a newspaper that tended to support the Reform agenda. Justitia was appalled by Krauskopf's lectures and condemned him for preaching Darwinism and pantheism to young minds, introducing them to ideas antithetical to Judaism. Like the "orthodox" opposition, Justitia called for Krauskopf to remove himself from the "Jewish" pulpit.[74]

Another negative response came from Benjamin Peixotto, editor of the *Menorah Monthly*, established in 1886 as the organ of the B'nai B'rith. Founded in 1843 in the United States, the B'nai B'rith operated as a secular, fraternal organization to provide a variety of cultural, social, and financial supports for Jews. In the early 1880s, a serious tension developed between B'nai B'rith and Reform leaders, especially Rabbis Wise and Hirsch. The tension involved the unwillingness of the B'nai B'rith leadership to provide money to support a struggling Hebrew Union College. It also involved a dispute about whether a secular or a religious Jewish organization should be the unofficial voice of American Jewry. Reform rabbis denigrated the frivolous socializing they thought went on at meetings as well as the ritual elements of the fraternal order. B'nai B'rith leaders, in turn, condemned the Reform focus on morality as narrow and a religious definition of Jewishness as parochial. It was this growing tension that spurred B'nai B'rith leaders to press for a publication of their own.[75]

While the *Menorah Monthly* aimed to provide equal space and opportunity to both sides of any controversy, Peixotto did not conceal his "pleasure" in publishing letters and articles that attacked radical Reform Judaism. In July, 1887, he introduced a letter by the Galveston lawyer Leo Levi, a strong critic of Reform, by noting that "*evolutionists* have had the field for some years" and it was time to hear the other side.[76] While Levi's letter discussed Jewish life and practice, Peixotto's statement reflected the fact that Reform leaders were using the language of "evolution" to describe and defend their version of Judaism.

Peixotto then turned to the topic of organic evolution. Krauskopf's lectures on evolution and Judaism, he proclaimed, were "as far away from the principles of Judaism as they were wanting in scientific knowledge." However, it was from the standpoint of science, in particular, that "their superficiality deserved exposure." To accomplish this, Peixotto published a three-part article by a Methodist clergyman, Thomas Mitchell, who had been drawn to Krauskopf's book through references made by the preacher Henry Ward Beecher. Mitchell was no mild critic of organic evolution. He argued that no compelling evidence existed for organic evolution, for the antiquity of human beings, or even for the antiquity of the earth. He adhered to an extremely literalist interpretation of Genesis, arguing that God had created the world as we know it in six 24-hour days. Mitchell's refutation of evolution received a favorable response by at least one Jewish reader, whose letter Peixotto published. This person, "whose profound learning has not infrequently embellished these pages," applauded Mitchell's criticism of Krauskopf. Krauskopf's views on evolution were "unripe trash" and reflected an excessive amount of "hutzpa." At the same time, the correspondent concluded that he was "not prepared sufficiently to say whether Darwin or Agassiz has hit upon the truth," preferring to remain undecided rather than judge the matter prematurely.[77]

In addition to offering a scientific critique of evolution, Mitchell also argued that the biblical text should be taken literally and as such controverted the fact of evolution. It was this point that especially concerned Rabbi Sonneschein, who wrote a rebuttal that quickly appeared in the *Menorah Monthly*. Sonneschein believed the Genesis account of creation to be wrong in its details and Darwin to be right. He pointed to Jewish authorities who believed that Genesis was not an infallible text on science or philosophy and that the "six days" need not be interpreted literally.[78] Mitchell responded to Sonneschein but, pressed for space, Peixotto could not publish it in *Menorah Monthly*.

However, Mitchell's response soon appeared in the pages of *The American Hebrew*. While they had warmly welcomed Mitchell's original article, the editors of *The American Hebrew* still supported evolution. In the same issue as Mitchell's critique of Sonneschein, they published a laudatory review of a biography of Darwin. At the same time, they also expressed their opposition to Krauskopf's views about evolution. The editors noted that Mitchell was "in error" for thinking that American rabbis were "generally evolutionists." "The Rabbis of America," they continued, "are not as a general rule, Evolutionists; at least, not in the sense of Rev. Joseph Krauskopf's definition of Evolution."[79] The editors did not explain Krauskopf's definition nor their specific objections to it. But certainly they opposed Krauskopf's *use* of evolution to support his version of radical Reform. The willingness of the editors to publish Mitchell's rebuttal, while strongly disagreeing with his views, illustrates how the discussion of organic evolution had become intertwined with the polemics about the future course of Judaism in America.

Into the 1890s

After the Krauskopf controversy, it appears that traditionalist Jews gave little attention to evolutionary theory. Solis-Cohen was the exception. He continued to write in support of evolution and the compatibility of evolution with Judaism. He endorsed evolutionary theory in a series of book reviews for the *American Hebrew* of works by George Romanes and Thomas Huxley and argued that Judaism, a religion of moral and ritual practices, had nothing to fear from the fact of evolution.[80] At the same time, Solis-Cohen was concerned about the absence of discussion of evolution or science by traditional Jews. In a letter to Morais in 1894, he suggested that some rabbis at the Jewish Theological Seminary should receive training in the sciences sufficient to allow them to address critical issues in the relationship between science and Judaism. The past two decades had been a time for battle with the Reform movement, Solis-Cohen recalled, a task that had required the tools of Jewish knowledge. Now, traditionalist leaders required a familiarity with modern science in order to address the concerns of a new and inquisitive generation of Jewish youth.[81] Morais' response is not known, but he certainly did not begin to publish his views on these matters. Indeed, the absence of public discussion about evolution by traditionalist Jews makes it difficult to determine whether there was any growth in support for evolution during the 1890s.

Unlike traditionalists, Reform Jewish leaders continued to discuss evolution in sermons and lectures in the early 1890s. And by this time, it appears that all Reform rabbis, except Wise, accepted evolutionary theory.[82] Hahn now supported evolution, although there is no indication of what prompted his conversion. In the early 1890s, he gave a series of lectures endorsing evolutionary theory and drawing parallels between evolution in the organic world and in the history of the Jewish religion.[83]

Most significantly, Reform Jewish support for evolution swelled as recently ordained rabbis began to take up pulpits around the country. Even during Wise's presidency of Hebrew Union College, which lasted until his death in 1900, the radical Reform of Kohler and Hirsch had a greater influence in shaping the ideology and intellectual commitments of graduates. Three of these new rabbis – David Philipson, Rudolf Grossman, and Henry Berkowitz – publicly supported evolutionary theory. They were joined by two immigrants who became leading voices in the Reform community in the 1890s: the rabbi J. Leonard Levy (London) and Jacob Voorsanger (Amsterdam).[84] These individuals accepted evolution as a matter of course and applied it to their understanding of Reform Judaism. Influenced by the critical study of biblical texts and the theology of progressive revelation, they shared a conviction that organic evolution, understood to be a teleological, progressive development, was consistent with and supportive of the core of Reform Judaism.[85]

Conclusions

By the 1890s, evolutionary theory was widely accepted among Reform Jews, and rabbis commonly appealed to evolution as justification for Reform Judaism. However, this was a new situation. During the 1860s and early 1870s, Reform Jews overwhelmingly rejected organic evolution. Only with Kohler's 1874 sermon did vocal support for evolution begin to grow in the Reform community. Others soon joined him, but opposition to evolution continued. The key period appears to have been the 1880s, when support for evolution swelled among Reform Jews. This reflected the growing strength of radical Reform, which embraced the theology of progressive revelation and the findings of biblical criticism. By the early 1890s, the transition was complete. Acceptance of evolution emerged to its prominence in the American Reform community as part and parcel of the rise of radical Reform Judaism.

The pattern of response among traditionalist Jews was different, although the paucity of their responses, compared to Reform Jews, makes it difficult to draw certain conclusions. Like their Reform contemporaries, traditionalist Jews opposed the evolution of species in the 1860s and 1870s. One cautious voice of support appeared in the mid-1870s, but unequivocal support for evolution emerged only in the 1880s in the pages of *The American Hebrew*. These younger Jews, more educated in the sciences than their older colleagues, accepted evolutionary theory and used evolution to buttress traditional Judaism. However, their acceptance of evolution did not initiate a trend toward acceptance of evolution among traditionalists; at least no such trend is evident in published materials. The traditionalist Jewish community, into the 1890s, continued to have both strong supporters and strong opponents of evolution.

In terms of the specific arguments offered, American Jews differed little from their Protestant contemporaries. Opponents of evolution typically emphasized the connections with materialism, whether they thought them to be logical or historically contingent. But Jewish opponents, like their Protestant contemporaries, also offered other sorts of arguments: evolutionary theory was bad science; it had not received unanimous support among scientists; it undermined the authority of the biblical text; and it challenged central tenets of religious faith. Reform and traditionalist supporters of evolution responded with a set of arguments to counter each of these claims. They emphasized the near-unanimous acceptance of evolution by scientists, argued that the authority of the Bible need not depend on the veracity of the opening chapters of Genesis, and disengaged evolution from materialism by advancing theistic versions of evolutionary theory. Like Protestant supporters of evolution, these Jews maintained that the origin of matter, life, mind, and conscience might require the special creative act of God and that God operated purposefully in nature and history to bring about His desired ends.

However, in terms of overall patterns, the dissimilarities begin to emerge. In his extensive analysis of Protestant responses to Darwinism, the historian Jon Roberts has argued that after 1875, when the vast majority of scientists had come to accept evolution, theology supplanted science as the focus of discussions about evolution. Protestant intellectuals generally agreed "that the evolutionary hypothesis was irreconcilable with the prevailing formulations of numerous Christian doctrines." Those who opposed evolution used this "as sufficient grounds" for the rejection of evolution. Others argued that this condition "made theological restatement imperative." This then became the task of the many books and articles published by Protestants during the last quarter of the nineteenth century.[86]

The situation in the American Jewish community was different. While Jewish opponents of evolution during the late 1870s and 1880s condemned evolution because it contradicted particular Jewish beliefs, they continued to emphasize scientific arguments as well as the claim that evolution encouraged materialism. More striking, the acceptance of evolution among Jews did not induce a major transformation in theology. Protestant supporters of evolution modified natural theology, emphasized God's immanence, viewed revelation as progressive, and became more receptive to higher biblical criticism. American Jews reacted differently. Traditionalists who accepted evolution wrote almost nothing on theological topics and expressed little interest in higher biblical criticism. In Reform Judaism, the reformulation of theology and the practice of biblical criticism had begun in Germany decades prior to the 1870s.[87] Moreover, American traditionalists as well as Reform Jews drew on progressive, historical interpretations of Judaism that had been developed earlier by their German co-religionists. The acceptance of evolution by the scientific community in the 1870s might have affected individuals, moving them toward the liberal end of the theological spectrum.

However, the spectrum itself – the range of theological options available to Jews in the 1870s and 1880s – had been established earlier in the century.

Rather than transform theology, evolutionary ideas and language were used by Reform and traditionalist Jews to reinforce already established positions. In some cases, the rejection of organic evolution was intimately connected with opposition to liberal religious views that had been linked to the new science. Indeed, disagreements about evolution among American Jews were often debates about the nature of Judaism; these debates had a particular urgency in the 1870s and 1880s. In the 1870s, support for organic evolution by Kohler, Ellinger, Mayer, and Sonneschein involved support for the more radical Reform ideology, including acceptance of a theology of progressive revelation and the findings of biblical criticism. In contrast, Wise's opposition to organic evolution was linked to a different theology of Reform Judaism. In the 1880s, public controversy erupted between Reform and traditionalist Jews about what the acceptance of organic evolution meant for the validity of their respective versions of Judaism. Hirsch argued that the law of evolution, applied to Judaism, entailed significant change in ceremonial and ritual practice, in order to adapt to the new era. The traditionalist editors of *The American Hebrew*, in contrast, invoked the gradualness of evolutionary change and the law of "survival of the fittest" to argue that the continuation of long-established ceremony and ritual was necessary for the survival of Judaism and the Jewish people. The science of evolution was a cultural resource used by various Jews in their struggle to determine the future of American Jewry.

This essay has offered an overview and analysis of American Jewish responses to evolution from the 1860s through the 1880s. However, a comprehensive understanding of the American scene must await further research on European Jewish responses to Darwin and evolutionary theory. A comparative analysis will provide a means to further examine the relative importance of various factors shaping both the substance and the chronology of Jewish responses. In addition, such investigations will also help to illuminate American Jewish responses more directly, since many American rabbis and lay leaders during the 1870s and 1880s were either immigrants from Britain or German-speaking countries or received their rabbinic training or higher education there. The responses of Eastern European and Russian Jews will also be important as the story of American Jewish responses to evolutionary theory continues into the 1890s and beyond. The massive immigration of Eastern European and Russian Jews to the United States that began in the 1880s included a large number of religiously orthodox as well as many varieties of secular Jews, some of whom were quite receptive to evolutionary theory.[88]

As the nineteenth century drew to a close, the American Jewish community dramatically increased both in numbers and in varieties of Jewish identity. An examination of responses to evolution by American Jews will have to occur in light of this new situation.

Notes

1 Kaufmann Kohler, "Science and Religion," *Jewish Times*, 20 February, 1874, p. 821.

2 Jonathan Sarna, "New Light on the Pittsburgh Platform," *American Jewish History*, 76 (1987):358–68, especially p. 363; Leon Jick, *The Americanization of the Synagogue, 1820–1870* (Hanover, N.H.: University Press of New England, 1976), pp. 191–92; and Michael Meyer, *Response to Modernity: A History of the Reform Movement in Judaism* (Oxford: Oxford University Press, 1988), p. 262.

3 On the Christian uses of evolutionary theory, see Naomi Cohen, "The Challenges of Darwinism and Biblical Criticism to American Judaism," *Modern Judaism*, 4 (1984):121–57, especially pp. 139–40; and Egal Feldman, *Dual Destinies: The Jewish Encounter with Protestant America* (Urbana: University of Illinois Press, 1990), pp. 146–48. On Adler, see Benny Kraut, *From Reform Judaism to Ethical Culture: The Religious Evolution of Felix Adler* (Cincinnati: Hebrew Union College Press), 1978, pp. 52–57.

4 Cohen, "Challenges of Darwinism," p. 151; also see Marc Swetlitz, "Responses of American Reform Rabbis to Evolutionary Theory, 1864–1888," in *The Interaction of Scientific and Judaic Cultures in Modern Times*, eds. Yakov Rabkin and Ira Robinson (Lewiston, N.Y.: Edwin Mellen, 1995), pp. 103–25.

5 Cohen, "Challenges of Darwinism," pp. 121–57; Meyer, *Response to Modernity*, pp. 273–74; Marc Lee Raphael, *Profiles in American Judaism: The Reform, Conservative, Orthodox, and Reconstructionist Traditions in Historical Perspective* (San Francisco: Harper & Row, 1984), pp. 21–24; and Joseph Blau, "An American-Jewish View of the Evolution Controversy," *Hebrew Union College Annual*, 20 (1947):617–34.

6 Swetlitz, "Responses of American Reform Rabbis." In this paper, I analyzed the views of the most determined Reform opponent of evolution, Rabbi Isaac Mayer Wise.

7 Of the thirty serials I systematically examined, the following twenty-three contained references of one sort or another to Darwin and evolution: *American Hebrew* (New York), *American Israelite* (Cincinnati), *Asmonean* (New York), *Deborah* (Cincinnati), *Hebrew* (San Francisco), *Hebrew Leader* (New York), *Hebrew Review* (Cincinnati), *Hebrew Standard* (New York), *Jewish Advance* (Chicago), *Jewish Chronicle* (Baltimore), *Jewish Exponent* (Philadelphia), *Jewish Free Press* (St. Louis), *Jewish Messenger* (New York), *Jewish Progress* (San Francisco), *Jewish Record* (Philadelphia), *Jewish Reformer* (New York), *Jewish Tidings* (Rochester), *Jewish Times* (New York), *Maccabean* (Chicago), *Menorah* (New York), *New Era* (New York), *Occident* (Chicago), and *Zeitgeist* (Milwaukee). The seven serials that did not contain any references to Darwin or evolution include the following: *Boston Hebrew Observer*,

Hebrew Observer (San Francisco), *Jewish Herald* (New York), *Jewish Record* (New York), *Jewish South* (Atlanta), *Occident and American Jewish Advocate* (Philadelphia), and *Sinai* (Baltimore). The invaluable guide to pre-twentieth-century American Jewish literature is Robert Singerman, *Judaica Americana: A Bibliography of Publications to 1900*, 2 vols. (New York: Greenwood, 1990).

8 In the sketch of American Jewish religious life in the next four paragraphs, I have drawn freely from several sources: Jeffrey Gurock, "Resistors and Accomodators: Varieties of Orthodox Rabbis in America, 1886–1983," *American Jewish Archives*, 35 (1983):100–87; Raphael, *Profiles in American Judaism*; Meyer, *Response to Modernity*, pp. 225–95; Robert E. Fierstein, *A Different Spirit: The Jewish Theological Seminary of America, 1886–1902* (New York: Jewish Theological Seminary of America, 1990); Hasia R. Diner, *A Time for Gathering: The Second Migration, 1820–1880* (Baltimore: John Hopkins University Press, 1992), pp. 114–41, 206–13; and Gerald Sorin, *A Time for Building: The Third Migration, 1880–1920* (Baltimore: Johns Hopkins University Press, 1992), pp. 170–90.

9 The idea of accomodation is from Gurock, "Resistors and Accomodators," pp. 100–9. This concept, I suggest, applies just as well to Reform rabbis, although Gurock did not do so himself.

10 Historians of American Jewish history have much debated to what extent the twentieth-century distinction between "Conservative" and "Orthodox" applies to belief and practice in the nineteenth century. "Conservative" and "Orthodox" were used as self-descriptive labels in the 1880s. However, so were "traditionalist" and "positive historical," and some rabbis used these terms interchangeably. For discussions of these issues, see Jacob J. Petuchowski, "One Hundred Years of American Conservative Judaism: An Essay Review," *American Jewish History*, 80 (1991):546–65, especially p. 557; Abraham Karp, "The Conservative Rabbi – 'Dissatisfied But Not Unhappy,' " *American Jewish Archives*, 35 (1983):188–270, especially pp. 194–98; Raphael, *Profiles in American Judaism*, pp. 81–88; and Gurock, "Resistors and Accommodators," p. 162, note 6.

11 Gurock, "Resistors and Accommodators," p. 105.

12 It is the standard view that Reform Jews outnumbered traditionalist Jews during the 1870s and 1880s, although given the overlap in practice between moderate Reform and some traditionalists and the paucity of numerical data, it is impossible to offer a more precise characterization of the relative size of these two groups.

13 For reactions of Eastern European Jews to the founding of the Seminary, see Fierstein, *A Different Spirit*, pp. 51–52.

14 Review of *Man in Genesis and Geology*, by Joseph P. Thompson, *Jewish Messenger*, 18 October, 1869, p. 6.

15 Dr. Boynton, "Dr. Boynton on Genesis I," *American Israelite*, 17 February, 1860, pp. 260–61; Dr. Mayer, "The Origin and Education of Men," ibid., 3 May, 1867, pp. 4–5; A. R. Abbot, "Religion and Science," ibid., 15 October, 1869, p. 9, 22 October, 1869, pp. 9–10; E. C., "Science and Faith," ibid., 17 December, 1869, p. 6; and Isaac Mayer Wise, "Brevities," ibid., 24 January, 1868, p. 4.

Although prior to 3 July, 1874, Wise's weekly was called *The Israelite*, all references in this article will be cited as *The American Israelite*. Also see David Einhorn, "Dogmatical Difference between Judaism and Christianity," *Jewish Times*, 25 June, 1869, pp. 2–3. Einhorn edited a monthly periodical, *Sinai*, between 1856 and 1863, but it did not carry any articles that addressed evolution.

16 Jon H. Roberts, *Darwinism and the Divine in America: Protestant Intellectuals and Organic Evolution, 1859–1900* (Madison: University of Wisconsin Press, 1988), pp. 40–41.

17 For example, Sabato Morais, "On Avoiding Abstruse Topics in the Pulpit" [n.d.], microfilm #206, American Jewish Archives, Cincinnati, Ohio. This sermon is not easy to date, but a comparison of Morais' handwriting with dated sermons suggests that it was written during the 1880s.

18 "Philadelphia," *American Israelite*, 4 August, 1882, p. 38; and "Angelo Heilprin," *American Hebrew*, 11 August, 1893, pp. 468–69. Little has been written on American Jewish scientists in the nineteenth century. See Louis Gershenfeld, *The Jew in Science* (Philadelphia: Jewish Publication Society of America, 1934), pp. 192–93, 198–99; Yakov M. Rabkin, "Science," in *Jewish-American History and Culture: An Encyclopedia*, eds. Jack Fischel and Sanford Pinsker (New York: Garland, 1992), pp. 565–69; and Jacob Rader Marcus, *United States Jewry, 1776–1985*, 4 vols. (Detroit: Wayne State University Press, 1989–93), vol. 4, pp. 352–54.

19 See below for more on this physician. On American Jewish physicians, see Abram Kanof, "Medicine," in *Jewish-American History and Culture*, pp. 381–86; Marcus, *United States Jewry*, vol. 3, pp. 354–58.

20 In the 1860s, two of five articles published by Wise in the *American Israelite* that mentioned evolution derived from non-Jewish sources; Boynton, "Dr. Boynton on Genesis I," and Abbot, "Religion and Science" (see note 15 for full citations). Boynton's talk was summarized by Wise, and Wise's comments strongly suggest that Boynton was not Jewish. Abbot's sermon was reprinted from the *Universalist Quarterly* of Boston.

21 Roberts, *Darwinism and the Divine in America*, pp. 32–63.

22 "Theology of Nature," *Asmonean*, 16 July, 1852, p. 103; S. Solis, "The Spirit of the Beautiful," *Occident*, September, 1852, pp. 339–40; Isaac Leeser, "A Survey of the Field," ibid., April, 1856, pp. 1–7; Abraham De Sola, "Address of the Rev. Abm. De Sola, LL.D., to the Graduating Class at McGill College on May 3, 1864," ibid., August, 1864, pp. 216–24; H. K., "Physico-Theological Evidence of the Existence of a God," *American Israelite*, 19 June, 1857, p. 397; De Sola, "The Montreal Historical Society and Rev. Dr. De Sola," ibid., 13 March, 1863, pp. 281–82; Chautaubriand, "The Existence of God Demonstrated by the Works of Nature," ibid., 2 October, 1863, p. 109, 9 October, 1863, pp. 114–15, 16 October, 1863, p. 125, 23 October, 1863, p. 133; Wise, "Nature and Revelation," ibid., 16 October, 1863, pp. 124–25; Myer S. Isaacs, "Life," ibid., 27 August, 1858, p. 5; S. M. Isaacs, "Our Religion," ibid., 23 January, 1863, p. 5; Josiah, "A Jewish Free School," ibid., 17 July, 1863, pp. 19–20; Leeser, "The Object of the Creation," in *Discourses on the Jewish*

Religion, 10 vols. (Philadelphia: Sherman and Co., 1867), vol. 5, pp. 338–58; Adolph Loewy, "God in Nature," *Jewish Times*, 7 May, 1869, pp. 4–5, 28 May, 1869, pp. 6–7, 11 June, 1869, pp. 6–7, 16 June, 1869, p. 5; Moritz Ellinger, "Alexander von Humboldt," ibid., 17 September, 1869, pp. 8–9; and Conrad Jacoby, "Gratitude," *Hebrew*, 15 October, 1869, p. 4.

23 Bachya ibn Pakuda, "Nature Proclaims A God," in *Sabbath Readings* (London: Jewish Association for the Diffusion of Religious Knowledge, 1860–61), pp. 4–5; Bachya, "From Nature to Nature's God," in *Sabbath Readings*, pp. 41–46; Bachya, "Nature Proclaims a God," *Occident*, April, 1861, pp. 21–22; Bachya, "From Nature to Nature's God," ibid., April, 1863, pp. 9–15; Bachya, "Nature Proclaims a God," *Jewish Messenger*, 12 April, 1861, p. 4; Bachya, "From Nature to Nature's God," ibid., 17 April, 1863, p. 4; Bachya, "From Nature to Nature's God," *American Israelite*, 20 March, 1863, p. 290, 27 March, 1863, p. 301; Bachya, "Nature Proclaims a God," *Hebrew*, 19 March, 1869, p. 4; and Bachya, "From Nature to Nature's God," ibid., 10 February, 1871, p. 4, 17 February, 1871, p. 4.

24 Frederick Gregory, "The Impact of Darwinian Evolution on Protestant Theology," in *God and Nature: Historical Essays on the Encounter Between Christianity and Science*, eds. David C. Lindberg and Ronald L. Numbers (Berkeley: University of California Press, 1986), pp. 385–87; and Gregory, *Nature Lost? Natural Science and the German Theological Traditions of the Nineteenth Century* (Cambridge: Harvard University Press, 1992).

25 Meyer, *Response to Modernity*, pp. 64–66, passim.

26 Bernhard Felsenthal, "Progress of Jewish Ideas," *Jewish Times*, 17 June, 1870, pp. 244–46; reprinted as "Our Pulpit," *Jewish Chronicle*, 2 December, 1870, pp. 6–7, and as "Jewish Ideas Conquer the World," in *The American Jewish Pulpit* (Cincinnati: Bloch and Co., 1881), pp. 68–75. Also see "Ueber Das jüdisch-religiöse Leben: Bibel und Wissenschaft," *Sinai*, 3 (1858):1130–32; and Victor Leifson Ludlow, "Bernhard Felsenthal: Quest for Zion," Ph.D. diss., Brandeis University, 1979.

27 Wise, "Soph Dabar, Lecture XII: Darwin's *Descent of Man* Reviewed Anatomically," *American Israelite*, 22 January, 1875, p. 4; and Wise, *The Cosmic God* (Cincinnati: Office of American Israelite & Deborah, 1876), p. 55.

28 Adolph Kessler, "The Origin and Antiquity of Man," *Jewish Times*, 3 March, 1871, p. 9, 10 March, 1871, pp. 21–22; quotation from 10 March, 1871, p. 22. For Kessler's theology, see "Religion," ibid., 7 October, 1870, p. 505; "The Personal God Philosophically Demonstrated," ibid., 30 June, 1871, pp. 278–79.

29 Ellinger, "The Descent of Man," *Jewish Times*, 5 May, 1871, p. 153; "Darwin's Book on the Descent of Man," ibid., 16 June, 1871, pp. 243–44 [written by the London correspondent of the *Augsburger Allgemeine Zeitung*]; Adolph Huebsch, "Rev. Dr. Huebsch's Second Thursday Night Lecture," ibid., 5 December, 1873, pp. 644–45; Adolph Kohut, "Alexander Von Humboldt and the Bible," *American Israelite*, 13 October, 1871, p. 6 [selections from Kohut, *Alexander von Humboldt und das Judentums* (Leipzig: F. W. Pardubitz, 1871)]; M. R. Miller, "Contrast Between the Fruits of Infidelity and the Fruits of

the Bible," ibid., 12 January, 1872, pp. 5–6 [Miller was a Christian cleric who often wrote for the *American Israelite*]; J. H. M. Chumaceiro, "Darwinism," ibid., 19 January, 1872, p. 6; Wise, "Varieties," ibid., 16 February, 1872, p. 8; "Dr. Bree on the Darwinian Theory," ibid., 1 November, 1872, pp. 5–6; Michael Henry, "A Jewish View of Darwinism," *Jewish Messenger*, 5 January, 1872, p. 2 [originally published as "Darwinism," *Jewish Chronicle*, 15 December, 1871, p. 8]; and Abraham Treuenfels, "Darwinism and Religion," ibid., 4 April, 1873, pp. 4–5, 11 April, 1873, pp. 5–6 [selections from Treuenfels, *Die Darwin'sche Theorie in ihrem Verhältniss zur Religion* (Magdeburg: W. Simon's Buchhandlung, 1872)].

30 Ellinger, "Descent of Man"; "Darwin's Book on the Descent of Man"; and Henry, "Jewish View of Darwinism."

31 Kohut, "Alexander Von Humboldt and the Bible"; Chumaceiro, "Darwinism"; Henry, "Jewish View of Darwinism"; and Huebsch, "Rev. Dr. Huebsch's Second Thursday Night Lecture."

32 Miller, "Contrast Between the Fruits of Infidelity"; and Treuenfels, "Darwinism and Religion."

33 Solomon H. Sonneschein, "Anti-Büchnerisches," *Deborah*, 4 April, 1873, p. 2; and Sonneschein, "Aphorismen," ibid., 18 April, 1873, p. 2. See Kessler, "Origin and Antiquity of Man," p. 9; Henry, "Jewish View of Darwinism"; and Treuenfels, "Darwinism and Religion."

34 Isidor Walz, "The Modern View of Nature," *New Era*, 1 (1870–71):22–27; and Raphael D'C. Lewin, "Science, Arts, and Facts," ibid., 2 (1871–72):277–78, 434.

35 Kohler, "Das neue Wissen und der alte Glaube," *Jewish Times*, 6 February, 1874, pp. 797–99; Kohler, "Science and Religion," pp. 820–21; and Kohler, *Das neue Wissen und der alte Glaube!* (Chicago: Rubovits, 1874). Kohler's sermon was originally delivered on 18 January, 1874.

36 Ellinger, "Theology and Evolution," *Jewish Times*, 23 January, 1874, p. 760; Sonneschein, "The Theory of Evolution Manifested 450 Years before Darwin by a Jewish Philosopher," *American Israelite*, 13 March, 1874, p. 4; Sonneschein, "Judaism and Darwinism," *Jewish Messenger*, 20 March, 1874, pp. 4–5; and Jacob Mayer, "The Theory of Evolution and the Bible," *American Israelite*, 31 July, 1874, p. 5.

37 Kraut, "Judaism Triumphant: Isaac Mayer Wise on Unitarianism and Liberal Christianity," *American Jewish Studies Review*, 7–8 (1982–83):179–230; Swetlitz, "Responses of American Reform Rabbis."

38 Wise, "Lectures Introductory to the Philosophy of Mind," *American Israelite*, 6 March, 1874, pp. 4–5; Mayer and Wise, "The Theory of Evolution and the Bible," ibid., 31 July, 1874, p. 5.

39 Wise, "Soph Dabar," *American Israelite*, 16 October, 1874, p. 4, 23 October, 1874, p. 4, 6 November, 1874, p. 4, 13 November, 1874, pp. 4–5, 20 November, 1874, p. 4, 27 November, 1874, p. 4, 4 December, 1874, p. 4, 11 December, 1874, p. 4, 1 January, 1875, p. 4, 8 January, 1875, p. 4, 15 January, 1875, p. 4, 22 January, 1875, p. 4, 5 February, 1875, p. 4, 12 February, 1875, p. 4, 19 February, 1875, p. 4, 5 March, 1875, p. 4, 12 March, 1875, p. 4, 19 March, 1875, p. 4,

26 March, 1875, p. 4, 2 April, 1875, p. 4, 9 April, 1875, p. 4, 16 April, 1875, p. 4; and Wise, *The Cosmic God*. For more discussion of Wise's *Cosmic God*, see James G. Heller, *Isaac M. Wise: His Life, Work and Thought* (New York: Union of American Hebrew Congregations, 1965).

40 Wise, "The World of My Books," in *Critical Studies in American Jewish History*, ed. Jacob Rader Marcus, 3 vols. (Cincinnati: American Jewish Archives, 1971), vol. 1, pp. 175–78.

41 Wise, "Soph Dabar," 15 January, 1875, p. 4, 22 January, 1875, p. 4, 12 February, 1875, p. 4; and Wise, *Cosmic God*, pp. 47–61, 108–13, quotations on p. 51.

42 Charles Darwin, *The Descent of Man* [1871] (Princeton: Princeton University Press, 1981), p. 68; Wise, "Soph Dabar," 6 November, 1875, p. 4, 5 February, 1875, p. 4; and Wise, *Cosmic God*, pp. 42, 62–69.

43 Wise, "Soph Dabar," 12 February, 1875, p. 4, 2 April, 1875, p. 4; and Wise, *Cosmic God*, pp. 114–18, 157–64, quotations on pp. 114, 159, 163, 164.

44 Wise, "An Abstract of Our Cleveland Sermon," *American Israelite*, 26 October, 1866, p. 4; Wise, "The Philosophy of Providence," ibid., 13 February, 1874, p. 4; Wise, "The Mosaic Account of Creation," ibid., 25 December, 1874, p. 4; Wise [editorial], ibid., 10 December, 1880, p. 188; Wise [editorial], ibid., 31 October, 1884; and Wise, "Reformed Judaism" [1871], in *Selected Writings of Isaac Mayer Wise*, eds. David Philipson and Louis Grossman (New York: Arno Press, 1969), pp. 260–62. On Wise's views on revelation and biblical criticism, see Meyer, *Response to Modernity*, pp. 240–41.

45 Wise, "The Morals of Darwinism," *American Israelite*, 15 October, 1875, p. 6; Wise, "The Book of Genesis and the Theology of History," ibid., 5 November, 1875, p. 4; Wise, "Darwinism, Evolution and Spirit," ibid., 17 December, 1875, p. 4; Wise, "The Fourth Theory on the Origin of Species," ibid., 24 December, 1875, p. 4; Wise, "Revelation," ibid., 23 February, 1877, p. 4; Wise, "Darwinism Chastised," ibid., 26 October, 1877, p. 4; ibid., 27 February, 1880, p. 4; Wise, "The Negative Position of Judaism," ibid., 26 March, 1880, p. 4; Wise, "The Bible of the Patriarchs," ibid., 22 October, 1880, p. 132; Wise, "Skeptical Atheism," ibid., 3 March, 1882, pp. 284–85; and ibid., 28 April, 1882, p. 348.

46 Huebsch, "Evolution or Creation: The End of Perfection Only Found in God" [4 October, 1876], in *Rev. Dr. Adolph Huebsch* (New York: privately printed, 1885), pp. 132–35; Huebsch, "Man as a Winged Creature" [18 August, 1877], ibid., pp. 162–65; and Huebsch, "Scripture and the Scientists" [5 October, 1880], ibid., pp. 238–40.

47 Aaron Hahn, "The Jewish Theology in the Age of Criticism," *American Israelite*, 29 April, 1881, p. 340; Hahn, "History of the Arguments for the Existence of God," ibid., 31 August, 1883, p. 5, 5 October, 1883, p. 4; Hahn, "System of Jewish Theology," ibid., 21 November, 1884, p. 4; and Hahn, *History of the Arguments for the Existence of God* (Cincinnati: Bloch, 1885), pp. 60–68, 101–102.

48 Samuel M. Isaacs, "Tyndall's Address," *Jewish Messenger*, 11 September, 1874, p. 4; Isaacs, "Problematic Evolution," ibid., 3 November, 1876, p. 4; Isaacs, "Virchow and Haeckel," ibid., 25 January, 1878, p. 4; "The Troubles of

Scientists," *Hebrew Leader*, 1 January, 1876, p. 4; "Whence You Came?," ibid., 26 March, 1876, p. 4; "God in History," ibid., 21 June, 1878, p. 4; Abraham S. Cohen, "Man the Child of God," ibid., 25 March, 1881, p. 4; Cohen, "Darwin and His Theory of Evolution," ibid., 5 May, 1882, p. 4; and Alexander Kohut, "Science and Judaism," *Jewish Messenger*, 7 May, 1886, p. 6, 14 May, 1886, p. 6.

49 "The Troubles of Scientists," *Hebrew Leader*, 26 March, 1876, p. 4.

50 Morais, "A Few Words of Caution to Youth, Predicated on Tyndall's Address, Wherein He Advances Materialistic Views" [n.d.], microfilm #207, American Jewish Archives, and Morais, "On Avoiding Abstruse Topics in the Pulpit."

51 De Sola, "Scripture Zoology," *Occident*, April, 1857, pp. 38–40; De Sola, "Lecture on Scripture Botany," ibid., May, 1858, pp. 90–93; De Sola, "Address of the Rev. Abm. De Sola, LL.D."; De Sola, Review of *Archaia*, by J. W. Dawson, *Jewish Messenger*, 25 May, 1860, p. 4, 15 June, 1860, p. 4, 22 June, 1860, p. 4, 29 June, 1860, p. 4, 6 July, 1860, p. 4, 13 July, 1860, p. 4, 3 August, 1860, p. 4; De Sola, "The Mosaic Cosmogony," ibid., 11 March, 1870, pp. 4–5, 18 March, 1870, p. 4, 25 March, 1870, p. 4; De Sola, "The Montreal Historical Society and Rev. Dr. De Sola," *American Israelite*, 13 March, 1863, pp. 281–82; De Sola, "The Study of Natural History," ibid., 25 March, 1870, pp. 6–7; and Gerald Tulchinsky, *Taking Root: The Origins of the Canadian Jewish Community* (Hanover, Conn.: University Press of New England, 1993), pp. 40–60.

52 De Sola, Review of The Origin of the World, by J. W. Dawson, *Jewish Messenger*, 21 December, 1877, p. 4, 28 December, 1877, p. 4, 11 January, 1888, pp. 4–5.

53 Roberts, *Darwinism and the Divine in America*, p. 219.

54 Morais, "A Few Words of Caution to Youth" and "On Avoiding Abstruse Topics in the Pulpit."

55 Alfred T. Jones, "The Bible and Science," *Jewish Record*, 6 October, 1876, p. 4; and Joseph Le Conte, "Evolution, in Its Relation to Materialism," ibid., 1 March, 1878, p. 8.

56 Jonathan D. Sarna, *JPS: The Americanization of Jewish Culture* (Philadelphia: Jewish Publication Society, 1989), pp. 14–15; Sarna, "The Making of an American Jewish Culture," in *When Philadelphia Was the Capital of Jewish America*, ed. Murray Friedman (Philadelphia: Balch Institute, 1993), pp. 145–55; Philip Rosen, "Dr. Solomon Solis-Cohen and the Philadelphia Group," ibid., pp. 106–125; Moshe Davis, *The Emergence of Conservative Judaism* (Philadelphia: Jewish Publication Society, 1963), pp. 169, 349–54; Frederick de Sola Mendes, "Open Answer to Rev. I. M. Wise," *American Hebrew*, 16 May, 1884, pp. 4–5; and Philip Cowen, *Memories of An American Jew* (New York: International Press, 1932), pp. 40–61.

57 Review of *Natural Science and Religion*, by Asa Gray, *American Hebrew*, 9 April, 1880, p. 87; Henry Pereira Mendes, "Creation and Science," ibid., 15 October, 1880, pp. 103–104.

58 Sonneschein, "St. Louis, 10. December," *Deborah*, 17 December, 1875, p. 2; Sonneschein, "Milwaukee, 2. January," ibid., 14 January, 1876, pp. 2–3;

Sonneschein, "Die Disputation," ibid., 18 February, 1876, pp. 2–3; Sonneschein, "The Debate," *American Israelite*, 25 February, 1876, pp. 4–5; Max Lilienthal, "Religion or No Religion," ibid., 25 September, 1874, pp. 4–5; Lilienthal, "Philosophy, Judaism, and Reform," ibid., 6 October, 1876, pp. 4–5; and Lilienthal, "Has Judaism a Future?," ibid., 3 October, 1879, pp. 4–5.

59 Kohler, "Adam; or Man in Creation," *Jewish Messenger*, 24 October, 1879; Kohler, "Old and Modern Judaism," in *Hebrew Review*, 1 (1880–81):97–116; Kohler, "Darwin und das Judenthum," *Der Zeitgeist*, 11 May, 1882, pp. 154–56; Morris M. Coh, "History and Religion, Article II," *Jewish Record*, 26 March, 1880, p. 3; Cohn, "Something About Evolution," ibid., 18 November, 1881, p. 3; Emil G. Hirsch, "Dedication Sermon," ibid., 22 April, 1881, p. 3; Hirsch, "The Crossing of the Jordan," *Jewish Advance*, 10 September, 1880, pp. 9–10; Hirsch, "Evolution in Religion," ibid., 6 May, 1881, p. 5; Hirsch, "Darwin und der Darwinismus," *Zeitgeist*, 22 June, 1882, pp. 201–204; Hirsch, *Darwin and Darwinism* (Chicago: Occident, 1883); Artesian, "Our Denver Letter," *Jewish Tribune*, 5 October, 1883, pp. 339–40; Emanuel Schreiber, "Dr. Schreiber's Theories," *Occident*, 19 October, 1883, p. 4; David Stern, "Rev. Dr. Stern's Sunday Lecture: The Theory of Evolution," *Jewish Record*, 25 January, 1884, p. 4; Stern, "The Struggle for Existence," ibid., 29 February, 1884, pp. 4–5; and Bostonia, "Boston Brieflets," *American Israelite*, 29 May, 1885, p. 6.

60 Kohler, "Old and Modern Judaism," p. 100.

61 Hirsch, "Evolution in Religion," p. 5.

62 Fierstein, *A Different Spirit*, pp. 19–20.

63 Ismar Schorsch, "Ideology and History in the Age of Emancipation," in *The Structure of Jewish History, and Other Essays* (New York: Jewish Theological Seminary, 1975), pp. 23–31; Meyer, *Response to Modernity*, pp. 84–89; Petuchowski, "One Hundred Years of American Conservative Judaism," pp. 554–56.

64 Hirsch, "Ungluekliche Parallelen," *Zeitgeist*, 2 March, 1882, p. 73; Hirsch, "Radicalism," *American Hebrew*, 27 January, 1882, p. 122; and Hirsch, "Evolution in Religion," ibid., 10 March, 1882. Hirsch did not immediately answer the *American Hebrew* editors, but in 1886, he argued that evolution from lower to higher clearly involved the "dropping [of] this or that tool which the earlier life needed." The human arm and the bird wing had historical connections, but they were certainly not the same; so too with Judaism. Ceremonies served as tools in crafting Judaism's spirit, so when circumstances changed, old ceremonies had to be dropped in favor of new ones. See Hirsch, "Historical Judaism," *Jewish Reformer*, 5 March, 1886, pp. 8–9.

65 Cyrus L. Sulzberger, "The Evolution of Religions," *Jewish Tribune*, 8 February, 1883, p. 84; Sulzberger, "Chicago Philosophy," *American Hebrew*, 13 November, 1885, pp. 2–3; Solomon Solis-Cohen, "Jewish Separatism," ibid., 19 February, 1886, pp. 20–21, 26 February, 1886, pp. 34–35; and ibid., 16 April, 1886, p. 145.

66 Fierstein, *A Different Spirit*, pp. 20–25.

67 "New Jews and Judaism," *American Hebrew*, 14 March, 1884, p. 66; "Chicago Philosophy," ibid., 13 November, 1885, pp. 2–3; Solis-Cohen, "A Step Backward," ibid., 27 November, 1885, pp. 40–41; and Solis-Cohen, "Jewish Separatism." The quotation is from "New Jews and Judaism."

68 John M. Efron, *Defenders of the Race: Jewish Doctors and Race Science in Fin-De-Siècle Europe* (New Haven: Yale University Press, 1994), pp. 66–67, 82–84; Efron, "Science, Religion, and the Modernization of the Jews," presented at the History of Science Society Annual Meeting, New Orleans, 15 October, 1994; and Alan M. Kraut, *Silent Travelers: Germs, Genes, and the "Immigrant Menace"* (Baltimore: Johns Hopkins University Press, 1994), pp. 147–49.

69 Joseph Krauskopf, "Evolution and Judaism," *American Israelite*, 5 November, 1886, p. 5, 12 November, 1886, p. 5, 19 November, 1886, p. 5, 26 November, 1886, p. 5, 3 December, 1886, pp. 7 and 10, 10 December, 1886, pp. 4–5, 17 December, 1886, pp. 1–2, 24 December, 1886, pp. 1–2, 7 January, 1887, p. 5, 14 January, 1887, pp. 4–5, 21 January, 1887, p. 5, 28 January, 1887, p. 5, 4 February, 1887, pp. 4–5, 18 February, 1887, p. 5, 24 February, 1887, p. 5, 4 March, 1887, pp. 4–5; and Krauskopf, *Evolution and Judaism* (Kansas City: Berkowitz, 1887).

70 Krauskopf, "Evolution and Judaism," 21 January, 1887, p. 5; and Krauskopf, *Evolution and Judaism*, pp. 220–28.

71 Krauskopf, "Evolution and Judaism," 10 December, 1886, p. 5; and Krauskopf, *Evolution and Judaism*, p. 103.

72 Krauskopf, "Evolution and Judaism," 18 February, 1887, p. 5; and Krauskopf, *Evolution and Judaism*, pp. 276, 278, 286.

73 Krauskopf, "Evolution and Judaism," 24 February, 1887, p. 5; and Krauskopf, *Evolution and Judaism*, pp. 291–310.

74 Justitia, "Kansas City, April 24," *Jewish Free Press*, 29 April, 1887, p. 2.

75 Deborah Dash Moore, *B'nai B'rith and the Challenge of Ethnic Leadership* (Albany: SUNY Press, 1987), pp. 43–51; Edward E. Grusd, *B'nai B'rith: The Story of a Covenant* (New York: Appleton-Century, 1966), pp. 92–96.

76 Benjamin F. Peixotto, "Editor's Sanctum," *Menorah*, 3 (1887):73.

77 Peixotto, "Editor's Sanctum," *Menorah*, 3 (1887):296–300; Thomas Mitchell, "Evolution and Judaism," ibid., 3 (1887):112–18, 179–82, 338–44.

78 Sonneschein, "Evolution and Judaism: A Brief Reply to Professor Thomas Mitchell," *Menorah*, 3 (1887):424–27, 4 (1888):70–72.

79 "Literary," *American Hebrew*, 29 July, 1887, p. 185; Thomas Mitchell, "Evolution and Judaism," ibid., 2 March, 1888, pp. 51–52, 9 March, 1888, pp. 67–69; Review of *Life of Darwin*, by G. I. Bettany, *American Hebrew*, 2 March, 1888, pp. 52–53; and ibid., 2 March, 1888, p. 51.

80 Solis-Cohen, Review of *Mental Evolution in Man*, by George John Romanes, *American Hebrew*, 29 November, 1889, pp. 106–7, 3 January, 1990, pp. 212–13; Solis-Cohen, Review of *Darwin and After Darwin*, by George John Romanes, ibid., 19 August, 1892, p. 513, 26 August, 1892, pp. 544–45; Solis-Cohen, Review of *Essays Upon Some Controverted Questions*, by Thomas

Henry Huxley, ibid., 2 October, 1892, pp. 78–79; and Solis-Cohen, Review of *Darwiniana*, by Thomas Henry Huxley, ibid., 2 February, 1894, p. 402.

81 Letter, Solis-Cohen to Morais, 29 October, 1894, microfilm #207, American Jewish Archives.

82 Wise, "On the First Chapter of Genesis," *American Israelite*, 21 October, 1887, p. 4, 11 November, 1887, p. 4, 22 June, 1888, p. 4, 20 February, 1890, p. 4; and Wise, "Darwinism Denied," ibid., 21 November, 1889, p. 5, 1 March, 1900, p. 4.

83 Hahn, "The Great Science of Evolution" [11 December, 1892], *A Course of Lectures by Aaron Hahn*, 1 (1892–93):6; Hahn, "The Great Problems of Life" [1 January, 1893], ibid., 1 (1892–93):9; and Hahn, "The Religion of the Future" [5 March, 1893], ibid., 2 (1893):6.

84 David Philipson, "The Conflict between Science and Religion," *American Israelite*, 19 March, 1891, p. 4; Rudolf Grossman, "Evolution and Religion," 3 February, 1893, MS #494, Box 1, Folder 1893, Rudolf Grossman Papers, American Jewish Archives; Grossman, "Darwin and Lincoln," 15 February, 1895, MS #494, Box 1, Folder 1894, Rudolf Grossman Papers; Henry Berkowitz, "Judaism and the Gospel of Modern Thought" [2 November, 1894], in *Pulpit Message* [bound collection of sermons in the Hebrew Union College Library, Cincinnati, Ohio]; J. Leonard Levy, "Reliance on Science" [19 November, 1893], in *Sunday Lectures* (Philadelphia: Oscar Klonower, 1893–94), Ser. VII; Jacob Voorsanger, "The Rise of Man," *Emanu-El*, 10 April, 1896, pp. 8–10, 17 April, 1896, pp. 9–12; and Voorsanger, "Evolutionary Tendencies," ibid., 28 October, 1898, pp. 6–8.

85 The historian Naomi Cohen has argued that Reform supporters, in the 1890s, responded to a polemical challenge by Christian theologians – that evolution, applied to the history of religions, supported the superiority of Christianity over Judaism; see Cohen, "The Challenges of Darwinism and Biblical Criticism to American Judaism," pp. 139–43. My research, which focused on sermons, editorials, and articles on science or evolution, has not uncovered any concern for this challenge. Such responses, however, may more likely be found in the literature of Jewish-Christian polemics. This topic deserves more exploration to discover how seriously this challenge was viewed by Jews and whether such concerns arose before the 1890s.

86 Roberts, *Darwinism and the Divine in America*, p. 87.

87 In 1885, Morais accused Kohler of using Darwin, Spencer, and Dutch biblical critics Kuenen and Oort to destroy Judaism; Kohler replied that his critical tools had been formulated "independently from the Dutch or the Darwinian schools, and long before they were started!"; see Morais, "Morais to Kohler," *American Hebrew*, 26 June, 1885, p. 99, 3 July, 1885, p. 115; and Kohler, "Kohler to Morais," ibid., 3 July, 1885, p. 120.

88 The following examine responses to evolution by nineteenth-century European Jews: Shalom Rosenberg, "Introduction to the Thought of Rav Kook," in *The World of Rav Kook's Thought* (Jerusalem: Avi Chai, 1991), pp. 91–97; Mordechai Breuer, *Modernity Within Tradition: The Social History of Orthodox*

Jewry in Imperial Germany, trans. Elizabeth Petuchowski (New York: Columbia University Press, 1992), pp. 203–14; Mark Pittenger, *American Socialists and Evolutionary Thought, 1870–1920* (Madison: University of Wisconsin Press, 1993), pp. 103–10; Lois Dubin, "The Reconciliation of Darwin and Torah in *Pe'er Ha-adam* of Vittorio Hayim Castiogloni," *Italia Judaica IV: Gli Ebrei Nell'Italia ebrei nell unita* (Rome: Ministero per i beni culturali e ambientali, 1993), pp. 273–84; David Cesarani, *The Jewish Chronicle and Anglo-Jewry, 1841–1991* (Cambridge: Cambridge University Press, 1994), p. 59; Efron, *Defenders of the Race*, p. 69, passim; and Ralph Colp and David Kohn, "'A Real Curiosity': Charles Darwin Reflects on a Communication From Rabbi Naphtali Levy," in "Science and Religion in Modern Western Thought," Bernard Lightman and Bernard Zelechow (eds.), *European Legacy*, 1 (1996):1671–1776.

9

Black responses to Darwinism, 1859–1915

ERIC D. ANDERSON

American blacks, it would seem, had good reason to be suspicious of Charles Darwin and his theories of human origins. According to one scholar, Darwin's evolutionary hypothesis was by 1900 "the chief scientific authority for racists" in the United States. "His emphasis upon physical differences between races and his theory of natural selection – in fact the whole idea that racial characteristics result from evolution – became cornerstones of scientific racism." Darwin's influence was "pervasive," according to this interpretation, although popularizers of racist ideas seldom quoted him directly.[1]

If Darwinism was indeed the chief prop of racism in late-nineteenth-century America, one would expect a strong protest against Darwinian evolution by blacks, especially from the educated elite among African Americans. Yet, in fact, there was no strong, sustained black reaction either for or against Darwin and his theories. The story of black responses to Darwinism is complex and marked by unexpected turns. Darwin looms larger in white writings about "the future of the Negro"[2] than in the speeches and writings of important black Americans.

The two most influential black secular leaders of the years between Emancipation and the 1920s, Frederick Douglass and Booker T. Washington, did not find Darwin's ideas particularly threatening. Though neither man publicly endorsed Darwin, each employed a vague concept of evolution in his theory of progress.

In 1854, five years before the publication of *The Origin of Species*, Douglass had given a lengthy address devoted to "The Claims of the Negro Ethnologically Considered." Vigorously affirming the humanity of the Negro to an audience at Western Reserve College, he exclaimed:

> Away ... with all the scientific moonshine that would connect men with monkeys; that would have the world believe that humanity, instead of resting on its own characteristic pedestal – gloriously independent – is a

> sort of sliding scale, making one extreme brother to the ou-rang-ou-tang, and the other to angels, and all the rest intermediates!

He then turned to the issue of the unity of mankind, attacking Josiah Nott, Louis Agassiz, and other prominent "polygenists," who argued that Negroes and whites belonged to scientifically distinct species. "The credit of the Bible is at stake," said Douglass. St. Paul's declaration "that God hath made of one blood all nations of men," as well as the "whole account of creation, given in early scriptures," would have to be reinterpreted or "be overthrown altogether" if the polygenists' position was correct.[3]

Despite his rejection of an evolutionary link between man and beast, Douglass himself employed some Lamarckian evolutionary concepts in this address.[4] Built upon the inheritance of acquired characteristics, evolutionary change could be very rapid, he thought. He remembered seeing the common people of Ireland and noting "that these people lacked only a black skin and wooly hair, to complete their likeness to the plantation negro." "The open, uneducated mouth – the long, gaunt arm – the badly formed foot and ankle – the shuffling gait – the retreating forehead and vacant expression . . . all reminded me of the plantation and my own cruelly abused people." These facts suggested to him that "the condition of men" explained "their various appearances." The Negro's color, for example, could be fully explained by the "vertical sun," damp soil, and "enervating miasma" of the Niger, the Gambia, and the Senegal regions. The "gaunt, wiry, ape like appearance of some of the genuine negroes" could be explained by "the vicissitudes of barbarism."[5]

There is no evidence that either *The Origin of Species* or *The Descent of Man* had a major impact on the thinking of Frederick Douglass. After 1854 his religious views shifted away from conventional Protestantism to a more skeptical, secular position, but new scientific ideas do not appear to have been central in this shift. Rather his commitment to social reform led him to criticize the actions of Christian churches and doubt the direct intervention of divine providence.[6]

"I do not know that I am an evolutionist," he declared in an 1883 speech explaining "the philosophy of reform." He hoped "that each generation shall be an improvement on its predecessor," and "to this extent" he was an evolutionist. No longer fearful of animal ancestory for man, Douglass stated: "I certainly have more patience with those who trace mankind upward from a low condition, even from the lower animals, then [sic] with those that start him at a high point of perfection and conduct him to a level with the brutes. I have no sympathy with a theory that starts man in heaven and stops him in hell."[7]

Unlike modern scholars who have found "the military metaphor" for the relationship between science and religion to be misleading,[8]

Douglass saw science at war with superstition. The telescope teaches the insignificance of the earth, refuting "the old theory for which the church proposed to murder Galileo." Thoughtful people "are compelled to admit that the Genesis by Moses is less trustworthy as to the time of creating the heavens and earth than are the rocks and stars," he said.[9]

Booker T. Washington could be described as an "evolutionist" or even a "social Darwinist," although both terms are misleading without careful qualification.[10] Certainly his education at Hampton Institute in the 1870s was both weak in scientific content and clear in its repudiation of Charles Darwin. One of the textbooks assigned to all students at Hampton during Washington's study there was Mark Hopkins' *Outline Study of Man*, an answer to Darwin's theory written by the distinguished American educator.[11]

The themes of Washington's mature ideology – the necessity of struggle and self-help, and the futility of "artificial" or "forced" development – sound Darwinian (or perhaps Spencerian). But it is possible to read too much into Washington's use of evolutionary concepts. He meant little more by "evolution" than "progress," and his political and social ideas were not dependent upon a particular scientific theory.

When a correspondent asked Washington's opinion of Darwin, he received a very revealing answer: "I am not a student of biology and cannot answer with any definiteness the questions you ask." The issue of evolution, Washington wrote, "is very largely a question for experts . . . ." For him the importance of Darwin's ideas lay outside the technical details of biology. Washington affirmed those "general ideas of the progress of mankind, which have been associated with the name of Darwin in most of our minds," without taking "much stock in the theories people have advanced . . . as to the way in which this progress has come about."[12] Unless one keeps this view of evolution in mind, Washington's use of evolutionary metaphors in discussing American race relations makes no sense. For example, when Washington told a black audience in 1904 that Negroes were "a child race – very largely an undeveloped race," he was not conceding the inherent inferiority of blacks nor commenting on a specific biological theory. His point was that blacks were making rapid progress, more rapid progress, indeed, than any other "race of people in history under similar circumstances."[13]

By the time of Washington's death in 1915, William Edward Burghart Du Bois was rapidly becoming one of the most powerful spokesmen for black America. As editor of *The Crisis*, a crusading journal affiliated with the National Association for the Advancement of Colored People, Du Bois was in the habit of fearlessly, eloquently attacking all forms of Negrophobic thought and action, yet he had remarkably little to say about Darwin or evolution. Even in his earlier role as an academic, the

Fisk-, Harvard-, and Berlin-trained intellectual said less about evolution than one might expect.

In an 1897 address to a group of Negro intellectuals, for example, Du Bois offered extended comments on the concept of race, but mentioned Darwin only in passing. Invoking "the hard limits of natural law," Du Bois declared that "the history of the world is the history, not of individuals, but of groups, not of nations, but of races, and he who ignores or seeks to override the race idea in human history ignores and overrides the central thought of all history." Modern science's racial classification, he noted in passing, "is nothing more than an acknowledgment that, so far as physical characteristics are concerned, the differences between men do not explain all the differences of their history." Darwin's recognition (as Du Bois paraphrased it) "that great as is the physical unlikeness of the various races of men, their likenesses are greater" was, Du Bois asserted, the basis of "the whole scientific doctrine of human brotherhood." But the main point of his speech had been to repudiate radical black assimilationism in favor of a racial chauvinism that cherished "race identity" for the sake of the distinctive contributions "the Negro people" would make to civilization.[14]

Du Bois more directly engaged Darwin in a 1909 speech entitled "The Evolution of the Race Problem." Speaking at the National Negro Conference which launched the NAACP, Du Bois reflected on the coincidence that John Brown's "martyrdom" occurred in the same year that Darwin published *The Origin of Species*. Since 1859 advocates of Brown's principles of equal rights and human brotherhood had been on the defensive, Du Bois noted, as influential thinkers advanced arguments for the "social utility" of war and "human degradation," and the inevitability of inequality among classes and races. Du Bois did not blame Darwin or other scientists for these ideas. The problem was that "the splendid scientific work of Darwin, Weismann, Galton and others" had been "popularly interpreted" as meaning that "no philanthropy can or ought to eliminate" the essential inequality among "men and races of men," that the weaker individuals and groups were doomed to be overwhelmed by the strong.[15]

Without challenging evolution in any way, Du Bois skillfully dismantled the "supposed scientific sanction" for racism:

> First, assuming that there are certain stocks of human beings whose elimination the best welfare of the world demands; it is certainly questionable if these stocks include the majority of mankind and it is indefensible and monstrous to pretend that we know today with any reasonable certainty which these stocks are.

Second, "the present distribution" of advanced culture was not "a fair index of the distribution of human ability." He granted that the Sudan in 1909 was, in cultural terms, a thousand years behind France – just as "the valley of the Thames" had once been "miserably backward" compared to the civilization flourishing on "the banks of the Tiber." Such cultural disparities were better explained by "climate, human contact, facilities of communication, and what we call accident" than by innate racial characteristics.[16]

Rightly understood, the science of "the age of Darwin" expanded "possible human achievement," argued Du Bois. "Freedom has come to mean not individual caprice or aberration but social self-realization in an endless chain of selves . . . ." Moreover, he asserted, cleverly reversing racist conventional wisdom, "the present hegemony of the white races" threatened to promote "by brute force" the "survival of some of the worst stocks of mankind," putting in power "the slums of white society" over nonwhite groups in Asia and Africa and leaving "intelligent, property-holding, efficient Negroes of the South" to the "absolute mercy" of the ignorant supporters of white demagogues such as Tillman and Vardaman.[17]

Du Bois even used evolutionary language to mock white southern hysteria about "social equality." He argued that "free racial contact" would not lead to "contamination of blood" and lowering of culture. The result, instead, would be "survival of the fittest by peaceful personal and social selection," allowing "the best to rise to their rightful place."[18]

This important speech is an isolated statement, however, and not a central theme of Du Bois's propaganda of racial defense. Nor did other black Americans make these ideas central to their thought. Though they used Darwinian terms and metaphors, they did so casually, almost offhandedly – a fact that complicates the task of assessing the impact of Darwinism on this community. For example, a reader of George Washington Williams' *History of the Negro Race in America from 1619 to 1880* might be impressed by this historian's explanation of Negro death rates in the first years of freedom: "The law of the survival of the fittest is impartial and inexorable." Yet Williams was not talking about natural selection at all. He immediately equated the "law" of survival of the fittest with another law: "the soul that sinneth shall surely die." Blacks crowding into large cities to test their freedom were simply reaping the consequences of ignoring "sanitary laws," breathing the "poisonous atmosphere" of urban areas, and living "imprudent" lives of idleness.[19]

Williams, in fact, began his two-volume *History* with an implicit rejection of human evolution. In his first chapter, entitled "The Unity of Mankind," he appealed repeatedly to the book of Genesis to refute "the absurd charge that the Negro does not belong to the human family." All human beings were descended from Adam and Eve and the remnant of eight people who survived the Flood. The story of the Tower of Babel "furnishes reasonable proof" that not only did "all mankind" once share the same language, but also "they were of one blood." In the New Testament, the words of the Apostle Paul ("an inspired writer") were equally clear: God "hath made of one blood all nations of men for to dwell on the face of the earth . . . ."[20]

Another black writer, attorney and political activist D. Augustus Straker, redefined "survival of the fittest" in an even more startling way. Writing in *The Colored American Magazine* in 1901, Straker argued that "survival of the fittest" as applied to society did not mean "oppression or power over weakness, but an equal opportunity in competition." The problem was that blacks were prevented from participating in "human competition for the development of true manhood ripening into the image of the Divine Creator," he wrote. "Survival of the fittest is not allowed, for in competition only can such a result be achieved."[21] Straker had a point, for the emerging system of segregation was clearly designed to *prevent* certain forms of interracial competition. But his use of the phrase "survival of the fittest" reveals little about his opinion of Darwin or scientific theories about development.

Even black religious leaders seldom directly attacked Darwinism or competing theories of evolution, preferring, instead, to make traditional, biblically based appeals for the "brotherhood of man." Francis Grimké, the eminent Presbyterian divine, mentioned Darwin only once in all of his published sermons, and that was in an 1897 sermon on marriage. Grimké called Darwin a "distinguished naturalist who was careful never to make a statement until he had first carefully verified it" and cited him merely to show that marriage was an aid to longevity![21] In his private reflections, however, written in retirement and in the midst of the fundamentalist controversies of the 1920s, Grimké condemned Darwin and evolution directly:

> There is no evidence anywhere to show that man was ever anything but what he is now, human, pure and simple. The Bible teaches in Genesis that all forms of life on the globe were created; and that each form was to propagate itself after its kind. There is not the scintilla of evidence, despite the pretensions of Science, to show that this is not the case.[22]

Pointing to the absence of "missing links," he concluded that man was "not the result of evolution" from "some inferior animal." If it were otherwise, he added, we should see signs of continuing evolution in man over the thousands of years of recorded history: "How is it that he has not grown into something higher?"[23]

Without directly referring to Darwin, Bishop Lucius H. Holsey of the Colored Methodist Episcopal denomination made a similar point in a sermon entitled "The Fatherhood of God and the Brotherhood of Man." "Human nature is found by experience as well as by history and philosophy to be the same in quality and essentials, in all ages, states and conditions," he said. The defining characteristics of human beings had not changed from "the Garden of Eden to the present day," and no man was superior to another by nature. "Man is not his own creator," Holsey declared on another occasion. "No affinity of co-operative forces, controlling precipitant elements, could produce and round into the masterly parts displayed in the constitution of man. That which did not previously exist, could not, of itself, begin to be."[24]

One of the few black clergymen to discuss "the latest conclusions of physical science" in a more direct fashion was Theophilus G. Steward, a prominent preacher and prolific writer in the African Methodist Episcopal denomination. In several articles and a book entitled *Genesis Re-read*, Steward attempted to reconcile belief in divine providence – and the validity of the Bible – with the recognition that development was an important part of the natural order. "The Mosaic record," he believed, "while surely teaching a direct creation, also allows for orderly development." The six days of creation "may have been six periods of vast duration," Steward wrote. "It does not affect the result whether the creation was completed in six days or six millions of years." Steward sought to separate "the cosmogony of Moses" from the underlying theological message of Genesis.[25]

At the same time, he saw shortcomings in Darwinism. "A good hypothesis" as far as it went, the theory traced progress, but was "utterly unable to account for it."

> Mr. Darwin believed that life was "originally breathed by the Creator into few forms, or into one," but that all which follows has been developed under the action of two principles, viz., variation and natural selection. But Darwinism does not account for this impulse of variation nor of selection.[26]

Steward's views do not fit the mold of late-twentieth-century creationism. He affirmed the special creation of man ("real science tells us that the first link of the line ending in man has not yet been found") and the Mosaic authorship of Genesis, yet rejected as an anachronism the idea that Noah's flood covered the whole earth. This does not mean that his work was considered shockingly heterodox in 1885, for, as Ronald L. Numbers has demonstrated, nineteenth-century creationism was more diversified than its present-day descendant. *Genesis Re-read* was advertised in the official publication of the A. M. E. denomination as "Orthodox, but Sparkling and Progressive" and praised in a review by one of Steward's most prominent clerical colleagues. For several years it was used as a textbook at Payne Theological Seminary at Wilberforce University.[27]

Still, Steward did not represent the majority of black clergymen, even those with seminary training. A reviewer in the *A. M. E. Church Review*, a quarterly theological journal, thought the author of *Genesis Re-read* conceded too much in his attempt "to fight unbelievers on their own ground." If the Bible contained errors, how could it be inspired? "If Moses was not inspired when he wrote the first chapter of Genesis, he was not inspired when he wrote the tenth, or the twelfth or the fiftieth."[28]

Even as they rejected evolution (including Darwin's version), black Christian intellectuals had a vision of progress that made them open to theological change. Indeed, their confidence in the Bible was stronger than their commitment to any particular theological hypothesis. They believed, moreover, that science itself could be expected to change.

A. M. E. Bishop Benjamin T. Tanner is a good illustration of both attitudes. "Can evolution be accepted as true and the character of the Bible be maintained?" he asked in an 1892 article. "We frankly answer: No." But evolution would not overthrow the Bible. The scriptures, he wrote, had faced apparently "insurmountable" obstacles before:

> Take, for instance the question of the *age* of the world. The time was when men were perfectly well satisfied that the Bible and Bishop Usher [sic] were at one. But we know better now; and without in the least affecting the faith we had in the character of the Bible. So also was it with the *creative* days. The men are yet living who in their youth would have deemed it the gravest heresy to think other than that six literal days of twenty four hours each, alone, were occupied in the work of creation. But we know better now, nor is our faith in the character of the Bible disturbed.[29]

Tanner approached science with equal confidence. Science was "eminently progressive," he reminded his readers, and likely to change the conclusions now presented. A century ago, scientists were "altogether as dogmatic" about the theories they then advanced "as they are today upon the theory of evolution." In another century or so, scientists were bound to discard evolution. "The all-necessary 'missing link' will never come to light; and for the reason, like the stone of the alchemists, it has no existence."[30]

This conservative bishop found false applications of scripture more threatening to the status of black Americans than any theory of evolution. When three white Methodists suggested that "the Negro race" might not be descendants of Noah, he issued a strong reply, denouncing his colleagues' uncertainty as an "infidel remark." This issue, unlike the age of the earth or the days of creation, was worth a fight. "The Negro is a man," he concluded. "He is of Adam. He is of Noah. The Negro is a brother, and will be until science can demonstrate the Bible to be no more than a fable . . . ."[31]

Although a recent scholar has described Bishop Tanner as one of the leading "defenders of orthodoxy" among black Protestants, he carefully discriminated among his theological opponents. An essay on "The Negro and the Flood" is particularly revealing. Tanner was so determined to refute the claim that Negroes lacked "a tradition of the Deluge" – a claim with disturbing polygenist implications – that he appropriated Steward's "liberal" notions to back up his traditional interpretation. He began by arguing that some white groups lacked a Flood tradition and some Africans had such legends – and vigorously defending the idea of a universal flood. Tanner then asked, "but suppose that Moses and Christ and Peter were mistaken" and "learned men" including Theophilus G. Steward were correct in denying the possibility of a worldwide flood? "What then, as it relates to the Negro and the Flood?" In that case, he concluded, Steward's alternate explanation would "serve most potently for an excuse" if it should be proven that there was no African deluge tradition. No matter which interpretation one adopted, whether Tanner's or Steward's, there was no valid reason to hypothesize separate origins for black and white.[32]

The battle lines of the 1920s (or the 1990s) simply cannot be read back into late-nineteenth-century discussions of theology among the black elite. The clergyman who declared in 1892 that science had already "confuted the fallacious reasonings of Darwin and teachers of evolution" may have perceived no contradiction in the statement of an African scholar two years earlier: "Science is not, *per se*, an enemy to religion. It is time men begin to discriminate between the teachings of the Scriptures and the teachings of men on the Scriptures."[33] The minister who doubted "the possibility of natural science being able to prove or disprove creation as an act" may well have been amused if he read black Episcopal bishop James Theodore Holly's ironic description of the call of Abraham as "selection" and the "subsequent history of this patriarch and that of his descendants" as an illustration of "the struggle for life" and "survival of the fittest."[34]

Typical black ministers declined, however, to discuss evolution as directly as Steward or Tanner. Such a muted response is not entirely surprising. There was good reason to dismiss Darwin or completely ignore him in a poorly educated, largely rural community struggling with the practical issues of poverty, proscription, and demoralization. Though Grimké's elite congregation might have been willing, perhaps, to consider aspects of what has been called "the spiritual crisis of the Gilded Age," certainly most black Christians would find other matters more pressing.[35] Kelly Miller, the black mathematician and philosopher, exaggerated only a little when he wrote early in the twentieth century: "The Negro has no serious controversy over Scriptural interpretation. He is never tried for heresy. He does not wrangle over questions of the higher criticism." Unlike white Christians, black believers had a "fundamental agreement" as to "the intepretation and value of Bible teaching," Miller observed.[36]

It is more surprising that an African-American writer with some claim to scientific training, writing on a subject closely related to "the descent of man," could also ignore Darwin. Martin R. Delany, experienced as a physician, explorer, Civil War officer, and Reconstruction-era politician, wrote an 1879 tract directly addressing some of the same questions engaged in Darwin's work: *Principia of Ethnology: The Origin of Races and Color*. But instead of responding to Darwin, Delany built much of his 95-page tract around a question posed by George Douglas Campbell, the Duke of Argyll, in *Primeval Man* (1869). The Duke, an ornithologist and a theistic evolutionist, had written "the question is not the rise of Kingdoms, but the origin of Races .... When and How did they begin?"[37]

Delany began by rejecting Darwinism out of hand. "The Mosaic or Biblical Record" was the basis of his enquiry, "without an allusion to the Development theory," he wrote. "In treating on the Unity of Races, as descended from one parentage, we shall make no apology for a liberal use of Creation as learned from Bible," Delany added. "Upon this subject ethnologists and able historians frequently seem at sea, without chart or compass, with disabled helm, floating on the bosom of chance, hoping to touch some point of safety; but with trusty helm and well-set compass, we have no fears with regard to a direct and speedy arrival, into the haven."[38]

Delany assumed that the Duke of Argyll also completely rejected the "Darwinian development theory," though in fact this was not the case. The Duke insisted that man was the result of special creation, but accepted the idea that most of God's creative work was accomplished through the nonmiraculous processes of evolution, with a significant role for natural selection. Delany misunderstood, apparently, a statement he quoted from this aristocratic Calvinist: "It [Darwin's theory] is not in itself inconsistent with the Theistic argument, or with belief in the ultimate agency and directing power of a Creative Mind."[39]

The origin of races and color, according to Delany, lay in the differing complexions of Noah's three sons, Shem, Ham, and Japheth. (The Original Man – as well as Noah and his wife – had been red skinned, "resembling the lightest of the pure-blooded North American Indians," argued Delany, citing linguistic speculation on the Hebrew word Adam.) At the time of the Tower of Babel, when languages had been confused and the people scattered, God used the differing complexions of these men and their families to accentuate the new differences created by language. Working naturally, without recourse to miracles, "the all-wise Creator" enabled "each individual of any one of the now three grand divisions of the new tongues ... to identify the other without speaking." For the first time, people noticed skin color differences that until then "would have no more been noticed as a mark of distinction, than the variation in the color of the hair of those that are white, mark them among themselves as distinct peoples."[40]

This point established to his satisfaction, Delany moved on to other subjects. The rest of *Principia of Ethnology* is devoted to a discussion of biological aspects of skin color, the historical achievements of "the black race," and an extensive comment on ancient Egypt and Ethiopia, complete with hieroglyphics and black pyramid builders. There are few implications for evolution in this lengthy digression, except in his dogmatic claim "that the races as such, especially white and black, are indestructible; that miscegenation as popularly understood – the running out of two races, or several, into a *new race* – cannot take place." He explained that "a general intermarriage of any two distinct races" would simply result in the minority group being entirely absorbed by the majority – not in the creation of a new race.[41]

The most extended black philosophical response to the issues of evolution appeared in an idiosyncratic, meandering tome entitled *The African Abroad, or His Evolution in Western Civilization* (1913). It was written by William H. Ferris, holder of a Yale bachelor's degree and a master's degree from Harvard, a man trained for the ministry and variously described as an "itinerant philosopher" and a "free-lance journalist and lecturer." The historian David Lewis has characterized *The African Abroad* as "a disjointed display of nineteenth-century learning, in which history fulfilled the Hegelian world spirit, the Anglo-Saxon race was the 'advance guard of civilization,' and the 'Negrosaxon' was endowed with uplifting 'spiritual and emotional qualities.'" Ferris was the sort of black intellectual, observes recent scholar Wilson J. Moses, who was "ridiculed by Tuskegee" and "jeered at in the popular literature of the day," an undisciplined and untidy man motivated "by a sincere love of knowledge" and "the life of the mind." He was a member of the American Negro Academy, a supporter of the Niagara Movement, and, later, an advocate of Marcus Garvey's authoritarian pan-African movement.[42]

Though Ferris averred "The universe needs a God back of it to explain it," he was no fundamentalist. He recognized that life on earth was much older than the six thousand years allowed by some biblical literalists, dating the activities of "the cave men," for example, to "five hundred thousand years ago."[43] He accepted evolution "as a working hypothesis with a high degree of probability." "The principle of survival of the fittest holds universal sway in nature," he wrote. "And by the survival of the fit, Nature does not mean the intellectually, aesthetically and ethically fit, but those who can best adapt themselves to their physical environment." At the same time, he found in evolution's "upward trend through millions of years" evidence that "a divine plan, a divine idea was being realized through evolution." "We cannot understand evolution," wrote Ferris, shifting to the language of Hegelian idealism, "save as the method of the world ground [World Ground, a page later] in creating beings and manifesting himself."[44]

Ferris's ruminations on evolution were less a response to the biologist Darwin, who was mentioned only in passing as one of "the famous men of the nineteenth century," than to the philosopher Herbert Spencer. Ferris referred to the "Spencerian theory of universal evolution" and saw Spencer's agnosticism as the primary challenge to his philosophy. "We have passed forever beyond the age of atheism," wrote Ferris, "when men could say, 'There is no God,' and we are now in the age of Spencerian agnosticism, when men say that . . . . the human mind is impotent to grapple with transcendent realities."[45] In short, *The African Abroad* offers no systematic critique of Darwin's idea of the role of natural selection in evolutionary change.

All of the black responses to Darwin – ranging from Douglass's vague appropriation and certain clergymen's sublime neglect to Delany's peremptory rejection – are remarkably muted if Darwin's ideas were central to American racist thought. Is it possible, one wonders, that black leaders (excluding, perhaps, Du Bois) missed the full implications of Darwinism? Or if they saw the social mischief inherent in his ideas, did they lack the resources to combat a widely accepted viewpoint?

Historians are on shaky ground, of course, when they attempt to explain what did not happen – the dogs that did not bark, so to speak. Although some mysteries may require an analysis of silence, most historical problems are solved with less risky methods. In this case, the mystery evaporates when placed in the appropriate context. Quite simply, the idea that black Americans somehow *ought* to have responded more strongly to Darwinism overlooks three important facts.

First of all, other scientific theories were more threatening to black Americans than Darwin's interpretation of evolution. The most direct intellectual threat to their status and rights was the polygenist theory, which denied a common ancestor for all human beings. Such thinking was not merely a curiosity of the pre-Darwinian era, immediately discredited in the years after 1859. As the historian of anthropology

George W. Stocking, Jr., has demonstrated, polygenist thought persisted (in one form or another) well after *The Origin of Species* and *The Descent of Man*, sometimes even combining with "nominal Darwinism."[46]

Black spokesmen had the option of taking an agnostic posture toward Darwin, but they were forced to respond with vigor to the claims of Josiah Nott, Louis Agassiz, George R. Glidden, and other scientists who rejected the unity of mankind. Frederick Douglass, for example, could say "I do not know that I am an evolutionist," but he repeatedly attacked the polygenists. In an 1861 speech, he pointed to the creative powers common to all humans. "Some of our most learned naturalists, archeologists and Ethnol[o]gists, have professed some difficulty in settling upon a fixed, certain, and definite line separating the lowest man from the highest animal," he noted with some asperity.

> To all such I commend the fact that man is everywhere, a picture making animal. The rudest of men have some idea of tracing definite lines, and of imitating the forms and colors of objects about them .... The rule I believe is without an exception and may be safely commended to the Notts and Gliddens who are just now puzzled with the question as to whether the African slave should be treated as a man or an ox.[47]

The same imperative lay behind George Washington Williams' decision to begin the history of the American Negroes with a chapter on "the unity of mankind." As he noted in the first sentence of his history, "During the last half-century, many writers on ethnology, anthropology, and slavery have strenuously striven to place the Negro outside the human family ...."[48] Whatever problems Darwinism might raise about the history of life on earth or the historicity of Genesis, the theory at least affirmed that all human beings shared a common ancestor.

But to focus exclusively on theory is to miss a second important fact about the black responses to Darwinism. Spokesmen for black Americans, whether journalists, academics, or clergymen, did not have the luxury of ignoring pressing practical issues. In the late nineteenth century and early twentieth century, as American race relations deteriorated, such leaders were like firemen in an arson-mad neighborhood: they rushed from one crisis to another, responding as well as they could to new outbreaks of distortion or hostility.

When, for example, the white statistician Frederick L. Hoffman published a book in 1896 entitled *Race Traits and Tendencies of the American Negro*, thoughtful blacks were appalled – and eager to rebut the anti-Negro conclusions of the book. "The most influential discussion of the race question to appear in the late nineteenth century," *Race Traits* put the prestige of science behind the proposition that Negroes were headed for extinction. The disproportionately high mortality rate among blacks, Hoffman argued, was not caused by unfavorable environmental conditions but was the result of "race characteristics," such as "inferior organisms and constitutional weaknesses," as well as a group propensity toward sexual promiscuity and crime. White philanthropy since Emancipation had done nothing to arrest black retrogression, according to Hoffman, who advocated a complete end to counterproductive "modern attempts of superior races to lift inferior races to their own elevated position."[49]

It would have been possible to criticize *Race Traits* by attacking the book's scientific and philosophical assumptions. Modern scholars, after all, have labelled Hoffman as a "philosopher of racial Darwinism," and "the leading edge of neo-Darwinist extremism."[50] But the most significant black critic of Hoffman chose, instead, the narrowest and most practical basis for debate: Hoffman had his figures wrong.

The first "occasional paper" issued by the American Negro Academy (a select group of black intellectuals that included W. E. B. Du Bois, Alexander Crummell, and Francis Grimké) was a 36-page demolition of *Race Traits* written by Kelly Miller.[51] In a chapter-by-chapter commentary on Hoffman's work, Miller showed the German-born statistician repeatedly misrepresenting facts and taking unwarranted logical leaps to reach his conclusions. For example, Hoffman's claim that black population growth was faltering (as a result of diminished Negro vitality) was based on a simple oversight – the role of immigration, not natural increase, in relative white population increases. Without immigration, the white American population was increasing more slowly than the black population, Miller found. To blame the relative decline of the black population on "race traits" was ridiculous. "It would be as legitimate to attribute the decline of the Yankee element as a numerical factor in the large New England centers to the race degeneracy of the Puritan, while ignoring the proper cause – the influx of the Celt," observed Miller. Hoffman claimed to be open-minded, but he insisted that heredity rather than poor medical care or bad living conditions caused a high Negro death rate in the cities. Miller noted that "this high death rate of the American Negro" was equivalent to the white death rate in many places. The death rate for Baltimore blacks was lower than for whites in Munich, he discovered, and the death rate in Strasburg was about the same as that of New Orleans Negroes. "If race traits are playing such havoc with the Negroes in America, what direful agent of death, may we ask the author, is at work in the cities of his own fatherland?"[52]

Miller's *A Review of Hoffman's Race Traits and Tendencies of the American Negro* has virtually nothing to say about the social implications of evolution, the appropriateness of applying natural selection to human beings, or the meaning of "the struggle for life." Instead the reader faces a relentless parade of denials: the Negro is not constitutionally prone to tuberculosis; the mulatto is not "morally inferior" to the pure-blooded black; "race traits" do not account for black criminality; "rape is not peculiarly characteristic of the Negro"; the effectiveness of education for blacks is not in doubt; Negroes are not hopelessly inefficient workers who have "not yet learned the first element of Anglo-Saxon thrift."[53] These denials were more urgent to Miller and his colleagues than a purely academic discussion of Hoffman's scientific assumptions, for if Hoffman could be shown to be illogical or ill informed, his prophecy of black doom would be discredited immediately.

There was one other important reason for the apparently limited black response to Darwinism. Contrary to the claims of some modern scholars, Darwin's ideas were not the primary impetus behind racist thought. His ideas may have been "pervasive," but neither his doctrine of natural selection nor his personal philosophy was inherently racist.[54] Indeed, some aspects of Darwin's theory of evolution could be turned against advocates of racism, as Du Bois demonstrated.

The very pervasiveness of evolutionary language and vaguely Darwinian ideas in the late-nineteenth and early-twentieth centuries created confusion and contradiction. As the historian Gertrude Himmelfarb has pointed out, "almost every variety of belief" either claimed Darwin or was blamed on him. Karl Marx admired Darwin, and sent him an inscribed copy of *Das Kapital*. *The Origin of Species*, Marx believed, provided "a basis in natural science for the class struggle in history." At the same time, reformers attacked advocates of laissez faire and the limited state as the true "social Darwinists." Darwinism "was condemned by some as an aristocratic doctrine designed to glorify power and greatness, and by others, like Nietzsche, as a middle-class doctrine appealing to the mediocre and submissive." Evolutionary thought was at the same time the root of imperialism and a source of pacifist inspiration. No wonder, then, that some racists claimed Darwinian authority for their ideas, while others, such as certain southern antievolutionists in the 1920s, warned that "evolution made the Negro as good as a white man."[55]

When Darwinism is carefully defined, it cannot be construed to offer anything but the most "dubious support for racist doctrine." As Cynthia Russett has argued, a theory that discredits the idea of "separate and distinct origins of races" and does not assert the existence of "fixed and unchanging racial characteristics" is an insufficient foundation for racism.

Before the Mendelian revolution in genetics, she writes, "Darwinism could only provide a set of rather vague though undeniably suggestive phrases and ideas to stengthen the ideology of race."[56]

The militant racists of the period might invoke evolutionary ideas in their behalf, but the true sources of their beliefs were often older than *The Origin of Species*. Many of the key ideas of scientific racism, such as the value of race "purity" and the biological risks of miscegenation, the adaptibility of specific races to specific climates, and the allegedly different patterns of black and white mental development, were remnants of pre-Darwinian debates about the unity of mankind and can be fairly described as "polygenist thinking."[57]

One black writer who found Darwin a useful ally in the fight against the racial extremists of the early twentieth century was W. O. Thompson, a little-known essayist. Writing in 1906 in *The Voice of the Negro*, Thompson went well beyond the typical vague references to evolution and "survival of the fittest." Faced with the exceptionally harsh Negrophobic pronouncements of novelist Thomas Dixon and politician Tom Watson, among others, Thompson reminded his readers just what biological inferiority meant. If "fitness" is defined as success at survival, and inferiority entails extinction, "how is it, if the Negro is inferior [that] he has been enabled to remain for such long periods in an uncivilized state and not become extinct?" A truly inferior group, in biological terms, would have died out long ago.[58]

Unlike most popular writers on evolution, Thompson did more than merely invoke Darwin's name. He presented two extended quotations from the great scientist, commenting on mental and physical similarities among "the most distinct races of men" and discussing the role of imitation in human development. (Antiblack fanatics, Thompson complained, were in the illogical position of claiming both that "imitation is inferiority" and "the lack of imitation is inferiority.")[59]

For Thompson's purposes, the Bible and Darwin were in agreement. "Both the biblical account of creation and the theory of evolution must be discounted if the Negro is not regarded as one of the human species," he wrote. "For both accounts agree that all races descended from a single primitive stock," he stated, in a continuation of the old battle against polygenist ideas.[60]

The range of black responses to Darwinism suggests not only the difficulty in measuring the "impact of Darwinism" – as popularizers and nonspecialists appropriate, translate, and distort scientific ideas – but also raises questions about the nature of Darwinism itself. If the chief victims of racism did not interpret Darwin's theory of evolution as the primary source of their oppression, perhaps historians should reconsider certain convenient, time-honored generalizations about evolution, race, and society.

Notes

1 I. A. Newby, *Jim Crow's Defense: Anti-Negro Thought in America, 1900–1930* (Baton Rouge: Louisiana State University Press, 1965), pp. 12–13.

2 See George M. Fredrickson, *The Black Image in the White Mind: The Debate on Afro-American Character and Destiny, 1817–1914* (New York: Harper Torchbooks, 1972), pp. 228–55.

3 John W. Blassingame and John R. McKivigan, eds., *The Frederick Douglass Papers*, 5 vols. (New Haven, Conn.: Yale University Press, 1979–92), 1st ser., *Speeches, Debates and Interviews*, vol. 2, pp. 497, 501–02; 503–05.

4 Waldo E. Martin, Jr., *The Mind of Frederick Douglass* (Chapel Hill: University of North Carolina Press, 1984), pp. 248–49.

5 Blassingame and McKivigan, eds., *The Frederick Douglass Papers*, 1st ser., vol. 2, pp. 520–21.

6 Martin, *The Mind of Frederick Douglass*, pp. 175–82.

7 Blassingame and McKivigan, eds., *The Frederick Douglass Papers*, 1st ser., vol. 5, pp. 124, 129.

8 James R. Moore, *The Post-Darwinian Controversies: A Study of the Protestant Struggle to Come to Terms with Darwin in Great Britain and America, 1870–1900* (Cambridge: Cambridge University Press, 1979), pp. 19–49. Also see David C. Lindberg and Ronald L. Numbers, "Beyond War and Peace: A Reappraisal of the Encounter between Christianity and Science," *Church History*, 55 (1986): 338–54.

9 Blassingame and McKivigan, eds., *The Frederick Douglass Papers*, 1st ser., vol. 5, pp. 130–31, 133.

10 See William Toll, *The Resurgence of Race: Black Social Theory from Reconstruction to the Pan-African Conferences* (Philadelphia: Temple University Press, 1979), pp. 3–4, 69.

11 Louis R. Harlan and Raymond W. Smock, eds., *The Booker T. Washington Papers*, 13 vols. (Urbana: University of Illinois Press, 1972–1984), vol. 2, p. 61.

12 Letter, Booker T. Washington To Elmer Kneale, 29 November, 1911, in Harlan and Smock, eds., *The Booker T. Washington Papers*, vol. 11, p. 378.

13 *The Booker T. Washington Papers*, vol. 7, p. 470.

14 W. E. B. Du Bois, "The Conservation of Races," in *W. E. B. Du Bois Speaks: Speeches and Addresses, 1890–1919*, ed. Philip S. Foner (New York: Pathfinder, 1970), pp. 73–85.

15 Du Bois, "The Evolution of the Race Problem," in Foner, ed., *W. E. B. Du Bois Speaks*, pp. 202–03.

16 Ibid., pp. 204–05.

17 Ibid., p. 205. In a 1908 editorial paragraph in the short-lived magazine *The Horizon: A Journal of the Color Line*, Du Bois commented: "We have long since learned . . . that the 'fit' in survival is not necessarily the Best or the most Decent – it may be simply the most Impudent, or the biggest Thief or Liar." Herbert Aptheker, ed., *Selections from The Horizon* (White Plains, N.Y.: Kraus-Thomson, 1985), p. 49.

18 Du Bois, "The Evolution of the Race Problem," p. 207.

19 George W. Williams, *History of the Negro Race in America from 1619 to 1880*, 2 vols. (New York: Putnam, 1883), vol. 2, pp. 416–17.

20 Ibid., vol. 1, pp. 1–3, 4–5.
21 D. Agustus Straker, "Greater Opportunity for Civic Development of Mankind," *Colored American Magazine*, 2 (1901): 183–92.
22 Carter G. Woodson, ed., *The Works of Francis J. Grimké*, 4 vols. (Washington, D.C.: The Associated Publishers, 1942), vol. 2, p. 155.
23 Woodson, ed., *The Works of Francis J. Grimké*, vol. 3, pp. 185–86. See also p. 397.
24 L. H. Holsey, *Autobiography, Sermons, Addresses, and Essays* (Atlanta: Franklin Printing, 1898), pp. 57–59, 226.
25 William Seraile, *Voice of Dissent: Theophilus Gould Steward (1843–1924) and Black America* (Brooklyn: Carlson, 1991), pp. 88–90; Theophilus Gould Steward, *Genesis Re-read; or, The Latest Conclusions of Physical Science Viewed in Their Relation to the Mosaic Record* (Philadelphia: A. M. E. Book Rooms, 1885), pp. iii, 87, 89–90.
26 Steward, *Genesis Re-read*, pp. 113–14.
27 Ibid., pp. 52–75, 144, 154; Ronald L. Numbers, *The Creationists* (New York: Knopf, 1992), pp. 3–19; Levi J. Coppin, "Two New Books of Rare Merit," *Christian Recorder*, 10 December, 1885; advertisement, *Christian Recorder*, 17 December, 1885; and Frank N. Schubert, "Theophilus Gould Steward," in *Dictionary of American Negro Biography*, ed. Rayford W. Logan and Michael R. Winston (New York: Norton, 1982), pp. 570–71.
28 Book review of *Genesis Re-read* in *AME Church Review*, 2 (1886):235–37.
29 Benjamin T. Turner, "The Origin of Man," *AME Church Review*, 4 (1887):204.
30 Ibid., p. 211.
31 Turner, *The Descent of the Negro: Reply to Rev. Drs. J. H. Vincent, J. M. Freeman and J. L. Hurlbut, Editors of Sunday School Journal for Teachers and Young People, Methodist Episcopal Church* (Philadelphia: 1898), pp. 12, 23.
32 Moses Nathaniel Moore, "Orishatukeh Faduma and the New Theology," *Church History*, 63 (1994):65; and Benjamin T. Tanner, "The Negro and the Flood," *AME Church Review*, 4 (1892):439–45.
33 D. B. Williams, "The Harmony between the Bible and Science Concerning Primitive Man," *AME Church Review*, 9 (1892):19; and Orishatukeh Faduma, "Thoughts for the Times; or, The New Theology," *AME Church Review*, 7 (1890):142.
34 James W. Lavatt, "Creation of the World Opposed to the Theory of Natural Philosophy," *AME Church Review*, 8 (1891):68; and James Theodore Holly, "Biblical Inspiration," *AME Church Review*, 12 (1895):182. For a different view, see Moore, "Orishatukeh Faduma and the New Theology."
35 Paul A. Carter, *The Spiritual Crisis of the Gilded Age* (DeKalb: Northern Illinois University Press, 1971), pp. viii–ix, recognizes that the crisis of faith that hit certain upper-class and middle-class groups in the late nineteenth century was not relevant to lower- and lower-middle-class rural Christians.
36 Kelly Miller, *Radicals and Conservatives: And Other Essays on the Negro in America* (New York: Schocken Books, 1968), p. 158. (This is a reprint of Miller's 1908 work *Race Adjustment*.)
37 Martin R. Delany, *Principia of Ethnology: The Origin of Races and Color, with an Archeological Compendium of Ethiopian and Egyptian Civilization from Years*

of Careful Examination and Enquiry (Philadelphia: Harper & Brother, 1879), p. 7. On the Duke of Argyll, see Moore, *The Post-Darwinian Controversies*, pp. 221–22, 231–32, 306–07; and Peter J. Bowler, *Evolution: The History of an Idea* (Berkeley and Los Angeles: University of California Press, 1984), pp. 212–13, 223–24.

38 Delany, *Principia of Ethnology*, p. 9.
39 Ibid., pp. 15–16.
40 Ibid., pp. 11–19.
41 Ibid., pp. 41–95.
42 David Levering Lewis, *W. E. B. Du Bois: Biography of a Race* (New York: Henry Holt, 1993), pp. 275, 466; and Wilson Jeremiah Moses, *The Golden Age of Black Nationalism, 1850–1925* (Hamden, Conn.: Archon Books, 1978), pp. 75, 211–14.
43 William H. Ferris, *The African Abroad, or His Evolution in Western Civilization, Tracing His Development under Caucasian Milieu* (New Haven, Conn.: Tuttle, Morehouse & Taylor, 1913), pp. 6, 22.
44 Ibid., pp. 47, 55, 57, 58.
45 Ibid., pp. 8, 13–14.
46 George W. Stocking, Jr., *Race, Culture, and Evolution: Essays in the History of Anthropology*, rev. ed. (Chicago: University of Chicago Press, 1982), pp. 46–47.
47 Blassingame and McKivigan, eds., *The Frederick Douglass Papers*, 1st ser., vol. 3, p. 459.
48 Williams, *History of the Negro Race*, vol. 1, p. 1.
49 Fredrickson, *The Black Image in the White Mind*, pp. 249–51; and John S. Haller, Jr., *Outcasts from Evolution: Scientific Attitudes of Racial Inferiority, 1859–1900* (Urbana: University of Illinois Press, 1971), pp. 60–68.
50 Fredrickson, *The Black Image in the White Mind*, p. 250; and Robert C. Bannister, *Social Darwinism: Science and Myth in Anglo-American Social Thought* (Philadelphia: Temple University Press, 1979), p. 190.
51 Alfred A. Moss, Jr., *The American Negro Academy: Voice of the Talented Tenth* (Baton Rouge: Louisiana State University Press, 1981).
52 Miller, *A Review of Hoffman's Race Traits and Tendencies of the American Negro* (Washington, D.C.: American Negro Academy, 1897) pp. 7, 10–13.
53 Ibid., pp. 15–18, 21–22, 25–31, 33–34.
54 According to one scholar, the case that Darwin himself was a racist "rests largely on guilt by association and scattered quotations." See Bannister, *Social Darwinism*, pp. 183–86; cf. Carl N. Degler, *In Search of Human Nature: The Decline and Revival of Darwinism in American Social Thought* (New York: Oxford University Press, 1991), pp. 14–16.
55 Gertrude Himmelfarb, *Darwin and the Darwinian Revolution* (New York: Doubleday, 1962), pp. 420–23, 431; Adrian Desmond and James Moore, *Darwin* (New York: Warner Books, 1992), p. 601; Robert Nisbet, *History of the Idea of Progress* (New York: Basic Books, 1980), pp. 258–59; Bannister, *Social Darwinism*, pp. 114–36; and W. J. Cash, *The Mind of the South* (New York: Vintage Books, 1941), p. 347.
56 Cynthia Eagle Russett, *Darwin in America: The Intellectual Response, 1865–1912* (San Francisco: W. H. Freeman, 1976), p. 92; and Bannister, *Social Darwinism*, pp. 184–86. Darwin remains a whipping boy for some academics. For a

particularly inaccurate example, see Sandra Van Dyk, "The Evolution of Race Theory: A Perspective," *Journal of Black Studies*, 24 (1993):77: "The rise of Darwinism brought with it 'scientific' theories to explain race. According to those who built their ideas on Darwin's theories, African people are primitive, have smaller brains, and are a distinct, inferior species."

57 Stocking, *Race, Culture, and Evolution*, pp. 46–55.

58 W. O. Thompson, "The Negro: The Racial Inferiority Argument in the Light of Science and History," *The Voice of the Negro*, 3 (1906):507, 511.

59 Ibid., pp. 507–09.

60 Ibid., p. 509.

10

"The irrepressible woman question": women's responses to evolutionary ideology

SALLY GREGORY KOHLSTEDT AND MARK R. JORGENSEN

Darwinian evolutionary theory intensified debate about the origin and dimensions of sexual difference in species, not only in scientific and medical literature but also within wider intellectual circles whose members quickly entered a multidimensioned public debate about "the woman question."[1] Charles Darwin himself contributed directly to the discussion by his introduction of the mechanism of sexual selection in *Origin of Species* (1859) and through his widely read *Descent of Man and Selection in Relation to Sex* (1871), where he most fully elaborated his ideas about women's nature.[2] Other scientists, social scientists, and popularizers – mostly men – appropriated the evolutionary model into their discussions about women's nature in the last half of the nineteenth century and presumed that their personal observations represented biological determinations. The prescriptive sexual categories provided by scientists and physicians working within the evolutionary framework became dominant in a culture increasingly preoccupied with scientific explanations. Educated women in the nineteenth century were on the periphery of conversations about Darwinian theories for lack of institutional and professional forums, but they were hardly disinterested in evolutionary arguments, particularly as such concepts related to individual women and to women's collective circumstances. A few women, however, took up Darwinism directly, drawing on their own experience to extend or challenge theories and sometimes to position their advocacy of women's rights. They took up these arguments because the emphasis on evolutionary change and the mechanisms of sexual selection provided an opportunity to rethink "the woman question" in their own terms.

The authors thank Barbara Brookes of the University of Otago and Rene Burmeister, then at the University of Minnesota, as well as the editors of this volume and the anonymous referees for their thoughtful reading of the essay in manuscript form.

The singular designator, woman, of course glosses over virtually all variations among women and underscores the increasingly bipolar view of sexual identity in the nineteenth century.[3] Women concerned about women's political, economic, and social rights were especially outspoken in response to the assertions about sex and gender by scientists who used scientific arguments to buttress the social status quo. This paper looks closely at several women whose work directly addressed Darwinist formulations and who redesigned his arguments in ways that fitted their personal observations about the roles that women did and potentially could play in contemporary life.

Darwinian ideas related to women in many ways. This paper considers those aspects of Darwinian thought that bore most directly on what was termed "woman's nature" and on resulting assumptions about behavior, with particular reference to the responses of several women in the decades that followed *The Descent of Man*.[4] There was no simple, singular Darwinism. Likewise, responses to scientific arguments about sexual difference within the context of biological evolution demonstrate the multiple ways in which women and those interested in issues of gender chose to modify, extend, or react to the Darwinian arguments. Some women who opposed, and indeed some who supported, Darwinian ideas tended to argue on religious or other grounds.[5] Many remained quiet about the gendered discussion of their physical and intellectual traits by scientists and physicians but nonetheless created opportunities in defiance of the presumed limitations of their sex. This paper deals primarily with those who, in the context of late-nineteenth-century women's movements for greater educational opportunities and political rights, saw in Darwinism a concept requiring explication. Those who read Darwin closely and chose to comment did not often reject Darwin's overall theory, but typically they did seek to reconfigure the specific analysis or conclusions. Some challenged the assumptions about difference, some emphasized the principles of malleability and change, and some reconfigured the role of women to emphasize female choice in sexual selection. Women and gender are, of course, not synonymous and must not be conflated here, despite the fact that they were by most nineteenth-century commentators.[6]

Two American writers stand out in these discussions. Antoinette Brown Blackwell was among the first to take an interest in investigating specific aspects of Darwinian evolutionary theory as they related to sexuality and gender. A generation later, Charlotte Perkins Gilman would use Darwinism to speculate about the dimensions of a quite different future world. They, and others, understood that to argue in Darwinian terms was to engage a wider audience and, at the same time, to familiarize their women readers with concepts about gender that were common in scientific culture.

The Darwinians on women, 1859–1889

Darwin, certainly a man of his own time, carefully recorded his observations and thoughts about sex differences in his early notebooks and put sexual difference at the very core of his evolutionary theory.[7] Sexual selection was an essential part of his theory of natural selection and was described in *On the Origin of Species* (1859):

> [Sexual selection] depends, not on a struggle for existence, but on a struggle between the males for possession of the females; the result is not death to the unsuccessful competitor, but few or no offspring. Sexual selection is, therefore, less rigorous than natural selection. Generally, the most vigorous males, those which are best fitted for their places in nature, will leave most progeny. But in many cases, victory will depend not on general vigour, but on having special weapons, confined to the male sex.[8]

Because *Origin of Species* concentrated on the theory of natural selection to explain survival among species, it alluded to but did not as thoroughly account for the differences among individuals within species. Darwin explicated his theories on sexual difference more than a decade later in *Descent of Man*, a two-volume work written to argue that evolution brought an increasing differentiation between the sexes of each species and to apply the theory of sexual selection to human experience.

Thus, in two long volumes, Darwin discussed sexual differences between "male" and "female" within species, relying heavily on personal observations and analogies among species. More than half of *Descent of Man* discussed the characteristics of each sex as he perceived them. Darwin argued that "sexual selection has apparently acted on both the male and female side, causing the two sexes of man [sic] to differ in mind and body."[9] That he not only stressed sex differences in an explicitly bipolar way but also expressed a consequential "inequality between the sexes" became the key point in subsequent scientific discussions and popular debate. Although the term "inferiority" was never used, most commentators have concluded from the context that Darwin intended a rank ordering. After all, he had written, "The chief distinction in the intellectual powers of the two sexes is shewn by man attaining to a higher eminence, in whatever he takes up, than woman can attain – whether requiring deep thought, reason, or imagination, or merely the use of the senses and hands."[10] The presumed superiority of masculine behaviors described by Darwin were, not surprisingly, those in congruence with the norms and gender characterizations of middle-class Victorian society. Darwin also argued that the increasing variability between the sexes in most species, and the greater variability of the male of each species, contributed to male collective superiority in specialized form and function.[11] Zoologist and former Darwin protégée George Romanes, who wrote in "Mental Differences between Men and Women" that the two were so different they might reasonably be classed as separate species, offered familiar behavior as evidence and described his conclusions as simply "a matter of universal recognition."[12] The highly visible public commentator Francis Galton, a dedicated eugenicist and Darwinist, entered the ranks of those arguing male mental superiority by providing examples of male achievement in *Hereditary Genius* (1869) and by publishing measurements made in his laboratory in London.[13]

Scholars have debated the extent to which Darwin's own ideas about females, as expressed in his scientific journals and publications, were formulated by personal experience and influenced by his contemporaries. Retrospective evaluations by biologist Ruth Hubbard and historian Greta Jones argue that Darwin's ideas were explicitly derived from his intellectual and social milieu, primarily Oxbridge-educated white men and his deferential wife, Emma; within his own circle and among many Victorian scientists, he found an evidently responsive audience.[14] Other scholars, particularly Eveleen Richards, argue that Darwin's ideas on the importance of sex differences and the behavior of each sex was constructed as much "by his commitment to a naturalistic or scientific explanation of human mental and moral characteristics as they were by his socially derived assumption of the innate inferiority and domesticity of women."[15] Gillian Beer notes more generally that Darwin's language reveals that "Darwin was a man, writing for men, aware of his own species-characteristics . . . ."[16]

Some of the broader lines of influence have been traced and assessed by scholars who document in detail the fervor, the contradictions in logic, and the angst of leading British and American philosophers, scientists, anthropologists, sociologists and others, almost entirely men, whose work was intended somehow to answer "the woman question."[17] The phenomena led Virginia Woolf to inquire, rhetorically, of her women's college audience a half-century later,

> Have you any notion how many books are written about women in the course of one year? Have you any notion how many are written by men? Are you aware that you are, perhaps, the most discussed animal in the universe? . . . Sex and its nature might well attract doctors and biologists;

> but what was surprising and difficult of explanation was the fact that sex – women, that is to say – also attracts agreeable essayists, light-fingered novelists, young men who have taken the M.A. degree; men who have taken no degree; men who have no apparent qualification save that they are not women.... It was a most strange phenomenon.... Women do not write books about men – a fact that I could not help welcoming with relief....[18]

This paper does not elaborate on the sources or motivations of Darwinian thinking about women, but rather concentrates on the enormous influence it exerted and the reactions it sparked, particularly among women themselves.

Public audiences glimpsed debates about evolution and the nature of sexual differences in popular publications and lectures. Arguably the work most available to educated American audiences was that of Herbert Spencer, first promulgated broadly in his serialized "The Study of Sociology" (later a book of the same name) that launched *Popular Science Monthly*, in 1872.[19] Some scholars have argued that Spencer established evolutionary science as the "master discourse defining sexuality, knowledge, and power in the second half of the nineteenth century."[20] Spencer, like the biologist Thomas Huxley, wrote about sexual differences with even more vivid language and emotion than Darwin, but both popularizers emphasized, in one way or another, the biological destiny of women to be maternal, instinctive, and lacking in the mental and physical equipment for public life.[21] Interestingly, Spencer deviated from Darwin in suggesting that as civilization evolves away from war-inclined militancy, there may be a "corresponding re-adjustment between the natures of men and women, tending in sundry respects to diminish their differences."[22] Among the most strident supporters of the sharp contrast between the natures of "man" and "woman" was Edward Drinker Cope, the American zoologist, vertebrate paleontologist, and neo-Lamarckian, who argued, using skeletal differences, that "nature defined woman's position to be mostly dependent." He was an outspoken critic of women's expanding public aspirations, particularly suffrage.[23] *Popular Science Monthly*, under editor Edward Youmans, provided a public forum for often impassioned opinions arguing the social importance of sexual differences, and it also published somewhat shorter responses from men and women contesting the arguments of Cope and others.[24] Darwin had suggested women's relative physical inferiority, and his themes of women's physical frailty (except for childbearing) were promulgated by physicians. The most frequently quoted were Edward H. Clarke, who offered a small number of cases as proof that serious education for girls was dangerous to their health, and Henry

Maudsley, who argued that women's minds broke down when they attempted to defy their nature by competing with men rather than serving them.[25]

The last forceful expression of the nineteenth-century literature on the evolution of innate sexual differences between male and female was produced by the Edinburgh biologist and later urban planner Patrick Geddes and his student J. Arthur Thomson. Their *Evolution of Sex* (1889), printed in Great Britain, the United States, and France, compiled information from a wide range of sources on evolutionary theory to underscore the supposed differences between the human sexes. Following others, they defined two types of physiological activity, using the term "anabolism" for femaleness, which involved energy conservation, and the term "katabolism" for maleness, which involved more active and aggressive use of energy. From these Geddes and Thomson derived a long list of what we might view as stereotypical descriptors of men and women that accorded women greater altruistic emotions, constancy, stability, common sense, and intuition; by contrast, men were more independent and courageous, eager and passionate, and had a wider range of experience that enlarged their brain and developed their intelligence. Here, too, the correspondence with Darwin's own descriptions is almost exact. Geddes and Thomson were apparently motivated to write in part by contemporary politics. They concluded a discussion about the evolution of these tendencies by observing, "the differences may be exaggerated or lessened, but to obliterate them it would be necessary to have all the evolution over again on a new basis. What was decided among the prehistoric Protozoa cannot be annulled by Act of Parliament."[26]

While Geddes and Thompson drew primarily on natural history, biology, and their own experience, the emerging disciplines of ethnology and anthropology extended the emphasis on difference. For many of the early anthropologists, sexual differences appeared closely connected to more general patterns of social organization. Elizabeth Fee has noted that those who engaged in social anthropology, influenced by biological evolutionists, concentrated on the wide range of variability in the organization and individual self-identifications among the groups. The data were duly recorded in ethnographic journals and coincided with, rather than challenged, Victorian notions about appropriate sex roles. In most cases anthropologists argued that the evolution from primitive to civilized societies had paralleled the movement toward patriarchal societies, monogamy, and clearly defined sex roles.[27] Fee argues that the predominately male anthropologists of the late nineteenth century used circular logic to force their contemporary ideas about women and about other peoples into what they claimed were scientific conclusions – there is less evidence of women theorizing to quite different conclusions, as would Ruth Benedict and Margaret Mead a generation or more later.[28] As long as men in professional areas of study relating to human behavior relied on their own personal experiences and the ideas of Darwin and at the same time dominated their disciplinary institutions so exclusively, there could be few women and few places for those women to discuss the ideas that affected their lives so immediately.[29] Even in the 1870s and 1880s, however, a distinctive line of argument began to be developed among some reform-minded women and men.

Women's responses to Darwin: an overview

What of women and the Darwinian explications of their "nature" in the late nineteenth century? Many were well aware of the impact of the new scientific discourse and its potential impact on their lives. Those of the middle class were, after all, sufficiently literate to read popular journals and well enough affiliated in local voluntary societies to be invited to literary and scientific lectures.[30] One of the compounding dilemmas for women has always been that, while they are differentiated by physical characteristics, they are implicated in sharing social class and intense personal relations, including intellectual conversations, with the very men who often exercise economic and political authority over them. At the same time, those women familiar with gender dimensions of nineteenth-century reform could be highly skeptical of those experts for whom "the female body" seemed so simplistic and for whom the confining terms of "maleness" and "femaleness" were often so rigorously prescribed.[31] Still, Antoinette Brown Blackwell expressed her self-consciousness candidly: "I do not underrate the charge of presumption which must attach to any woman who will attempt to control the great masters of science and of scientific influence. But there is no alternative!"[32]

These issues were highly personal and there was a complicated borderland between the presumed private lives of women and the more public concerns of men.[33] Caution characterized respondents. While evolution was taught in women's colleges, for example, their faculties were not noticeably outspoken on the issue of how Darwin dealt with sex differences. Systematic study in science for women, except at new women's colleges such as Mount Holyoke, Vassar, and Bryn Mawr and later in home-economics departments at state universities such as Iowa State Agricultural College, was not much encouraged before the end of the nineteenth century.[34] Professional associations limited women's participation, and professionals themselves tended to disregard women's writings, except for the popularized texts used in classrooms and for amateur activity. Scientists, perhaps science too, were perceived to be masculine. Most of the women who did become scientists, like most male scientists, chose to do traditional and relatively noncontroversial research within their subspecialties. Only changes in these circumstances and the opening of new social-science disciplines to women at the end of the century allowed access to laboratory resources and seemed to promise unprecedented levels of expert influence on questions of human behavior and intelligence.

Most outspoken were a few reform-minded and literary intellectuals, who, having broken with convention on other matters, spoke out on this issue with regard to women's education and opportunity. The heady opportunity to rethink social conventions that might be offered by this new scientific theory intrigued them, and they found that they had an audience among women and some reform-minded men. The *Woman's Journal* and *Woodhull and Claflin's Weekly* regularly printed articles on Darwin and related topics. The authors were, like their scientific and popularizing counterparts, highly individualistic. Most of the women who responded to the theories of evolution and particularly sexual selection focused not on the physical (read reproductive) capacities of women but on their mental capacity and their social roles. Women were, after all, debating with well-established experts who presumed that, as part of womankind, these women were less intelligent, unable to accomplish the highest types of activity, incapable of profound abstract thought, and lacking in originality.[35]

John Stuart Mill and Harriet Taylor had warned about the probable "pernicious effect" of arguments from nature, and certainly such arguments were used in a prescriptive way in medical and political arenas to constrain women's activities. Nonetheless, some women did take up the issue directly, responding to the challenge of evolutionary theory and natural selection. In France, for example, Clemence Royer translated *Origin of Species* and later *Descent of Man*, including her own enthusiastic introductory interpretations about evolution and women's position.[36] Some women were drawn to the idea of difference, and they tempered it with the ideal of equality or equivalence. Inevitably, those who wrote on the topic sought to interpret or challenge Darwinian ideas and theories in ways that matched their own realities and aspirations. For them, the questions were hardly abstract. The debate and its outcomes had significance for education, suffrage, and public policy.[37]

The challenge to Darwinians and the proposition of inferiority

Perhaps the earliest extended responses to Darwin's *Descent of Man* in the United States came in 1852 from Antoinette Brown Blackwell, who had been the first woman ordained into a mainline Protestant ministry and later served the Unitarian church. While at Oberlin College she had studied natural philosophy, and in the 1860s she found herself drawn to the works of Darwin and Spencer.[38] Moreover, she was a speaker and writer familiar with the tradition of women disseminating scientific ideas through popular textbooks such as those of Jane Marcet and Almira Hart Lincoln Phelps or explications such as those of Mary Sommerville.[39] In a volume of essays entitled *Studies in General Science*, published in 1869, Blackwell proved an enthusiastic advocate and explicator of such scientific ideas as conservation of matter and evolution. Her commentary on "The Struggle for Existence" concluded that a benevolent deity operated behind the apparent harshness of that concept of competition. Turning attention from the apparent ruthlessness of individual struggles, Blackwell argued for a more universal conception: "The Struggle for Existence, then, regarded in its whole scope, is but a perfected system of cooperations in which all sentient and unsentient forces mutually co-work in securing the highest ultimate good." Her extended commentary argued that science provided a critical means of understanding a complicated world and, optimistically, concluded that science would demonstrate ways for human action to promote social progress.[40] She welcomed the specificity of science as a way to determine answers to difficult questions and did not shrink from the possibility that her own ideas might be disproven; she did, however, expect that science would honor its own stated methods.

Over the next six years, Blackwell helped found the Association for the Advancement of Women (1873) and began contributing regularly to sister-in-law Lucy Stone's *Woman's Journal*.[41] Blackwell decided to respond quite explicitly to the writings of Darwin, Spencer, and Edward Clarke with regard to women and their evolution. A series of critical essays, some published earlier in periodicals, became her best known book, *Sexes Throughout Nature* (1875). At one level, the book is a critique of masculine language, examples, and metaphors. Blackwell takes the position that science has answers, but not necessarily do scientific men. She asserts that women's own experiences are of direct relevance, and that women need not accept casual observations of male authorities as data. Like Mill and Taylor, Blackwell reacted against biological determinism, particularly the emphasis on reproductive functions of women and causal links to women's supposedly limited intellectual capacity. Darwin was her starting point, but hardly the ultimate authority for Blackwell, in part because she realized that his theory had become the basis for most debate on "the woman question." She particularly took issue with him and others who held the "time honored assumption that the male is the normal type of his species" and the female only a "modification to a special end."[42] She astutely perceived that much of the content by the so-called theorists directed explicit challenges to the social movements for education, for women's suffrage, and for an expanded public role for women. Certainly none of those she challenged accepted an idea of equality between men and women, and all presumed male superiority. The techniques, as Blackwell noted, could differ with the same results: "Mr. Spencer scientifically *subtracts from the female*, and Mr. Darwin *adds to the male.*[43]

Blackwell's extended response to the Darwinians evaluated their evidence, arguments, and conclusions. The now former minister, admittedly influenced by biblical higher criticism, had come to accept scientific authority as a fundamental tool for understanding the natural world and its inhabitants. Her construction of evolutionary biology, however, could be made to fit with a social agenda to promote cooperation and balance within the human species.

Blackwell's own central thesis was presented clearly: "The sexes of each species of beings compared upon the same plane, from lowest to highest, are always true equivalents – equal but not identical in development and in relevant amounts of all normal force."[44] Using plants as an example, thus following Spencer and others in their easy analogies between human and other natural species, she suggested that early growth concentrates on the entire organism, and when distinctive sexual characteristics arise they are intended to balance forces for mutual production of a new, like organism.[45] Blackwell's emphasis was on complementarity, even the necessary partnership of male and female in their mutual goal of survival of the species. Sexual differences existed, but they were intended to provide means of collaboration for reproduction for the advancement of the entire species. When Blackwell turned to the human experience and the development of social communities, she stressed the importance of mental functioning for both male and female. She accepted the proposition that science provided answers to empirical questions but insisted that nature must be explored by experiment and fair trials of scientific theories. She pointed out that women's experiences had not been consistent with some male theorizing, particularly in terms of the education of girls. Using her own experiences as examples, both her studies at Oberlin College and her teaching experiences in New York and Michigan, she documented the capacities of women for academic accomplishment. Rather than accept the negative positions of Spencer and Clarke, she argued that there needed to be more data gathered, more studies undertaken, and even more chemical testing done in physiology before any conclusions could be drawn about the relative physical and mental capacities of men and women.[46]

For Blackwell, women had been "defrauded" by lack of opportunity, and thus community development was also constrained.[47] She echoed feminists from Mary Astell and Mary Wollstonecraft to Margaret Fuller in her insistence on seeing education and opportunity essential before any woman could be judged or the group of women assessed for their mental capacities. She also argued that man "could never reach his true proportions while she [woman] remained in any wise short of hers."[48]

Blackwell thus joined a line of theorists bringing science to bear on the questions of women's capabilities and rights. She would be cited and published by other women for the remainder of the century. There is little evidence that her ideas received much attention from Darwin, Spencer, Huxley or other well-known male authors whose commentaries reached a wider audience.[49] Blackwell actively sought such dialogue. She presented her arguments at meetings of the American Association for the Advancement of Science and published them in *Popular Science Monthly*, but she was disappointed that they had not been given much attention.[50] Others also joined the debate in *Popular Science Monthly*, but frustration was evident in the observation of Nina Morais in response to one article that "to cover ancient prejudice with the palladium of scientific argument is to unite the strength of conservatives and of progress in one attack."[51]

Through the contributions of Blackwell and others, the editors of the leading feminist periodical, *Woman's Journal*, alerted their readers to new scientific arguments concerning "the irrepressible woman question" and encouraged discussion and debate.[52] Blackwell was particularly skillful in her use of rhetoric and aware that her most attentive audience included women of intellect and a public concerned about women's rights and related reforms. Not all contributors were women; reformer Thomas Wentworth Higginson, who wrote a weekly column for the periodical from 1870 to 1884, took considerable interest in the arguments surrounding the work of Darwin and Spencer and also challenged notions of women's inferiority.[53]

Blackwell's published essays in the 1860s and 1870s had a diffident but determined quality as she and others formulated responses to evolutionists, naturalists, and anthropologists, as well as to physicians and educational administrators. The approach may have been a tactical decision. The message was one of skepticism about the simplistic notions being presented as science. They queried the studies of genius by men like Galton, for example, and pointed out how few opportunities for creativity were given to women. Nonetheless, there were certainly some women of genius as well. Suffragist Phebe Hanaford wrote *Daughters of America; or, Women of the Century* (1882) in part to put women back into the historical record and in part to identify women who had exceptional qualities like those attributed to men of genius.[54] Phrenology and craniology had already focused attention on the biological determination of racial and sexual differences, and evolutionary theory gave renewed energy to such investigations. A few women possessed the resources and interest to conduct precise measurements of the size and shape of the skull in what Elizabeth Fee called "the Baroque period of phrenology," and they carefully read and critiqued the data being published by men.[55] Contributors to the *Woman's Journal* continued to raise questions about efforts to predict behavior and mental capacity from narrowly focused physiological measurements.[56] Such authors challenged with sarcasm and wit the measurements of skulls and weights of brains, and asked what food consumption had to do with intelligence.[57]

Actions spoke even more eloquently, as the movements for education of girls and votes for women proceeded rapidly. With or without scientific sanction, middle-class women were gaining access to informal physical exercise regimens during the 1880s and 1890s that many believed would sponsor good health even as reformers increased access for girls and women at every educational level.[58] Many, such as physician Mary Putnam Jacobi, believed that the very success of women graduates would be a refutation of the charges of innate inferiority.[59]

The women's rights movement in fact built on the position and empirical data of Blackwell and others, sometimes turning the relative status of men and women upside down. Editor and suffragist Matilda Joslyn Gage, for example, looked at the data on proportions of male and female births and patterns of subsequent longevity to conclude before a women's rights convention in 1884, "from these hastily presented scientific facts it is manifest that woman possesses in a higher degree than man that adaptation to the conditions surrounding her which is everywhere accepted as evidence of superior vitality and higher physical rank in life."[60]

Darwinism in feminist advocacy

A number of political radicals and activist feminists accepted, on their own terms, the evolutionists' theme of women's distinctiveness and complementarity with men. After all, Darwin had avoided arguing for the absolute superiority of men – although his language and emphasis led prominent interpreters in that direction – and credited "woman" with unique and exclusive capabilities and roles that were fundamental to the reproduction of the species.[61] Being able to claim special qualities for women led certain women intellectuals not only to accept the idea of sexual divergence but also to promote the idea of female choice in sexual selection.[62] They joined reform-minded male scientific colleagues who queried biological determinism and who rejected the argument that humans could or should let nature take its own course, emphasizing instead the capacity to use mental capacities to shape political and economic alternatives.

Among political activists of the late nineteenth century who relied on scientific approaches in advocating their causes, the socialists may have been the most ardent as they sought to pursue Marxist biosocial evolutionary ideas. British and American socialists worked within the scientific framework their founders espoused; but only a few accepted unrestrained biological determinism as fundamental to social evolutionary theory. Women socialists seemed particularly intent on exploiting the cultural authority of Darwin and Spencer even as they argued that evolution proved the inevitability of, but not a singular direction for, change.[63]

Perhaps for that reason and despite the tensions and competition among socialist groups, the women read and discussed each other's ideas, casually crossing party lines. References to socialist but anti-Marxist authors such as Charlotte Perkins Gilman and Lester Frank Ward were common in *Socialist Woman*. These two American authors addressed broad audiences with their versions of evolutionary thinking about the nature of women and their role in society. With Lester Frank Ward's gynecocentric theory of evolution – namely, that in the economy of nature the female sex was primary and the male a secondary variant – and Edward Bellamy's reform Darwinism to inspire her, Gilman wrote her classic feminist treatise *Women and Economics: A Study in the Economic Relation between Men and Women as a Factor in Social Evolution* (1898).[64]

Charlotte Perkins Gilman had a difficult childhood that included only limited contact with her father, Frederick Beecher Perkins, whose family connections were reform-minded and whose own accomplishments included heading the Boston and San Francisco public libraries. What the self-educated young lecturer and writer found intriguing about evolution was its emphasis on process and on science as a way of comprehending human experience within natural order.[65] Gilman's *Women and Economics* was optimistic, assuming that through attention to actual natural capacities of men and women, the women's movement could help them utilize their characteristics cooperatively within the social organism.[66] Gilman accepted Geddes' and Thomson's *The Evolution of Sex* with their description of the divergence of character between men and women,[67] but for Gilman, traditional sex roles were a transitional phase and the unfortunate position of women a tragedy to be corrected. Her feminist version of "social Darwinism" became widely known among women progressives; *Women and Economics* was frequently reprinted and was translated into at least six languages.[68] Gilman was persuaded that while male traits of combat and display had at some earlier point been essential for social development, they had since become outmoded. Insisting that contemporary American culture was too "masculinist," Gilman sought to humanize it by reestablishing cooperation and nurture as essential qualities. She used the language of Darwinians to argue that organisms – social or individual – require the use of physical, intellectual, spiritual, and social powers. Subordination of any of them inevitably leads to underdevelopment. In her observations, men and women were both constrained by conventions and institutions, but women were the more oppressed. Contemporary women were also, in Gilman's estimation, soft, weak, ill-proportioned, and pathologically fixed on mothering as a result.[69] To assist them, she advocated contemporary reform – kitchenless houses and cooperative childcare – but she also sought more dramatic long-term social change.

Fiction and poetry offered Gilman another way to comment on contemporary society and to envision alternatives. The most dramatic vision was expressed in her utopian novel *Herland* (1915), as it explored the ways in which an all-woman society could operate for the greater good of the entire community.[70] Her dialogue between three men who ventured into the fictional Herland and the wise women assigned as their tutors made the point neatly. The men adamantly asserted that competition was fundamental to a successful society and a stimulus to industry. The women were puzzled, "Stimulus? To Industry? But don't you *like* to work?" When one of the visitors replied, "No man would work unless he had to," the women quickly asked, "Oh, no *man*! You mean that is one of your sex distinctions?" The women guides remained unconvinced by the subsequent discussion of the importance of competition and asked, "Do you mean, for instance, that no mother would work for her children without the stimulus of competition?"[71] The message of this novel and its sequel, *With Her in Ourland*, was to suggest human biological and intellectual potentials if the emphasis was on cooperation rather than competition. With women reasserting their evident evolved capacities and at the same time exploring powers that had been repressed, a new world was possible. For Gilman, "The whole feminine attitude toward life differs essentially from the masculine, because of [woman's] superior adaptation to the service of others, her rich fund of surplus energy for such service. Her philosophy ... and her conduct, based on natural impulses, justified by philosophy and ennobled by religion, will change our social economics at the very roots."[72] Gilman's belief in a linear view of evolutionary progress, incorporating the concept of the inheritability of acquired characteristics, had its inconsistencies, but her outspoken advocacy of women's capacities made her preeminent among feminist intellectuals.[73] She was not, however, the only woman writer and activist to take biological differences as a basis for envisioning women in a dominating role, leading society into a new social order.

Eliza Burt Gamble, a writer and educational activist, also inverted the hierarchy implied in *The Descent of Man* in her assessment of *The Evolution of Woman: An Inquiry into the Dogma of Her Inferiority to Man* (1894). After reading Darwin's evidence, Gamble concluded that the female of the human species was the primary unit of creation, freer than the male from imperfections. Men had less endurance, were more disposed toward organic diseases, and more likely to revert to former type, she argued; the female "thus represents a higher stage of development than the male."[74] Moreover, she suggested that morally as well as physically, women were superior because the characteristics assigned to "woman" were the source of social bonding. She insisted that female "constructive" virtues like altruism and compassion were, in fact, superior to male energy and competitiveness, which had become useless, an "actual hindrance to further progress."[75]

The South African novelist and socialist Olive Schreiner, like Gilman, anticipated that women might be mobilized into a movement that would take leadership in a new society. Schreiner moved in radical circles during her London sojourn in the 1880s before returning to her homeland to take up the cause of women's rights. Schreiner thought men ought to ponder the "man question" as thoroughly as they did the position of women.[76] Like Gilman and Gamble, she expected women to move with the biological force of evolution, but she did not cede place to manmade ideas about women's place and women's rights. Schreiner stressed cooperation, interdependence, and the complementarity of the sexes, and was critical of the established order in her fiction as well as her prose writings.[77] Some of these themes were developed as she met with "The Men and Women's Club" in London, but she did not publish an extended response to evolutionary theory until her volume on *Woman and Labour* appeared in 1911, which, like Gilman's work, was widely read and reprinted.[78]

The pervasiveness of these reconstituted "evolutionary" ideas indicates a loose network of intellectual theorizing among women stimulated by the Darwinian interpretations around them. The women mentioned here evidently read and sometimes cited one another.[79] The progressive ideas of women such as Gilman and Schreiner were carried throughout the European empires.[80] Historian Rosaleen Love points, for example, to Schreiner's influence on the evolutionary and feminist ideas of Louisa MacDonald, the influential, outspoken, and Edinburgh-educated principal of a women's college in Sydney, Australia.[81] Undoubtedly additional examples might be found among the women reformers of the late nineteenth century whose efforts led to women's suffrage and women's higher education, literally around the globe. The segregated and precarious careers of such women suggest that being committed to reform-oriented evolutionary ideas mitigated against their credibility in many circles and their entree into professional participation in traditional disciplines.[82]

Yet another dimension of women's response yields something of the fascination and the dilemma of literary figures influenced by Darwinian evolutionary thinking. In recent years a number of scholars have examined cases of direct influence of the Darwinists on fictional writers, men and women.[83] George Eliot provides one of the best examples, since she knew Spencer personally and her work has been explicitly analyzed in terms of changing evolutionary theory.[84] As Spencer came to emphasize women's primary role as mothers and to comment on the limitations of their intellect, Eliot responded by arguing that "diversification, not truth to type, is the creative principle" in the evolutionary process.[85] Eliot would not accept the idea that gender roles are predetermined by biology, although she explored the moral and biological imperatives at work on individuals and on the larger society.

In America, courtship plots in the short stories of author Kate Chopin have recently been viewed as studies in natural history according to the logic of sexual selection. Here, too, the author accepted the premise of sexual selection and female choice, but conceded nothing in terms of women's intelligence. Given Chopin's rejection of any idea of female inferiority, the idea of self-conscious sexual selection by females is, for her early short-story characters, very liberating. By the time Chopin wrote her only novel, *The Awakening*, however, her protagonist realizes that the idea of sexual selection flies in the face of the search for constant love; in the end she, like Chopin herself, commits suicide by walking into the sea.[86] For Gilman, as for Chopin, the attraction of Darwin's theory had been in the mechanism of sexual selection, wherein female choice in the selection was of critical importance; but for that mechanism to work well, Gilman had to create fictional circumstances in a novel, *Herland*, where parthenogenesis replaced heterosexual mating.

Among those who found evolution persuasive at some level, there was still considerable variation and ambiguity. Few, if any, of the discussants fully trusted the empirical techniques or the detailed results of Darwinian science. It seemed evident that men used their personal experiences, as the women did theirs, in order to frame arguments.

Moreover, the outcomes were not predictable because neither men nor women universally conformed to "type," although historical evidence confirms extended conversations internal to each gendered group, with some notable exceptions. Male scientists such as Lester Frank Ward and Otis T. Mason could advocate women's capacities and leadership, while some women certainly accepted the special and perhaps less-well-developed (because of limited education) "nature" of women, even as they promoted the idea that women had superior capacity for altruism and community building. The development of social-science disciplines and the increasingly evident limitations of cranial and other physiological measurements shifted the research direction of those concerned about scientific explanations of sexual difference that were at the foundation of Darwinian theory and discussions of evolution. Beyond the scope of this paper is the work of the first generation of women psychologists, including Mary Calkins, Helen Bradford Thompson Wooley, and Leta Stetter Hollingworth. According to historian Rosalind Rosenberg, their impact on their field was so significant that "by 1918 references to sex differences in intelligence were beginning to go out of style in psychology."[87]

Conclusions

There was never a simple or static Darwin; his ideas changed as did those of his elaborators and critics. Darwin's ideas appeared too sacrosanct for direct dismissal by the women studied here, although they did not hesitate to confront and even discard arguments by Spencer and Huxley, Clarke and Maudsley, Cope, Galton, and other proponents of Darwinism. Such biologists, paleontologists, physicians, and anthropologists seemed to the women to be intent on using anatomical measurements and somewhat casual physiological observations not to examine the issue but to prove men's superiority. That scientists chose to put this "woman question" near the forefront of their inquiries led Francis Power Cobbe to ponder in 1888, "of all the theories current concerning women, none is more curious than the theory that it is needful to make a theory about them."[88] The answer lies in the importance of both Darwin and the women's rights movements in the last half of the nineteenth century, and the efforts of those with authority to exert it to preserve their social order.

Women's responses, like those of men scientists, journalists, and social commentators, were multifaceted. Relatively few women had the education, the publication forums, or the social support to express their views in print – and, among those who did, all brought their experiences as women of the nineteenth century to bear on the issues. They had grown up with the idea of separate social spheres for men and women. Nonetheless, from the outset a few intrepid women intellectuals welcomed the prospect of a scientific method to dispute what they viewed as simply traditional and limiting views of women's capabilities. With perhaps a very few exceptions, women rejected the implied mental inferiority of their sex that emerged during the debates in the late nineteenth century. Blackwell and others jettisoned ideas about biological determinism in favor of those that emphasized the significance of change within evolutionary theory. Sympathetic male colleagues actively joined them in refuting the idea of male biological supremacy that dominated scientific writing in the 1870s and 1880s. Most of these women and men confined their role to that of critic, responding to and challenging what they viewed as the nonscientific methods or conclusions of those who postulated a fixed and narrow scope for women. Gilman, Gamble, and novelists such as Eliot and Chopin more fully embraced the arguments about sexual difference and sexual selection in order to frame a potentially different world order where women might even be viewed as the superior sex.

By the early twentieth century another set of researchers established a different set of empirical investigations that incorporated nurture as well as nature. The persistence of well-credentialed "new women" such as Calkins, Thompson, and Hollingworth challenged old positions, posed the issue of difference in a very different way, and made conceptual breakthroughs that became a critical part of the emerging social sciences. Women were not alone in this – indeed they often relied upon male mentors and allies – but they were moved by their personal experiences to take the quest particularly seriously.[89] This group, empowered by predecessors, colleagues, and professional success, argued quite boldly in their responses to any assumptions not based on sound empirical evidence.

Social conditions, economic necessity, political change, and other factors all contributed to the demise of the simplistic "woman question" and indeed the discussion of sexual selection as formulated by the late Victorians.[90] Certainly the active discussion prompted Blackwell, Gilman, and their successors to reformulate issues and methods in the social sciences emerging from the debates over evolutionary theory. Their challenges and alternative hypotheses in fact posed new sets of problems and generated new types of data in relationship to biological assumptions about sexual and gender identity. By the end of the century, there were significant challenges to Darwinian notions of "woman's nature" and quite empirical standards for testing individual and sex differences. Antoinette Brown Blackwell has the first and last word, writing rhetorically, "is it not quite time, then, for women to reconsider the ground work of these conclusions, if possibly the *savants* have furnished and pointed the weapons which can be effectively used for the overthrow of such grossly one-sided theories?"[91]

Notes

1 "The 'irrepressible woman question' is broader and more radical in every direction than most of us have been accustomed to think," wrote Antoinette Brown Blackwell in *The Sexes Throughout Nature* (New York: Putnam, 1875), p. 183. The *Oxford English Dictionary*, 2nd ed. (Oxford: Clarendon Press, 1989), vol. 20, pp. 486–87, suggests that the term was used by George Eliot in a letter in 1857 and then in a series of essays published as *The Woman Question in Europe*, by Theodore Stanton, ed. (New York: Putnam) in 1884; the usage varied, but typically referred to "women's nature" and, given such nature, the particular social, political, and economic roles for which women might (or might not) be suited. The question, closely linked to scientific inquiry, was also highly germane to legal and political debates.

2 On the assessment of women by scientific and philosophical writers before Darwin, see particularly Londa Schiebinger, *The Mind Has No Sex? Women in the Origins of Modern Science* (Cambridge: Harvard University Press, 1989) and Nancy Tuana, *The Less Noble Sex: Scientific, Religious, and Philosophical Conceptions of Woman's Nature* (Bloomington: Indiana University Press, 1993). Most standard accounts of Darwin provide only limited discussion about gender and have few references to the women who wrote about evolutionary theory. The Watson-Davis Prize-winning book by Adrian Desmond and James Moore, *Darwin* (New York: Warner, 1991) offers a particuarly thorough and up-to-date interpretation of Darwin himself.

3 A much larger discussion of cultural definitions of the body and also of sexuality during this period is important but beyond the scope of this paper; see, particularly, Thomas Laqueur, *Making Sex: Body and Gender from the Greeks to Freud* (Cambridge: Harvard University Press, 1990) and Catherine Gallagher and Thomas Laqueur, eds., *The Making of the Modern Body: Sexuality and the Social Body in the Nineteenth Century* (Berkeley: University of California Press, 1987). Both books contribute to the argument that in the late eighteenth and nineteenth centuries, the emphasis on sexual difference replaced an older formulation steeped in the Greek and Judeo-Christian tradition of "woman as a lesser man."

4 Because the terms Darwinism, Social Darwinism, evolution, and even biological determinism are used so variably and because primary sources cited in this paper tend to gloss over their distinction, we will use the terms as did many contemporaries and later historians, i.e., in a fairly general way that finds considerable continuities and similarities in argument among the scientists and their popularizers despite the sometimes heated debates among them. A number of essays in David Kohn, ed. *The Darwinian Heritage* (Princeton: Princeton University Press, 1985) discuss this issue, including

that by David Hull, "Darwinism as a Historical Entity: A Historiographic Proposal," pp. 773–812.

5 In those liberal religions such as Unitarianism and Universalism where women were involved as speakers, they joined their colleagues in encouraging the reading of Darwin, as pointed out in Cynthia Grant Tucker, *Prophetic Sisterhood: Liberal Women Ministers on the Frontier, 1880–1930* (Boston: Beacon Press, 1990), p. 20, passim.

6 On these historical and philosophical issues, see Sally Gregory Kohlstedt and Helen Longino, eds., "Women, Gender, and Science," *Osiris*, 12 (1997).

7 The literature is voluminous. For books that particularly relate to the themes of this paper, see Robert J. Richards, *Darwin and the Emergence of Evolutionary Theories of Mind and Behavior* (Chicago: University of Chicago Press, 1987), and Cynthia Eagle Russett, *Darwin in America: The Intellectual Response, 1865–1912* (San Francisco: Freeman, 1976). On social Darwinism see Robert C. Bannister, *Social Darwinism: Science and Myth in Anglo-American Social Thought* (Philadelphia: Temple University Press, 1979); Robert M. Young, "Darwinism is Social," in *The Darwinian Heritage*; and John Greene, "Darwin as a Social Evolutionist," *Journal of the History of Biology*, 10 (1977):1–27. None of these latter authors addresses the controversies around Darwin's definition of women.

8 Charles Darwin, *On the Origin of Species* Facsimile edition, ed. Ernst Mayr (Cambridge, Mass.: Harvard University Press, 1966), p. 88. This often quoted statement is also found in Eveleen Richards, "Darwin and the Descent of Women," in *The Wider Domain of Evolutionary Thought* eds. D. R. Oldroyd and I. Langham (Dordrecht: D. Reidel, 1983), p. 64.

9 *Descent of Man* (London: J. Murray, 1871; reprinted by Princeton: Princeton University Press, 1981), vol. 2, p. 402.

10 *Descent of Man*, vol. 2, p. 327. Darwin also wrote "man on an average is considerably taller, heavier, and stronger than woman, with squarer shoulders and more plainly-pronounced muscles.... Man is more courageous, pugnacious, and energetic than women, and has a more inventive genius." *Descent of Man*, vol. 2, p. 316. In a letter to Caroline Kennard, he was very explicit that he thought women "though generally superior to men in moral qualities, are inferior intellectually"; quoted in R. Richards, *Darwin and the Emergence of Evolutionary Theories*, p. 189.

11 Nearly all the literature that discusses Darwin and sexual selection in relationship to women emphasizes these passages and sees them as the arguments most referenced in contemporary literature through the end of the nineteenth century.

12 George Romanes, *Essays* (London: Longman, 1897), pp. 113–51; see p. 113; cited in Rosaleen Love, "Darwinism and Feminism: The 'Women Question' in the Life and Work of Olive Schreiner and Charlotte Perkins Gilman," in *Wider Domain of Evolutionary Thought*, p. 115. Darwin had loaned Romanes his notes and also wrote the postscript for *Mental Evolution in Animals* (New York: Appleton, 1884).

13 For a discussion of Galton see Flavia Alaya, "Victorian Science and the 'Genius' of Woman," *Journal of the History of Ideas*, 38 (1977):266–69.

14 Ruth Hubbard, "Have Only Men Evolved?" in *Women Look At Biology Looking at Women*, Hubbard, et al., eds. (Boston: G. K. Hall, 1979); Greta Jones, "The Social History of Darwin's *Descent of Man*," *Economy and Society*, 7 (1978):1–23; Carl Degler, *In Search of Human Nature: The Decline and Revival of Darwinism in American Social Thought* (New York: Oxford University Press, 1991); and Gertrude Himmelfarb, *Darwin and the Darwinian Revolution* (Gloucester, Mass.: Peter Smith, 1967).

15 E. Richards, "Darwin and the Descent of Women," and for a similar general theme, see R. Richards, *Darwin and the Emergence of Evolutionary Theories.*

16 Gillian Beer, "'The Face of Nature:' Anthropomorphic Elements in the Language of *The Origin of Species*," in *The Language of Nature: Critical Essays on Science and Literature*, ed. L. J. Jordanova (New Brunswick: Rutgers University Press, 1986), p. 221.

17 Among the earliest was Elizabeth Fee, "Science and the Woman Problem: Historical Perspectives," in *Sex Differences: Social and Biological Perspectives*, ed. Michael S. Teitelbaum (Garden City, N. Y.: Anchor Press, 1976), pp. 177–233; more recently was Russett, *Sexual Science: The Victorian Construction of Womanhood* (Cambridge: Harvard University Press, 1989). Others will be discussed in the text and notes that follow.

18 Virginia Woolf, *A Room of One's Own* (New York: Harcourt, Brace and World, 1957 [1929]), p. 26.

19 See especially Herbert Spencer's "The Status of Women" in *Principles of Sociology* I (New York: Appleton, 1888 [1877]), pp. 713–32. For a discussion of Spencer, Edward D. Cope, and George Romanes, see Susan Sleeth Mosedale, "Science Corrupted: Victorian Biologists Consider 'The Woman Question,'" *Journal of the History of Biology*, 11 (1978):1–55; on Spencer, see pp. 9–16.

20 See Nancy L. Paxton, *George Eliot and Herbert Spencer: Feminism, Evolution, and the Reconstruction of Gender* (Princeton: Princeton University Press, 1991). Paxton also traces ideas as Spencer moved from early liberal ideas about women to staunch antifeminism.

21 E. Richards, "Huxley and Woman's Place in Science: The Woman Question and Control of Victorian Anthropology," in *History, Humanity, and Evolution: Essays for John C. Greene*, ed. James R. Moore (Cambridge: Cambridge University Press, 1989), pp. 253–84.

22 Quoted in Mosedale, "Science Corrupted," p. 12.

23 Edward Drinker Cope, "The Relation of the Sexes to Government," *Popular Science Monthly*, 33 (1888):723, 726–27; Joseph LeConte weighed in with a similar opinion in "The Genesis of Sex," *Popular Science Monthly*, 16 (1879):178–79.

24 A particularly forceful statement on the superiority of men, based on arguments about nutritive capacities, was put forward by G. Delauney, "Equality and Inequality in Sex," *Popular Science Monthly*, 20 (1881):184–92. One woman apparently joined that side of the argument as well. An otherwise unidentified Miss M. A. Hardaker contributed a short essay on "Science and the Woman Question" in *Popular Science Monthly*, 20 (1882):557–84. The author argued that woman, with smaller organs, takes longer to eat a pound of bread than a man, and by analogy woman with a smaller brain processes

ideas more slowly; ultimately she will learn less. A forceful refutation came quickly in Nina Morais, "A Reply to Miss Hardaker on the Woman Question," *Popular Science Monthly*, 21 (1882):70–78. An earlier article by Hardaker on "Ethics of Sex" in *North American Review*, 131 (1880):62–74. was also challenged Nina Morais in a carefully constructed critique entitled "The Limitations of Sex," *North American Review*, 132 (1881):79–95. For a more general discussion, see Louise Mitchell Newman, *Men's Ideas/Women's Realities: Popular Science, 1870–1915* (New York: Pergamon, 1985); and Mosedale, "Science Corrupted," pp. 24–32.

25 Clarke was a formidable protagonist because he lectured widely and his book *Sex in Education; or, a Fair chance for the Girls* (Boston: J. R. Osgood, 1873) went through seventeen editions in thirteen years. See Joan Burstyn, *Victorian Education and the Ideal of Womanhood* (London: Croom and Helm, 1980); and Elaine Showalter, *The Female Malady: Women, Madness, and Culture in England, 1830–1980* (New York: Pantheon, 1985). Some scholars argue that many of these authors were motivated to write on the question in reaction to the movements to accord women more education and rights; see John Haller and Robin Haller, *The Physician and Sexuality in Victorian America* (Urbana: University of Illinois Press, 1974).

26 Patrick Geddes and J. Arthur Thompson, *The Evolution of Sex* (New York: Humboldt Press, 1889). The work of Geddes and Thompson is also discussed in Mosedale, "Science Corrupted," p. 37; Jill Conway, "Stereotypes of Femininity in a Theory of Sexual Evolution," *Victorian Studies*, 14 (1970):47–62; and Alaya, "Victorian Science," pp. 269–72. There were others, such as Havelock Ellis and Ellen Key, who continued to write about sex differences and sexuality and called themselves sexologists.

27 Fee, "The Sexual Politics of Victorian Social Anthropology," in *Clio's Consciousness Raised: New Perspectives on the History of Women*, eds. Mary S. Hartman and Lois Banner (New York: Harper and Row, 1974); and E. Richards, "Huxley and Woman's Place in Science."

28 A tantalizing possibility for analysis may be folklore. See Katherine D. Neustadt, "The Nature of Woman and the Development of American Folklore," *Women's Studies International Forum*, 9 (1986):227–34. Women were nearly a third of the membership of the American Folklore Society within a decade after its founding in 1888, and the group appears to have paid considerable attention to women.

29 The way in which circumstances and personal identities affected research in primatology is the center of Donna Haraway's *Primate Visions: Gender, Race, and Nature in the World of Modern Science* (New York: Routledge, 1989). Also see Sarah Blaffer Hardy, "Raising Darwin's Consciousness: Females and Evolutionary Theory," *Zygon*, 25 (1990):129–37.

30 The classic source on the patterns of inclusion of women scientists is Margaret Rossiter's *Women Scientists in America: Struggles and Strategies to 1940* (Baltimore: Johns Hopkins University Press, 1982).

31 For reactions to the simple polarities being assumed, see John D'Emilio and Estelle Freedman, *Intimate Matters: A History of Sexuality in America* (New York: Harper and Row, 1988).

32 Blackwell, *Sexes Throughout Nature*, p. 22.

33 For a discussion of these matters, see Ann Digby, "Victorian Values and Women in Public and Private," *Proceedings of the British Academy*, 78 (1992): 195–215.

34 Rossiter's, *Women Scientists in America* offers a detailed account of these developments. Also see *Charlotte Perkins Gilman: A Nonfictional Reader*, ed. Larry Ceplair (New York: Columbia University Press, 1991), pp. 27–28.

35 Sometimes the women authors revealed their own insecurities, as when Antoinette Brown Blackwell admitted her difficulties as a "woman" writer beset with many family responsibilities and asked her reader for "indulgence" for "many deficiencies and faults" that might appear in her book *Studies in General Science* (New York: Putnam, 1869), p. vii. Her self-consciousness as a woman writer is a reminder of the personal cost of domestic responsibility and of women's being told they are intellectually inferior.

36 See Joy Harvey, "Strangers to Each Other: Male and Female Relationships in the Life and Work of Clemence Royer," in *Uneasy Careers and Intimate Lives: Women in Science, 1789–1979* eds. Pnina Abir-Am and Dorinda Outram (New Brunswick: Rutgers University Press, 1989), pp. 147–71.

37 Love has noted that "Women, writing about women's nature, never forget for one moment they are women; men writing about women's nature frequently forget they are men"; see Love, "Darwinism and Feminism," pp. 113–31.

38 For background, see Elizabeth Cazden, *Antoinette Brown Blackwell: A Biography* (Old Westbury, N. Y.: Feminist Press, 1983) and Elizabeth Anne Munson, "Thwarted Nature and Perverted Wisdom: Antoinette Brown Blackwell's Critique of Evolutionary Theory," M. A. thesis, Pacific School of Theology at Berkeley, 1990.

39 M. Susan Lindee, "The American Career of Jane Marcet's *Conversations on Chemistry*, 1806–1853," *Isis*, 82 (1991):8–23; Marina Benjamin, "Elbow Room: Women Writers on Science, 1790–1840," in *Science and Sensibility: Gender and Scientific Inquiry, 1780–1840* (Oxford: Blackwell, 1991); and Kathryn A. Neeley, "Woman a Mediatrix: Women as Writers on Science and Technology in the Eighteenth and Nineteenth Centuries," *IEEE Transactions on Professional Communication*, 35 (1992):208–16.

40 Blackwell compiled *Studies in General Science* in an effort to share her own understanding of the relationship between science and religion, and she sent a copy to Charles Darwin, among others; see Cazden, *Antoinette Brown Blackwell*, p. 154.

41 Carol Lasser and Marlene Deahl Merrill, eds., *Friends and Sisters: Letters between Lucy Stone and Antoinette Brown Blackwell* (Urbana: University of Illinois Press, 1987), especially part 3.

42 Quoted in William Leach, *True Love and Perfect Union: The Feminist Reform of Sex and Society* (New York: Basic Books, 1980), p. 33.

43 Blackwell, *Sexes Throughout Nature*, p. 231.

44 Blackwell, *Sexes Throughout Nature*, p. 11.

45 Blackwell, *Sexes Throughout Nature*, pp. 43–46.

46 On her own experiences, see *Sexes Throughout Nature*, pp. 165–68 and 191–97; on her call for more scientific work, see "The Trial by Science," pp. 226–40.

Within the Association for the Advancement of Women, she joined Maria Mitchell in trying to encourage more women to study science; see Kohlstedt, "Maria Mitchell and the Advancement of Women in Science," in *Uneasy Careers and Intimate Lives*, pp. 129–46.

47 Blackwell, *Sexes Throughout Nature*, pp. 117–18.

48 Quoted in Marjorie Julia Spruill, "Sex, Science and the 'Woman Question': The *Woman's Journal* on Woman's Nature and Potential," M.A.T. thesis, Duke University, 1974, p. 22.

49 Blackwell was critical of both and found their reasoning going far beyond science; for example, she observed in *Sexes Throughout Nature*, p. 231, "Mr. Spencer, by modern scientific reasoning, has succeeded in grounding himself anew upon the moss-grown foundations of ancient dogma."

50 Munson, "Thwarted Nature and Perverted Wisdom," p. 34. Blackwell joined the AAAS in 1881 and presented papers in 1882 and 1884.

51 Morais, "Reply to Miss Hardaker," p. 70.

52 The *Woman's Journal* was founded by Lucy Stone and Henry Blackwell, Antoinette Brown Blackwell's in-laws, and represented the views of the broadly based branch of the suffrage movement, the American Woman Suffrage Association. See note 1 for quotation.

53 Spruill, "Sex, Science, and the 'Woman Question,'" pp. 15–18.

54 Phebe Ann Hanaford, *Daughters of America, or Woman of the Century* (Augusta, Maine: True and Co., 1882). In a chapter on women in science (p. 251), Hanaford optimistically predicted, "The time is fast approaching, when the question of sex will not be mentioned in relation to brain work."

55 At Karl Pearson's laboratories at University College, London in the 1890s, Alice Lee and others raised the issue of correlating physical size and mental capacity. She persuaded a number of prominent male anthropologists to be part of her study. The results ranked these men in terms of their cranial size and indeed seemed to insult some of them who had argued a correlation of size with intelligence. Several contested her results and it took until 1901 for her to receive her Ph.D. See Love, "Alice in Eugenics-Land: Feminism and Eugenics in the Scientific Careers of Alice Lee and Ethel Elderton," *Annals of Science*, 36 (1979):145–58.

56 Stephen Jay Gould, *Mismeasure of Man* (New York: Norton, 1987). See, for example, Blackwell, "The Savants on the Woman Question," *Woman's Journal*, 7 (1876):233; Blackwell, "Woman's Poor Brains," *Woman's Journal*, 24 (1893):321; and Anne Murray, "Respective Brains," *Woman's Journal*, 25 (1894): 131.

57 Marie Tedesco, "Science and Feminism: Conceptions of Female Intelligence and Their effect on American Feminism, 1859–1920" Ph.D. diss., Georgia State University, 1978, especially Chap. 8.

58 Martha H. Verbrugge, *Able-Bodied Womanhood: Personal Health and Social Change in Nineteenth Century Boston* (New York: Oxford University Press, 1988). Verbrugge's book suggests that most of the women educators and activists paid little attention to the scientific theories and concentrated on the practical results of their efforts.

59 Ruth Putnam, ed., *Life and Letters of Mary Putnam Jacobi* (New York: Putnam, 1925).

60 Susan B. Anthony and Ida Husted Harper, *History of Woman Suffrage* (published for Susan B. Anthony in Indianapolis: Hollenbeck Press, 1902), vol. 4, p. 30. This information may well have been derived from Blackwell's 1884 address at the Montreal meeting of the AAAS; see Munson, "Thwarted Nature and Perverted Wisdom," p. 249.

61 Tuana, *The Less Noble Sex*, pp. 36–39.

62 Ludmilla Jordanova, "Gender and the Historiography of Science," *British Journal for the History of Science*, 26 (1993):469–83, argues that "some historians of science and medicine have misconstrued the dichotomies around masculine and feminine as directly oppressive to women" (p. 277) and suggests that much more than labeling is involved.

63 The *Socialist Woman*, published from about 1907 to 1913 under editor Josephine Conger-Kaneko, recommended texts and a study series outline for self-education on evolutionary topics for their readers. These women drew on the vision of a prehistoric matriarchy, a period when women's unique traits had suffused altruism and egalitarianism into human relationships. Women's subsequent displacement by men and a partiarchal system intent on property ownership and competition had brought exploitation, warfare, and degradation to the mass of society. Thus the women socialists could imagine that as society evolved or was forced by revolution into socialism, that transformation would also bring changes in the relationship between the sexes. Although recent historians have elaborated on the often "unhappy marriage" of women and socialism, the socialist feminists, by accepting an evolutionary framework, could at least in theory put women into the center of the evolving order where their maternal characteristics would reshape the social order. For a general discussion, see Mark Pittenger, "Evolution, 'Women's Nature' and American Feminist Socialism, 1900–1915," *Radical History Review*, 36 (1986):47–61; also Lydia Sargent, ed., *Women and Revolution: A Discussion of the Unhappy Marriage of Marxism and Feminism* (Boston: South End Press, 1981).

64 Another important influence was Ward's Smithsonian Institution colleague, Otis Tufton Mason, who posited similar arguments in his *Woman's Share in Primitive Culture* (New York: Appleton, 1894).

65 Details of her extraordinary publication and lecture record are documented in Gray Scharnhorst, *Charlotte Perkins Gilman: A Bibliography* (Methchen, N. J.: Scarecrow Press, 1985). Lois N. Magner, "Women and the Scientific Idiom: Textual Episodes from Wollstonecraft, Fuller, Gilman, and Firestone," *Signs*, 4 (1978):61–80.

66 For a discussion of Gilman's evolutionary ideas, see Love, "Darwinism and Feminism," pp. 113–31 and Maureen L. Egan, "Evolutionary Theory in the Social Philosophy of Charlotte Perkins Gilman," *Hypatia*, 4 (1989):102–19. Gilman's own *The Living of Charlotte Perkins Gilman: An Autobiography* (New York: Appleton-Century, 1935) provides a personal assessment of influences that shaped her work.

67 She simply recorded the book as "very interesting" in her journal; see *The Diaries of Charlotte Perkins Gilman*, ed. Denise D. Knight (Charlotte: University of Virginia Press, 1994), pp. 452–53.
68 Mary A. Hill, *Charlotte Perkins Gilman: The Making of a Radical Feminist, 1860–96* (Philadelphia: Temple University Press, 1980), p. 267.
69 Gilman, *Woman and Economics*, p. 81.
70 Carol Farley Kessler, *Charlotte Perkins Gilman: Her Progress Toward Utopia with Selected Writings* (Syracuse: Syracuse University Press, 1995), pp. 70–76, 96.
71 Initially published in Gilman's monthly magazine, *The Forerunner*, the novel was subsequently edited and republished: *Herland*, ed. Ann Lane (New York: Parthenon, 1979), quotation from p. 60.
72 Quoted in Hill, *Charlotte Perkins Gilman*, p. 269.
73 See Lane, *To* Herland *and Beyond: The Life and Work of Charlotte Perkins Gilman* (New York: Pantheon, 1990), pp. 294–98.
74 Quoted in Degler, *In Search of Human Nature*, p. 109–10, from Eliza B. Gamble, *The Evolution of Woman: An Inquiry into the Dogma of Her Inferiority to Man* (New York: Putnam, 1894), p. 61; it was revised and published under the title *The Sexes in Science and History: An Inquiry into the Dogma of Woman's Inferiority to Man* (New York: Putnam, 1916). Also see "Eliza Burt Gamble," *The National Cyclopaedia*, 10 (1922):220–21.
75 Gamble, *Evolution of Woman*, p. 398.
76 See the discussion of Schreiner in Showalter, *Sexual Anarchy: Gender and Culture at the Fin de Siecle* (New York: Viking Press, 1990), pp. 48–58.
77 Catherine Mary Bedder, "New Woman, Old Science: Readings in late Victorian Fiction," Ph.D. diss., Cornell University, 1993, Chap. 4.
78 Ruth First and Ann Scott, *Olive Schreiner* (New York: Schocken, 1980); Schreiner published two English editions of *Women and Labour* by 1914 and it was translated into Dutch and German (see p. 373). Also see Judith R. Walkowitz, "Science, Feminism, and Romance: The Men and Women's Club, 1885-1889," *History Workshop*, 21 (1986):37–59; and Love, "Darwinism and Feminism," pp. 113–31.
79 Women's networks and close relationships in the nineteenth century have been much discussed, and they extended into the scientific and intellectual communities as well. See, for example, Toby A. Appel, "Physiology in American Women's Colleges: The Rise and Decline of a Female Subculture," *Isis*, 85 (1994):26–56.
80 A plea for more comparative and international work on precisely this period is in Judith Allen, "Contextualizing Late-Nineteenth-Century Feminism: Problems and Comparisons," *Journal of the Canadian Historical Association*, 1 (1990):17–36.
81 Love, "Darwinism and Feminism," pp. 125–27.
82 E. Richards, "Huxley and Women's Place," pp. 271–76. Eliza Lynn Linton became an outspoken antifeminist and biological determinist but was nonetheless knowingly affected by Huxley's exclusionary practices.
83 Woolf read and critiqued evolutionary writers, using them in explicit and indirect ways; see Elizabeth G. Lambert, "'Fish and Faith, She Reasoned': Evolutionary Discussion in *The Voyages Out*, *Mrs. Dalloway*, and *Between the*

Acts," Ph.D. diss., University of Massachusetts, 1991. The impact might be on men as well as women, as suggested in Karen Vollard Waters, "The Perfect Gentleman: Masculine Control in Victorian Men's Fiction, 1870–1901," Ph.D. diss., University of Maryland, 1993.

84 The literature on Eliot is extensive, and recent studies examine her interest in Darwin and friendship with Spencer and find considerable influence of evolutionary thinking in her novels. See Paxton, *George Eliot and Herbert Spencer*; Sally Shuttleworth, *George Eliot and Nineteenth-Century Science: The Make-Believe of a Beginning* (Cambridge: Cambridge University Press, 1984); Gillian Beer, *Darwin's Plots: Evolutionary Narrative in Darwin, George Eliot, and Nineteenth-Century Fiction* (London: Routledge and Kegan Paul, 1983); and Valerie A. Dodd, *George Eliot: An Intellectual Life* (New York: St. Martin's Press, 1990).

85 See Paxton, *George Elliott and Herbert Spencer*, p. 173; the quotation is from Middlemarch.

86 Kate Chopin, *The Awakening*, ed. Margaret Culley (New York: Norton, 1976 [1899]). Bert Bender, "The Teeth of Desire: *The Awakening* and *The Descent of Man*," *American Literature*, 63 (1991):459–73; and Bender, "Kate Chopin's Quarrel with Darwin," *Journal of American Studies*, 26 (1992):185–204.

87 Rosaland Rosenberg, *Beyond Separate Spheres: Intellectual Roots of Modern Feminism* (New Haven: Yale University Press, 1982), p. 105.

88 Quoted in Linda Birke, "'Life' as We Have Known It: Feminism and the Biology of Gender," in *Science and Sensibility: Gender and Scientific Enquiry. 1780–1945* ed. Marina Benjamin (Oxford: Basil Blackwell, 1991), p. 249. I thank Barbara Brooks for bringing this quotation to my attention.

89 A similar argument is made about women physicians in the late nineteenth century by Nancy M. Theriot, "Women's Voices in Nineteenth-Century Medical Discourse: A Step toward Deconstructing Science," *Signs*, 19 (1993):1–30.

90 The debate disappeared for nearly a century according to Malte Anderson, *Sexual Selection* (Princeton: Princeton University Press, 1994). Now it has been revitalized. On the one side are sociobiologists such as Edward O. Wilson, *On Human Nature* (Cambridge: Harvard University Press, 1978) and, on the other, biologists such as Ruth Bleier, ed., *Feminist Approaches to Science* (New York: Pergamon, 1986) as well as sociologists, historians, and philosophers, including Sandra Harding, Donna Haraway, and Helen Longino. Discussing the contemporary debate is beyond the scope of this paper. But see Sally Gregory Kohlstedt and Helen Longino, eds., *Women, Gender, and Science: New Directions*, vol. 12 of *Osiris* (Chicago: University of Chicago Press, 1997).

91 Blackwell, *Sexes Throughout Nature*, p. 185.

Index*

*Prepared by Richard Davidson.